I n the spirit of the Bob Woodward of the 1970s, Conrad Miller, MD has written an important, no, groundbreaking, book, impeccably researched, and which should be read by all Americans, no matter their political persuasion. 'The Most Important Issues Americans THINK They Know Enough About.....' will enlighten you as to the issues the mainstream media will not touch.

- - - Allan Weisbecker, author:
In Search of Captain Zero
(Best book of surf literature ever)

D r. Miller's ability to write "big thoughts" clearly is exceptional and that is one reason I enjoy reading his work -- because he assumes the reader is intelligent, but largely uninformed or misinformed -- which is surely true -- so he provides the 'backup information' necessary to make his points. This makes it a GREAT book not just for beginners, but also for experts -- his books are well-documented and informative, and draw unique conclusions from unique collections of indisputable facts. Because of his gift of clarity, I think every literate American has the education necessary to understand the complicated ideas he presents.

- - - Ace Hoffman, nuclear archivist

I0086461

"I see in the near future a crisis approaching that un-nerves me and causes me to tremble for the safety of my country. Corporations have been enthroned and an era of corruption in high places will follow, and the money power of the country will endeavor to prolong its reign by working upon the prejudices of the people until all wealth is aggregated in a few hands and the Republic is destroyed."

—Pres. Abraham Lincoln, 1864

———⊃○○○⊂———

"Enlightenment, peace, and joy will not be granted by someone else. The well is within us, and if we dig deeply in the present, the water will spring forth."

— Thich Nhat Hanh[1] 1991

---

1   <u>Peace Is Every Step</u> by Thich Nhat Hanh, page 41, Bantam Books, published 1991.

IF a site is ever approved (Yucca Mountain in Nevada rejected as unsafe by multitude of uncompromised scientists), 20,000 to 70,000 high level waste shipments across 43 states could occur over next 30 years — plus MORE shipments if we build new reactors. Each inadequately tested truck cask to carry up to 40 times long lasting radioactivity released by Hiroshima atomic bomb. Ten seconds exposure to breached cask contents fatal from 3 feet away. And what of ambush danger from terrorists? (See more in mid-Chapter One)

GENERAL HOWELL M. ESTES III

"The increasing reliance of US military forces upon space power combined with the explosive proliferation of global space capabilities makes a space vision essential. As stewards for military space, we must be prepared to exploit the advantages of the space medium. This Vision serves as a bridge in the evolution of military space into the 21st century and is the standard by which United States Space Command and its Components will measure progress into the future."

US Space Command--dominating the space dimension of military operations to protect US interests and investment. Integrating Space Forces into warfighting capabilities across the full spectrum of conflict

Page three of the 'Vision for 2020,' a 17 page internet document espousing U.S. Space Dominance, that _every_ American should scroll through and know about. Available at www.crestofthewave.com

THE MOST IMPORTANT ISSUES AMERICANS THINK THEY KNOW ENOUGH ABOUT...

by

Conrad Miller, M.D.

The Most Important Issues Americans THINK They Know Enough About

Paperback Lightning Source Third Edition

Copyright 2008 © Conrad Miller M.D. All rights reserved.
Printed in the United States of America on recycled paper.
Any properly footnoted quotation of up to 100 sequential words may be used without permission, as long as the total number of words does not exceed 250 words. For longer quotations or for a greater number of total words quoted, written permission from the publisher is required.

ISBN 13: 978-0-9753832-9-2
ISBN 10: 0-9753832-9-9

For information or individual orders, go to
http://www.crestofthewave.com
or write: Crest of the Wave
            P.O. Box 180
            Watermill, N.Y., 11976-0180, USA

This book is also available at
http://www.amazon.com • http://www.barnesandnoble.com
http://www.amazon.co.uk
or at your all important local book store

For libraries and businesses:
http://www.btol.com • http://www.bertrams.com
http://www.ingrambook.com • http://www.gardners.co.uk

This book is also available in ebook form from
http://www.amazon.com • http://www.powells.com
http://www.amazon.co.uk
Or by arrangement with Crest of the Wave
contact: http://www.crestofthewave.com

Cover illustration: http://www.bookcovers.com
Back cover illustration: adapted from page 15 of U.S. Air Force's 'Vision For 2020'

This book is dedicated to my beloved late wife Brenda, who tolerated, encouraged, and assisted my exhaustive efforts to complete and publish the work you are about to read.

I am missing her buoyance, love, energy, insight, strength, perception, protectiveness, verve for living, cool, grace, passion, companionship, courage, wisdom, spirituality, sense of humor... in addition to our mutual adoration... and probably will forever....

Acknowledgments

Thanks to the hundreds of people who have aided and abetted my experience over the past eight years. I know I've missed a few but this is a start:

All the People at NIRS - Diane D'Arrigo, Don Keesing, Mary Olson, Michael Mariotte; Cindy Folkers; Paul Gunter and Kevin Kamps (now at Beyond Nuclear); Karl Grossman, Bruce Gagnon, The Late John Gofman M.D., Ph. D., Ace Hoffman, Simon Billonness, Michael Colby, Philip Coyle III, Evan Douple, Helen Caldicott M.D., Vandana Shiva, Lisbeth Gronlund, Margaret Gundersen, Marion Fulk, George Lewis, Arjun Makhijani, Gary Oliver, David Wright, Kristin Bricker, Farrell Callahan, Ronnie Cummins, Scott Denman, Emilia Gonzalez Esteva, Richard Webb, Mary Roberts, Michael Hansen, Chuck Harder, Rusty Haynes, Steve Harrison, Winnetta G. Jackson, Kevin Kolbenheyer, John LaMontaine, Stephen Lendman, Patty Lovera, Nadia Martinez, Elizabeth McPartlan, Bill Manci, Greg Mello, Jerry Nussbaum, Hugh Sampson M.D., Jeff Smith, Judy Szela, Arpad Pusztai, Mark Ritchie, Anthony Salloum, Dave Schubert, Scott Sklar, Mike Skladany, Sergio Paone, Patrick Woodall, Lori Wallach, Paolo Berrino, Mary Bottari, Kathy Belyeu, Connie Folay, Margrete Strand, Diane and Marlene Halverson, Cary of Kauai's KUIC, Rick Dove, Brewster and Cathleen Kneen of the *Ram's Horn*, Dr. Samuel Epstein, Phil Lansing, Philip D. Lusk, Esther Maynard, Harvey Wasserman, my son Lutha - who helped me see the light toward publication ASAP, Amy Holman, Wenonah Hunter, Andrew Kadak, Ron Kimmel, Marv Lewis, Mary Gnetz, Yashina Giles, Kim, Allen, and Amber at bookcovers.com, Jeff Odefey, Joe Golato - the best chiropractor in the world, fathering me along the process of fighting/getting this book on out, David McIntyre of the NRC, Danila Oder, Kaye Stearman, Carmen Olmedo, Larry Bohlen, Chris Weiss, Christina Salvi, Robert D. Steele, Steve Suppan, Karen Hudson, Amarjit Sahota, Char Greenwald, Tony Brooks, Katherine Nadeau, Will Williams, Matt Bivens, Joe Cosgrove, Peter Cousins, Stephen Baker at the Center For Defense Information (CDI), Irwin Greenblatt, my sons Kyle McGough and Blake Leahy-Miller, Diane Hatz, Joe Cummins, Gunter Stotzky, Faith Campbell, Sebastian Belle, B.J. McManama, Peter Meisen, Dr. Rhoda Nussbaum, Geert Ritsema, Jon P. Corsiglia, Simon Ferrigno, Bruce Wagner, Andy Condey, Heather Mansfield, Allan Weisbecker, Mark Jackola, Jim Schultz, Doug Orton, Sara Johnson, Ron Stanchfield, Serena Diliberto, Phil Waters, Mark Worth, Robert Weissman, Ryan Zinn, Alicia Beshian, Bonnie Carr, Avram Chetron, Sidney Goodman, John Hammell, Eileen Shapiro Mantel, Ellen Miller, Kerria Williams, Dr. Marvin Resnikoff.

Permissions

Grateful acknowledgment is made to the following for specific permission to reprint previously published material:

- Granta publications for the Ivan Klima quote from 'The Powerful and the Powerless' from the book 'The Spirit of Prague'
- Red Wheel/Weiser/Conari Press for quotes from John Robbins from 'The Food Revolution'
- Vandana Shiva for the opening quote from her 'Biopiracy' book
- The Union of Concerned Scientists
- Institute for Agriculture and Trade Policy:
  www.iatp.org
- Organic Consumers Association:
  www.organicconsumers.org
- Kristin Bricker
- Stephen Lendman
- The Indypendent website
  www.indypendent.org
- Philip D. Lusk
- Elena Filatova website:
  www.elenafilatova.com
- Catherine and Brewster Kneen of the Ram's Horn
  www.ramshorn.ca
- Svetlana Alexievich Author: 'Voices From Chernobyl'
  & Dalkey Archive Press

Table of Contents

Introduction xiv

Chapter One:
  Bush's Nuclear Push 1
  PLUTONIUM, THE MOST TOXIC ELEMENT KNOWN TO MAN AND
      REPROCESSING NUCLEAR WASTE • 12
  THE USA IS 'THE PERSIAN GULF OF WIND' • 18
  ENERGY EFFICIENCY: NEGAWATTS, NOT MEGAWATTS • 25
  MEANWHILE, GERMANY LEADS THE WAY IN ALTERNATIVE ENERGY • 30
      CHERNOBYL • 37
  DECOMMISSIONING OR DISPOSING OF OVER-AGED NUCLEAR PLANTS • 52

Chapter Two:
  Your Food: Mutated, Irradiated, or
  Pragmatically Pure? 83
  OPENING DISCUSSION • 83
  CORN • 101
  SABOTAGING THE CONCEPT OF 'ORGANIC' FOODS • 120
  GENETICALLY ALTERING OUR FOODS • 123
  MILK AND RECOMBINANT BOVINE GROWTH HORMONE • 141
  FOOD IRRADIATION • 144
  FACTORY FARMING AND INDUSTRIAL AGRICULTURE • 170
  AQUACULTURE/FISH FARMS • 180
  DEAD ZONES, (GMO) CORN, FERTILIZERS AND ETHANOL • 201
  REALISTIC PROSPECTS AND MEASURES FOR IMPROVEMENT FOR U.S.
      AQUACULTURE • 205
  FOOD, SEEDS, PATENTS, YESTERDAY, AND THE FUTURE • 213
      SLOW FOOD • 244
  WHAT CAN I AND MY FAMILY EAT? • 254

Chapter Three:
  Star Wars and Space Dominance 267
  THE THEORETICAL BASICS OF MISSILE 'DEFENSE' • 301
  STRATCOM IN OMAHA, NEBRASKA, WHERE THE NEXT WAR WILL BE
      PLANNED, LAUNCHED AND CO-ORDINATED • 328
  THE USA'S NAZI-RIDDLED SPACE HISTORY • 332
  HONESTLY EVALUATING OUR SPACE PROGRAM, HONORING THE OUTER SPACE
      TREATY OF 1967, KEEPING OFFENSIVE WEAPONS OUT OF SPACE • 340

Chapter Four:
Surrendering U.S. Sovereignty To The World Trade Organization 347

PROTECTING THE USA AGAINST ASIAN BEETLES AND BORERS • 359
THE 'DOLPHIN-SAFE' TUNA STORY PLUS THE GUTTING OF OUR SEA TURTLE LAW • 364
THE BURMA/MYANMAR BOYCOTT ILLEGALIZATION STORY • 377
LURKS THE DANGER OF AN UNVOTED UPON ESTABLISHMENT OF A NORTH AMERICAN UNION (NAU) • 398
NEW HOPE WITH A DEMOCRATIC MAJORITY IN US CONGRESS? • 416

Chapter Five:
Radioactive Wastes In Your Dump, Air, Water, Utensils, Baby Stroller, Zippers, Anyone? 421

ACCEPTING TOXIC RADIOACTIVE EXPOSURE TO YOURSELF AND YOUR LOVED ONES • 436
WHAT CAN YOU DO? • 469
FINAL THOUGHTS CONCERNING RADIOACTIVE WASTE AND NUCLEAR POWER • 475

Chapter Six:
Preserve Your Dignity, American 479

Appendices 501
Endnotes 533

Introduction

Americans value their freedom and leadership in the world. But lately, according to a mid-2008 New York Times/CBS News poll, 81% of us think we are heading "pretty seriously" on the wrong track.[1] There's nothing like an economic crisis to shake our confidence in what we believe and what we think we know about today's most important issues.

In this third edition of my book, I've supplied some different angles on these issues. Whether we actually have a liberal media or a right wing media that is questionably fair and balanced, much of what you will read in the following pages may not be exactly how you are hearing our nation and

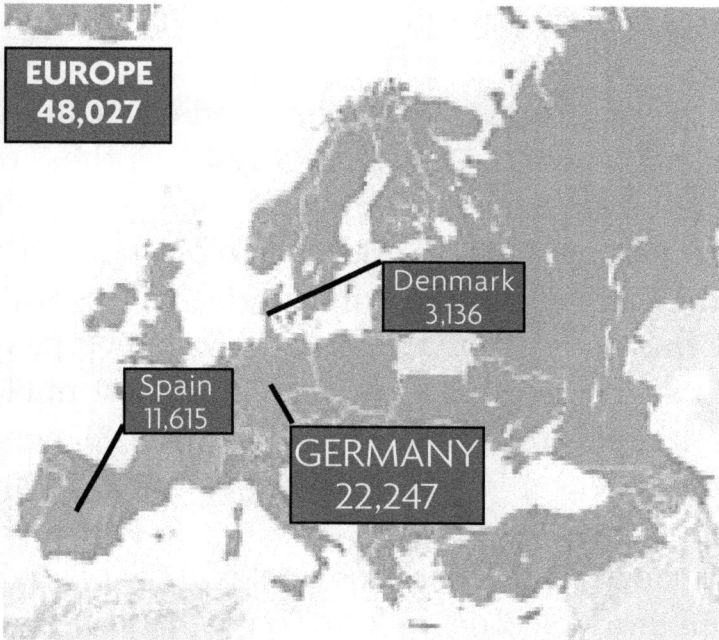

EUROPE
48,027

Denmark
3,136

Spain
11,615

GERMANY
22,247

The European Union's total wind power capacity is 56,535 megawatts. Germany leads the way with 22,247 megawatts. The average nuclear plant generates 1000 megawatts.[A] See www.ewea.org

our world being portrayed. Remember that basically five huge corporations control 75% of what Americans see and hear.[2]

Are you perplexed about the latest claim that 'nuclear is green?' Are you aware that T. Boone Pickens is investing $10 BILLION into what will be the biggest wind farm in the world, in Texas? That George Bush's home state now leads the USA in wind power megawattage? That the USA has been called 'The Persian Gulf of Wind?' That a prominent CalTech physicist tells us that we could provide all the USA's electricity with solar power from an 80-mile square area in one of our deserts?

Do you know that Germany is phasing out nuclear power? That Germany leads the world in both wind and solar power technology development today? That Dr. Alexey Yablokov, president of the Center for Russian Environmental Policy, informs us in his 2007 book that 300,000 people died so far as a result of the Chernobyl nuclear plant explosion and fire[3], not just the 31 people nuclear power proponents repeatedly claim in the media?

Though most Americans would eat organic food if given the choice, do we know enough about genetically modified patented crops to ensure their safety? Are you aware that you are probably in the majority in wanting genetically modified ingredients to be labelled? 87% of Americans polled want labelling[4], as they are already doing in Europe, Japan, Australia, even China. Do you know about a study showing test rats fed genetically modified potatoes had smaller brains, livers and testicles as compared to control rats fed non-genetically modified potatoes?[5] Have you heard that, if it is not stopped, sugar from genetically modified sugar beets will make up 50% of our non-organic sugar supply very shortly?[6] Have you been informed about grocery shelves in England being practically devoid of all and any genetically modified foods, because the English are acutely aware of these foods' ill affects?[7]

Are YOU Eating Monsanto's Genetically Engineered Sugar?

CONSUMER WARNING

If you are concerned about genetically engineered sugar and other foods in your shopping cart, you better act now! Here are some things you should know before you buy any foods or beverages from your local grocery store.

Kellogg's, Coca Cola, Kraft, and other large US sugar buyers, ignoring consumer concerns, will be sourcing their sugar from Monsanto's genetically engineered (GE) sugar beets, in stores in 2008. Like GE corn, soy, and canola, products containing GE sugar will not be labeled as such. Independent laboratory tests and industry disclosures indicate that upwards of 75% of all non-organic supermarket foods now "test positive" for the presence of GE ingredients.

Despite growing public concern over the safety of GE crops, like corn, soy, canola, and cotton, the US government refuses to require safety-testing or labeling for GE foods. Monsanto's RoundUp Ready GE sugar beets are designed to withstand massive doses of weed killers. In the last ten years since GE crops have become mainstream, the use of toxic pesticides has increased dramatically.

SCIENTISTS WARN THAT GE FOODS MAY

- Set off Allergies
- Increase Cancer Risks
- Produce Antibiotic Resistant pathogens
- Damage food quality
- Produce Dangerous Toxins
- Increase use of toxic pesticides
- Damage soil fertility
- Pollute adjoining farmlands
- Harm Monarch butterflies and beneficial insects such as ladybugs
- Create superpests, superweeds, and virulent new plant viruses.

TURN THIS LEAFLET OVER FOR GRASSROOTS ACTION GUIDELINES

WHAT YOU CAN DO TODAY TO AVOID MONSANTO'S GE SUGAR

No-K! BOYCOTT KELLOGG'S

TELL your grocer to boycott any product that isn't GE-free and to offer a full-line selection of certified organic foods, which prohibit Genetic Engineering.

JOIN the Organic Consumers Association and volunteer to help organize opposition to GE foods in your community. Keep informed on GE food and food safety issues at our web site: www.organicconsumers.org

CALL The major food processors and retailers and ask them if they commit to source GE-free sugar products. If they refuse tell them you will stop shopping at their store.

BUY certified organic foods from your local co-op, health food store, farmers market or through a CSA (Community Supported Agriculture program).

TAKE ACTION TODAY!

Call or E-mail (via their websites) the five companies listed below. Tell them you will not purchase their foods or beverages unless they can provide you with written assurance that their products will not contain genetically engineered sugar.

COMPANY	PHONE
KELLOGG'S · www.kelloggs.com	800-962-1413
MORNINGSTAR FARM www.seeveggiesdifferently.com	800-962-1413
KRAFT · www.kraftfoods.com	847-646-2000
M&M/MARS · www.mars.com	800-627-7852
COCA COLA · www.coca-cola.com	800-438-2653

FOR MORE INFORMATION:
ORGANIC CONSUMERS ASSOCIATION
6771 SOUTH SILVER HILL DRIVE
FINLAND MN 55603
PHONE: 218-226-4164
FAX: 218-226-4157
campaigns@organicconsumers.org
www.organicconsumers.org

Half of the non-organic sugar supply in the USA is threatened to become genetically modified sometime in 2008. You can help stop this travesty by calling the corporations listed above. More on this story: 2 one page letters to /from Kellogg's in the Appendix.

Page 16 of the U.S. Air Force's 'Vision for 2020' showing four 'fields' of warfare: ground, water, air and space. Viewable at many internet sites including www.gsinstitute.org

What about so-called 'missile defense?' Are you aware that our Star Wars generals actually want to weaponize space, preferably including nuclear weaponry, in defiance of the Outer Space Treaty of 1967 that the USA primarily authored to *prevent* an arms race in the heavens above? That the whole Star Wars 'missile defense/missile offense' boondoggle will be the 'largest industrial project in the history of the planet Earth?'[8]

Do you recall that there was no World Trade Organization (WTO) before the mid-1990's threatening the sovereignty of U.S. laws? Has anyone told you that 'free trade' might really be 'corporate managed trade?' That NAFTA (North American Free Trade Agreement) may soon lead to a North American Union (NAU), like the European Union (EU), being installed overshadowing our own USA democracy, and Canada's and Mexico's too? But no voting will be offered to our citizenry or state or federal governments to approve this, in contrast to each European nation and its citizens democratically voting to join or not become a member of the EU?[9]

What about large portions of our nuclear waste that we still do not know how to safely store, being deregulated to possibly end up de-monitored in our dumps, utensils and bra-clips? Do you know about EnergySolutions corporation wanting to import 20,000 tons of Italy's nuclear waste? That this could be the first step in making the USA the world's dumping ground for nuclear waste?[10]

As a physician, grandfather, naturalist, surfer and political observer, I am also very concerned about the course our nation is taking, and has been taking during the last few decades, to land us in the mess we find ourselves today. Hopefully, as you journey through the six chapters of this book, you will learn and be fascinated by the stories behind what I have mentioned above. In addition, I am sharing solutions with you for our problems, that you as an active citizen can help our nation implement with integrity and practicality.

Chapter 1:
Bush's Nuclear Push
(And Now McCain Wants USA 80% Nuclear For Our Electricity, Just Like France[a])

> This chapter is dedicated to Karl Grossman, a prolific American champion of common sense and justice.

Nuclear power is "safe and clean," our President who thinks God speaks to him personally, repeats to us over and over and over again. While God might have told him to invade Iraq,[1] the town of Godley in Illinois has been contaminated by vast nuclear waste spills so its citizens in our nation's heartland now have to drink bottled water. Exelon, who at first denied any such spills ever occurred from their Braidwood nuclear reactor and its pipings, eventually, after persistent local investigation, admitted, well, uh, yes, they did.

22 "tritium-laced" "unplanned releases," start-

---

a- France's actual percentage of electricity produced by nuclear power is closer to 75 percent. 'Concern Over French Nuclear Leaks,' BBC News, 24 July 2008, http://news.bbc.co.uk/2/hi/europe/7522712.stm

Godley, Illinois in the heartland of America, was a victim of multimillion gallon radioactive waste spills that were denied to have ever occurred for a decade by the Exelon nuclear corporation.

ing in <u>1996</u>, were not confessed to have happened, until <u>nine years later</u>![2]

MILLIONS of gallons of radioactive water meanwhile poisoned Godley's shallow sandy local wells. Unfortunately, while citizens pick up their family rations of bottled water today, provided guiltily by Exelon, they still live in fear of future nuclear waste spills.

Many Godleyans are also aware that tritium, an 'activation product' resulting from fissioning of uranium in their "cream of the crop"[3] nuclear reactor, can pass through their skin while they are showering or even washing their dishes.

According to the Grandfather of Health Physics, the late Karl Z. Morgan, tritium

"is the only radionuclide for which we assume as much is taken into the body via skin penetration as by inhalation. It is the MOST invasive of all radionuclides and distributes itself rather uniformly to all organs and all body tissues on a microCurie per gram basis. It presents a somatic, genetic and teratogenic [cancerous] risk. It cannot be separated from liquid waste by evaporation, a process used to concentrate most radionuclides [especially in nuclear reactors]."[4]

Though our President may claim that nuclear power is "safe and clean," or other Nuclear Energy Institute (NEI)– paid touters like Patrick Moore or Christy Todd Whitman might croak to us that nuclear is 'green,' Americans should know that there are more than 500 other radioactive elements or "radionuclides" besides the deadly tritium, that are produced in each of our 104 nuclear

Patrick Moore, public relations flak for the nuclear industry. Salaried via the lobbying organization of the nuclear industry, the Nuclear Energy Institute. Since 1986 has "spent more time working as a PR consultant to the logging, mining, biotech, nuclear and other industries than he did as an environmental activist." He was NOT a founder of Greenpeace, with which he was associated from 1971-1986, according to prominent Greenpeace senior advisor Harvey Wasserman.[†]

---

† 'Money is the Real Green Power: The Hoax of Eco-Friendly Nuclear Energy,' by Karl Grossman, Extra! Feb. 3, 2008. Mr. Wasserman is the coiner of the rejoinder "No Nukes."

plants every day. And, all of these are not perfectly contained within the walls or pipes or cooling ponds of these ultimately dangerous structures that all too often leak, or vent some amount of these radionuclides to our towns to contaminate our air and water.

What we should be worrying about is cancer, or the death or maiming of our fetuses, from these various radionuclides. Very very few of which were present in our environment in any mini-significant quantity before there were nuclear weapons or nuclear power.

Remember this: "nuclear power" is basically a building or plant where enriched uranium is fissioned or split apart, producing heat that turns water to steam, which rotates a turbine, which generates electricity. People think it is some mythical magical super-powerful irreplaceable entity. But that's all it is. Unfortunately, lethal radionuclides and radioactivity are produced that somehow must be contained and disposed of for a period of many centuries. We still do not know how to do this properly.

As Michael Keegan says: "Electricity is but the fleeting byproduct from nuclear power. The actual product is forever deadly radioactive waste."[6]

That's what is 'green' about nuclear power - - slimey, forever radioactive green.

Be aware that one of the reasons the USA never signed the Kyoto Protocol on climate change is that nuclear power has been excluded from being accepted as a 'clean' technology[7].

Also be aware that the _original_ function of today's nuclear power plant was to produce plutonium for our atomic bombs back in the 1940's. Then someone figured the plant could also generate electricity that would be "too cheap to meter." Which has not proven true. In fact, nuclear power is still so uneconomical, even today it requires massive subsidies from our federal government for our utilities to continue utilizing it. This includes the latest Energy Act of 2005 that could give $20.5 BILLION as subsidies to nuclear power![8] And if Joe Lieberman, the Senator from Connecticut, and his cronies have their way, you could make that $550 BILLION of your and my money in our fast falling economy where dollars are supposed to be scarce. That would be via the so-called 'Climate Security Act' and its sneaky not-yet-publicized _amendments_, which you should get all your friends and foes to call your Senators about via 202-224-3121, and tell them to vote against, with all its components, whenever the bill is re-introduced.[9]

Funny: if your house is in the worry-zone, located close to a nuclear plant, you <u>cannot</u> get any insurance against a nuclear accident. [Actually, you can't get insurance for a nuclear accident no matter where you live in the USA.] The insurance industry, in other words, considers liability for nuclear accidents a totally unacceptable risk. Yet, thanks to the Price-Anderson Act, recently re-approved by our Congress in 2005, the companies that deploy nuclear power plants <u>ARE</u> protected against nuclear accident expenses above $10 BILLION. If an accident occurs at one of the nuclear reactors around New York City, and contaminates the metropolitan area essentially forever, that liability would certainly be in the TRILLIONS of dollars, not a 'mere' $10 BILLION. AND <u>YOU THE U.S. TAXPAYER</u> WILL FOOT THE BILL FOR THE EXCESS ABOVE $10 BILLION, <u>NOT</u> THE NUCLEAR CORPORATIONS! (The cost incurred from the Chernobyl nuclear accident in the Ukraine has been estimated at $300 BILLION and rising....[10])

Latest reports in the New Scientist and the scientific press in general are finally linking increased numbers of cases of childhood leukemias and other cancers in areas surrounding those citizen-uninsureable nuclear plants to the nuclear

Nuclear plants originally were constructed to produce plutonium for atomic bombs. They have not proved to be 'too cheap to meter' in producing electricity, instead requiring multi-billion dollar subsidies for the industry to survive. Latest scientific studies are finally finding increased incidence of childhood cancers in the immediate areas surrounding nuclear plants.

plants themselves.[11] This is from common day-to-day operations, not from the ultimate worry of an explosion like the one that occurred in 1986 at Chernobyl. At least 300,000 people have died so far due to that infamous accident, mankind's worst industrial accident ever.[12] (More on Chernobyl later in the chapter.)

As an example of a dangerous radionuclide that can cause cancer in your child, let us talk about 'the canary in the coal mine' for *fuel failures*

7

in nuclear plants: strontium. When the shieldings of fuel rods or pellets of enriched uranium get so maximally hot that they fail or leak, strontium is the one radionuclide most easily detected of the more than 500 radionuclides that actually leak into the nuclear plant's water, floor drains and environment in general.

Strontium has a similar structure to the element calcium, which we hear on our TV's builds strong bones and teeth. However, unlike calcium, strontium is radioactive, emitting beta rays, essentially high speed electrons, flying out of those strontium atoms that don't belong in any of our children's bones. These rays or electrons hit the DNA in our child's bone cells, and bone marrow, where their blood cells are being made. Damage to the DNA strands occurs. Often this damage is spontaneously repaired by our miraculous body. Other times it is not. What can result are mutations or changes of the DNA. Sometimes these result in bizarre abnormal cancer cells. Or cells that are too damaged to survive.

In the bones, tumors can be produced that stay in one place and grow and grow until we notice them. Or leukemia can result, which is a cancer of the blood, with abnormal cancerous blood cells multiplying rapidly, or not enough cells surviving, or too many cells being killed off as a result of the beta rays' strikes.

Strontium can stay in the environment for a very long time. It is not like some bacteria we can kill immediately with boiling our water. Unnatural as it is in this Earthly world of ours, strontium's radioactivity hangs around for somewhere between 280-560 years. Half of it and its radioactivity decays away after 28 years. That is called its 'half life.' Over the following 28 years, another half of its remaining radioactivity will decay away, leaving ¼ of the original radioactivity active and dangerous. But it takes 10-20 'half lives' for strontium's 'hazardous life' to finally die out before we can stop worrying about it.

Too bad that today, nuclear spillage containing detectable strontium has been found in Westchester County in New York State, heading for the Hudson River that runs south alongside the island of Manhattan. Strontium has also been found in goats near the Millstone nuclear plant in Connecticut. Both of these contaminations should concern us all about the safety and cleanliness of nuclear power. The one in Westchester had strontium detected in three wells beside the

NY Yankees baseball announcer John Sterling shamelessly peddles the Indian Point nuclear reactor in Westchester county, just north of New York City, as having "zero greenhouse gases and reliable energy. Next to the Yankees infield, that's about as green as it gets." goes the ad Yankee fans have to listen to every day on their radio. Does Mr. Sterling know or care that Indian Point is currently leaking radioactive strontium outside the plant, already detected in a nuclear liquid plume underground headed for the Hudson River?

Indian Point reactor[13] that you hear John Sterling authoritatively advertise on New York Yankee radio. The Connecticut contamination has been claimed to have originated from the nearby New London, Connecticut submarine base where nuclear-powered submarines are made. However, many citizens of Joe Lieberman's home state angrily assert that their strontium actually is leaking from the aged Millstone nuclear plant complex[14], which activists have been trying to close for the last few decades.

If you could look inside a nuclear power plant, you would find an average of about 40,000 uranium fuel rods producing heat as a by-product of nuclear fission, or splitting of the uranium atom into its various radionuclides. Within each fuel rod are immensely potent 'pellets' of enriched uranium. Each pellet is about the size of the first bone of your pinky finger, and can put out energy equivalent to what three barrels of oil can. That is quite a condensed parcel of power! And there are about four MILLION pellets hopefully safely contained within all those fuel rods.

The other side of the story is that one of these pellets can kill you from a distance of forty feet away if you are exposed to it for only twenty seconds without proper shielding. That is, after the pellet has been active in the plant for a year or more, building up its radioactivity.[15] In addition, we are not living in a perfect world, so, Americans should know that, even without an accident or cataclysmic breach, minimal amounts of these nuclear pellets' various 500-plus radionuclides/ byproducts do weep through a protective cladding or covering into the surrounding cooling water.[16] Included here are decay or 'daughter products' of these radionuclides that can radioactively injure us if we are exposed to them due to imperfect containment during a plant's day-to-day operation.

Aren't we dealing with the Devil then, when we air-condition every room in our house should we feel too warm, or inefficiently heat our buildings in the winter to keep off the chill, if we use nuclear power to do so? Aren't there better alternatives to produce electricity? (We'll answer that in the pages ahead.)

Plutonium, The Most Toxic Element Known To Man And Reprocessing Nuclear Waste

Plutonium is the biggie radionuclide that we should be most afraid of and learn about. Why? Because plutonium-239 has a half-life of 24,000 years! That means we have to worry about it being a hazard to us and our descendants for 240,000 to 480,000 years. AND we should remember henceforth that only one MILLIONTH of a gram is the lung cancer-causing dose!

If you let me do the math, there are 454 grams in one pound [see helpful accompanying table on next page]. That means, by advancing our nuclear power insanity, we could produce 454 MILLION lung cancers in our own citizenry should some sort of accident occur releasing just one <u>pound</u> of plutonium! Hmmm, we only have about 300 MILLION Americans today....But, wait, this is a globalized world, isn't it? Why not think even more grandly?! <u>TWENTY</u> pounds of plutonium could kill ALL 6.8 billion human beings on this Earth via lung cancer if it is dispersed in small enough particles that could float all around our glorious planet! Yes, it may take 20-30 years for these lung cancers to develop; and, No, they will not have a label on them telling us that they came from inhaling some ultra-tiny, micron-sized plutonium particle into one of our lung sacs or 'alveoli.' But isn't nuclear power worth the risk?

Numbers to Help You

One trillion dollars >> $1,000,000,000,000
equals
one million million dollars
(or one million millionaires' money)

For plutonium ----> one microgram
equals a millionth of one gram
454 grams = one pound
one pound = 454,000,000 micrograms =
454 <u>million</u> micrograms
enough to give lung cancer to
454 <u>MILLION</u> people
@ 1 microgram per cancer

13

Oh, you think there can be another way? You know now that the original nuclear power plants were devised to produce plutonium for our first atomic bombs. What about the fact that twenty pounds of plutonium is enough for some terrorist or 'rogue state' to make an atomic bomb equal to the destructive power of the bombs we dropped onto Hiroshima and Nagasaki, Japan, to end World War II? Or that EACH of our 104 nuclear plants produces between 400-1000 POUNDS of plutonium every year?! Are you aware that George Bush wants to import nuclear waste from other countries to 'reprocess' it, and extract the plutonium (and uranium) for our use?

Would you think it advisable to consider, when all is said and done with "the dirtiest single step in the nuclear fuel chain"[17] - - meaning reprocessing - - where the nuclear waste will go? The act of reprocessing will have multiple NEW waste streams created from chopping up the fuel rods and dissolving them in hot acid. Radioactive gases and liquids resulting from the reactions will produce "significant 'routine' releases of radioactivity into the environment."[18] There could be explosions in the reprocessing plant of the solvents and volatile materials utilized, that could contaminate wide areas surrounding the reprocessing site. Eventually, the reprocessing plant itself will get too contaminated to operate, and have to be treated just like radioactive waste itself,[19] to be buried somewhere....probably in some poor area where the locals will not have the financial sustenance to fight such siting, as usually happens with most waste siting in our great country.

A 2001 report found that "80 percent of the collective radiation dose of the entire French nuclear power industry, and 90 percent of the radioactive emissions and discharges from the British nuclear power program, come from commercial waste reprocessing,"[20] which is allowed in these two countries. And their citizens pay the price.

"The British reprocessing center at Sellafield has discharged over 1000 pounds of plutonium into the sea, which has been detected in children's teeth throughout the British Isles... [This concen-

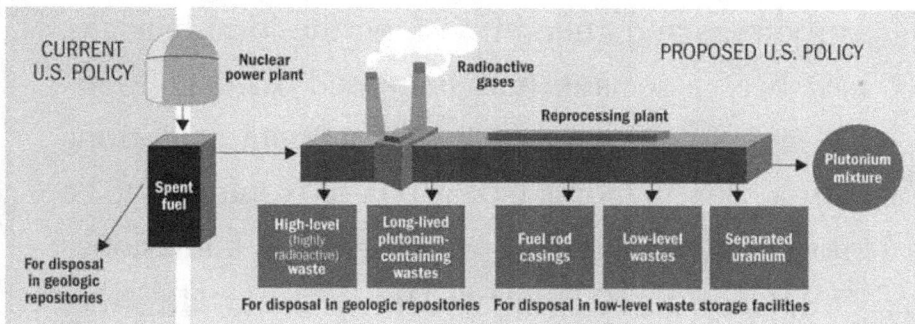

15

tration] in children's teeth decreases with distance from Sellafield, which indicates that the releases from the reprocessing facility are to blame. Radioactive contamination of the seafood supply has caused downstream governments from Ireland to Scandinavia to protest [these discharges]."[22]

One study "found that male Sellafield workers' exposures increase their children's risk of leukemia and non-Hodgkins lymphoma."[23] Not something you would like to have as a consequence of your daily labor.

Similarly, around the French reprocessing center at La Hague on that country's northern coast, the surrounding population also has increased incidence of childhood leukemia. The difference here is that this was associated with the radioactive pollution of the environment around the facility, not the daddies' radiation exposure.

And for you beach-goers, and seafood-lovers: "Consumption of local fish and shellfish, as well as mothers and children visiting the local beaches, have been associated with increased risk of contracting leukemia,"[24] around La Hague. Not something your local or national tourist boards would like to be broadcast over global or even European television.

Further stated: "elevated levels of certain child-

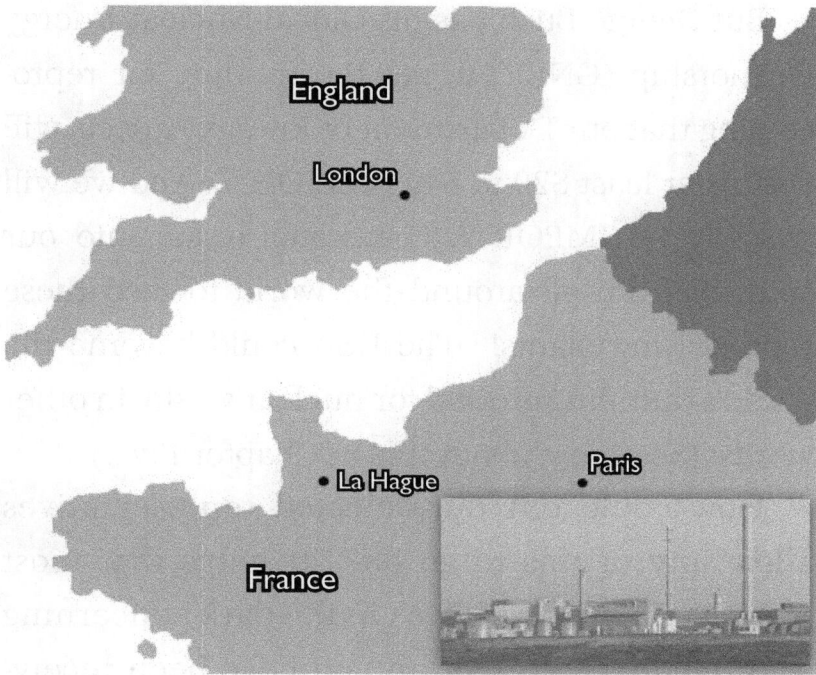

Scientists have found an increased risk of contracting leukemia from consuming local fish and shellfish, and even from visiting the beaches around the contaminating nuclear reprocessing plant in La Hague, France.

hood diseases and stillbirths are present around these currently operating reprocessing facilities in Europe,"[25] that you don't want to go near. Have you EVER heard about this?

At least President Gerald Ford did ban U.S. reprocessing of nuclear waste back in the 1970's after India produced their first nuclear bomb and exploded it, supplying the plutonium for the device via reprocessing — that they had claimed would be performed as a "peaceful use" of the technology.[26]

But George Bush has his Global Nuclear Energy Partnership (GNEP) to get the funding for reprocessing that our Department of Energy expects will cost us at least $20 to $40 BILLION.[27] And we will possibly be IMPORTING nuclear waste into our country from all around the world to feed these reprocessing plants! The USA could become the world's dumping ground for nuclear waste, in other words! (See more about this in Chapter Five.)

How could our duly elected representatives allow any of this to go on? It helps that most Americans live their lives in the dark concerning such nuclear information you have been receiving on these pages.

The USA Is 'The Persian Gulf of Wind'

Why doesn't our media repeat and hype up the wonderful news that our country is so blessed with such great wind power potential that is has been called "The Persian Gulf of Wind?" Yes, let it be known that the winds that blow through our states of North and South Dakota alone could supply 2/3rds of our nations' electricity. And George Bush's state of Texas could supply the other necessary 1/3rd ![28] In fact, just about every state has some wind power potential.

T. Boone Pickens is investing $10 billion into what will be the world's biggest windfarm in Texas.[29]

Texas now leads the USA in windpower production.[30] The Lone Star state already has enough wind turbines harvesting Texas breezes to power 1,000,000 homes![31]

Denmark produces greater than 20% of its electricity today from wind power.[32] Germany is phasing out nuclear power because it leads the world in windpower production (and is fast buying up much of the world's solar power technology[33]). Renewable energy supplied 14% of

T. Boone Pickens is investing $10 billion to build the world's biggest windfarm in Texas, the USA's #1 windpower state. The USA is now ranked Number Two in the world in windpower capacity, behind Number One, Germany.

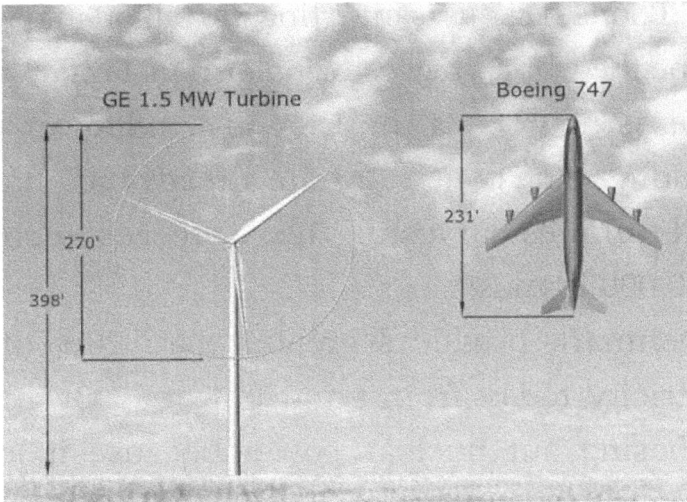

But <u>37</u> wind turbines like this 1.5 megawatt GE model could supply all of Kauai's homes with electricity.

Germany's electricity in 2007, with windpower producing 7.4% of German electricity. By 2020 Germany is projecting that 27% of its electricity will be generated by renewable forms of energy production.[34]

Just to show you how close we are to using wind power everywhere the wind blows today in the USA - - our great nation is now ranked as the world's Number Two windpower producer, by the way[35] - - let me tell you that our currently most popular selling larger wind turbine produces enough electricity to power 400 homes. It produces 1.5 megawatts and is made by General Electric, the same company that also is our major producer of nuclear power plants.[36] (The average nuclear power plant produces 1000 megawatts when it is functioning properly online.)

Putting this in very practical terms, let us take the island of Kauai in the state of Hawaii. With a population of 58,000 currently, taking an average home to house a family of four, one windmill to 400 homes with then 1600 people supplied by one windmill, only <u>37</u> windmills would be all that would be needed to power Kauai's homes' electricity needs.[37] And Kauai is a very windy island. Yet, that does not include the possibility of using solar power there, which surely should also occur in the very near future.

In fact, the way the smaller home wind turbines are made includes a solar component connection mounted on the turbine. Ron Kimmel of the American Wind Energy Association (AWEA) tells me that 90 percent of all home wind turbines sold have this connection available for solar system hook up, provided for the homeowner by the wind turbine salesperson. The most popular home wind turbine version currently costs about $55,000 minus rebates and the back-spinning of your electricity meter reducing your monthly bill by the amount of electricity generated above the needs of your home that you end up sending back into the grid system. The most commonly sold home wind turbine that can supply all your home's needs is rated to theoretically generate ten kilowatts of electricity (1000 kilowatts equals one megawatt).[38]

If only Ronald Reagan didn't take those solar panels off the White House roof back in 1981 that Jimmy Carter had placed there.....And cast our energy funding back then to nuclear power predominantly.... where would we be today, one has to wonder...?....

Yes, solar power. Cal Tech physicist and author David Goodstein tells us that an area in one of our deserts just 80 miles square could supply <u>all</u> of the USA's electricity needs. If we go whole hog, this could be accomplished within a decade, he states.[39] Imagine if we round up our manufacturing resources like our being-abandoned Michigan auto plants, as one example, and utilize these and our valuable workers to mass produce solar technology, now that the car production is going south and east. . .

And don't forget hydrogen as another alternative form of energy supply. Instead of propane tanks heating your home, or electrical wires supplying power flowing one way from a centralized utility company plant into your house, you may have your own hydrogen fuel cells on your property, storing and steadily contributing to your home's electricity current. Fuel cells do require some form of energy to produce the hydrogen to get a chemical reaction going that will result in water as the only waste product(s), in either liquid or vapor form (plus heat). That could mean having a wind turbine or some solar power apparatus as your primary energy source(s) if one wanted to engineer a most ecologically unpolluting system.

Electrolysis, or the simple splitting of water, into hydrogen and oxygen would be the ideal technology to produce the hydrogen for the fuel cells, via the energy forms just mentioned above.

A most attractive attribute of fuel cell technology is the extreme reliability. Typical downtime for minimal maintainance is less than one minute over a six year period![40]

As with other alternate forms of energy, if you have extra electricity running from these hydrogen fuel cells that you don't have to use right now, that you want to sell because of the needs of whatever market might be out there, or just to help balance your budget, you can send it off into the electricity grid system, saving/earning money as you do so. For today more than half of our fifty states have legally enstated 'net metering,' forcing utility com-

Home generation fuel cell units, as big as a suitcase or a refrigerator. Japanese government earmarking $309 million per year for fuel cell development; plans for 10 million homes ~1/4 all Japanese households - to be powered by fuel cells by 2020. Though may use natural gas as hydrogen source. See http://www.nextenergynews.com/news1/next-energy-news3.5b.html for more details.

panies to allow the consumer the right to lower her/his electricity bill by having her/his meter turn backwards when extra electricity is sent flowing back into the grid. Yes, we non-commercial homeowners could make modest profits by this activity.

However, during most of the latter half of the 20th century, utility companies successfully prevented such a thing from being allowed. This is one big factor that prevented especially wind power from taking off in America back in the 1970's.

Individual homes, or communities, or businesses could have their own wind turbine, solar energy devices, and/or hydrogen fuel cells integrated into a functional system to produce real energy independence. Though utility lobbyists tried to ward off the inevitable, please be informed that industries requiring uninterrupted electricity flow are already riding on the fuel-cell bandwagon, including banks, manufacturing plants, large commercial buildings, telecommunications, and soon, very soon, perhaps your own home may be included![41]

Many of us may have thought that President Bush got the message when he promised over one million dollars to fund hydrogen power research in his January 2003 State of the Union address. However, the reality is that instead of seeing a pollution-free hydrogen future, our nuclear-deranged President meant <u>nuclear power generated</u> hydrogen. And thanks to the Energy Act of 2005, $1.25 BILLION dollars has been carved into our bloated pro-nuclear budget for another ridiculous nuclear scam to be heaped upon the backs of the American taxpayer. That money will go to Idaho to construct a nuclear reactor there that might be coupled with the production of hydrogen.

Energy Efficiency: Negawatts, Not Megawatts

Of course, the most practical and least expensive way to lower our energy bills and needs, is the idea of 'Negawatts, not Megawatts,' otherwise known as energy efficiency. According to Amory Lovins, the physicist who might be my Energy Department Secretary if I were President: "each dollar of federal investment in energy efficiency has yielded over four dollars in economic benefits to the nation - - benefits in the form of new products, new jobs and energy cost savings to American businesses and households."[42] The array of new light bulbs, dimmers, tuneable ballasts, superwindows; plus turning off motors and electrically draining appliances efficiently, are just some ways to make us more energy secure.

With electricity being the "costliest form of energy...Each unit of saved electricity saves three or four units of fuel, chiefly coal, at the power plant. Saving electricity avoids much pollution, because power plants use one-third of all fuel and produce one-third of the resulting carbon dioxide (CO_2), one-third of the nitrogen oxides (NO_x), and two-thirds of the sulphur oxides (SO_x) [plus unacceptably unquantified cancer-causing radioactivity[43]]. Saving electricity therefore yields great environmental as well as economic leverage. And because saving electricity is cheaper than making it, pollution is avoided not at a cost but at a profit."[44]

Remember that only two percent of USA electricity is generated by oil, for all the confusing talk about our being "independent of foreign oil," that you continually hear. Most of our electricity is actually generated by coal [~50%], and about 19% is generated by nuclear power. 19% is also the amount generated currently by natural gas.[45] Less than two percent is generated by our renewable sources of the future: wind and solar, almost all of this percentage from wind right now.[46]

Also be aware that most of the uranium, about 80%, that we use in our nuclear plants also comes from foreign sources[47]. The debate about nuclear power should not mix apples and oranges. The USA uses 40% of the world's gasoline every year; nuclear power does NOT power your car, SUV or truck. Though hydrogen could, and does supply 50% of the power for the buses of Iceland, out there in the north Atlantic Ocean.

One problem with solar, wind, and any form of electricity production that has been very errantly presented to the American public is transmission and distribution. Although many responsible individuals may tell us that resistance in our grid cables limits efficient transmission to a few hundred miles from its point of generation, the truth is that DC transmission can efficiently travel 4000 miles; and AC transmission about half that. About 1/2 percent is lost per 100 miles of transmission. So if we went coast to coast, say 3000 miles, with our electricity, we might only lose 15% of it, which is only a nominal loss for the overall efficiency and profit that will result. Distribution losses are higher than transmission losses actually, which have to be dealt with no matter how far the electricity travels.

Yes, the DC current usually has to be converted to AC current at the distribution centers in the majority of cases.

The real big problem today is that our electricity grid is not very well interlocked between our three major grids (eastern, western, and Texas) and nine major power pools. This makes it very difficult to send power from Oregon to Georgia, or Ohio to Arizona. However, President Bush has done one excellent thing while in office: he has started the conversion of these various grids into a smoother, better conjoined interstate system linking our smaller transmission systems across the nation.[48] Meanwhile, our scientists are working on improving electricity transmission, for example, by developing 'superconductors,' minimizing the resistance to energy flow at everyday temperatures.[49]

Using your imagination outside the current box of reality, please picture that wind, solar, and hydrogen power do NOT have to be produced in some central 'farm' or utility-owned plant. They could be produced on your own land, or on your roof, or just outside of your little town on public land. Of course, just think about it: would the monopolistic utilities want THAT to happen, and then lose all their business with such de-centralized activity going on? Parlay that onto the stock merchants and amoral investors who only care about making some profit on the dollars they've piped into these utilities within our market economy...

But, when Justice arrives on some sweet train of Wisdom, tomorrow's practical forms of energy production that cause only minimum pollution, if any, shall arrive to power your home's needs, and that of our innovative businesses and industry. And perhaps our transportation too. It's coming. Once the above information gets out, and obstructive corrupted politicians of today are replaced with enlightened dedicated new faces who do CARE about us millions of Americans, we can be the ones to lead the world in proferring such ecologically favorable technology.

Meanwhile, Germany Leads The Way In Alternative Energy

What about you evening news watchers? Are you aware that Germany is phasing out nuclear power? How could that be? Ah, because they are installing at least a nuclear power plant's megawattage-worth of new wind generators EACH YEAR. Without the radioactive pollution, expense, and anxiety that goes along with the nuclear power option.

As of 2008, Germany has 22,247 megawatts of wind power[50], generating 7.4% of their electricity. The average nuclear power plant generates 1000 megawatts. Germany currently has 24 nuclear plants. (See windpower map in Introduction.)

There could be another reason for the German government to move in this denuclearizing direction. Have you ever heard about Gorloben? and the biggest deployment of police in Germany since the end of World War II. . . . ? . . .

No? Even with your hundreds of cable and DirecTV channels? Don't feel too left out. Most Americans missed the 1997 debacle that most of Europe was very aware of, when German authorities tried to ship just six (6) containers of nuclear waste to Gorloben, a small farm town south of Hamburg. The casks were going to be 'interim' stored above ground in a building that looked like 'a soft drink bottling warehouse.'[51] 30,000 police and $100 MILLION dollars had to be allocated to overcome adamant public protests and roadblocks that were occurring along a 300 mile transport route.

In one narrowed road, incensed citizens deeply imbedded a ten foot tall stainless steel cross to prevent the accompanying motorcade from proceeding toward its goal. People have to be pretty angry, and many fellow villagers must be very sympathetic, for something like that to happen.

Just think what will happen if and when attempts are made in America to transport 20,000 to 70,000 high-level nuclear waste shipments <u>thousands</u> of miles from all corners of the country to unsafe sites, or 'repositories,' like the one at Yucca Mountain in Nevada. Scientists have already said that Yucca, with its 33 earthquake faults, and more than 600 earthquakes that occurred within a 50 mile radius between 1976 and the 1990's, registering at least 2.5 on the Richter scale,[52] did not constitute an acceptable area to store nuclear waste. And the entire state of Nevada does not even have <u>one</u> nuclear plant.

Any of these shipments could be susceptible to a terrorist ambush or bazooka-ing that could blow the shipping casks apart. "Mobile Chernobyls," these radioactive transport loads have been called, after the world's worst nuclear/industrial accident ever at the Chernobyl nuclear plant in the Ukraine in 1986. Hundreds of our cities, and thousands of our smaller towns will be at risk along the myriad of routes these shipments will take. Once they might begin, expect them to go on for at least thirty years.

I don't think you will be very happy to find out that an accident breaching the inadequately tested casks, could kill your curious child with a mere ten seconds of exposure? Yes, if she stands but three feet away for those ten seconds, say on a dare, after the truck with its very hot cargo turns over, or the train derails, she could die within two weeks, from radiation sickness, where her hair falls out, her immune system implodes, she bleeds from many orifices, and dies an agonizing nuclear-powered-death.[53]

Additionally, those 'casks' that your trusted officials may claim are "safe" for nuclear waste transport have not been sufficiently tested at diesel fire temperatures, or adequate heights of descent for compromisability. You should know that the Nuclear Information and Resource Service (NIRS) and the World Information Service on Energy (WISE) jointly reported on March 17, 2006 in their Nuclear Monitor issue number 643 that "conservative estimates reveal that each truck cask on the highways would carry up to 40 times the long-lasting radioactivity released by the Hiroshima atomic bomb. Rail and barge casks, six times larger, would carry over 200 times the long-lasting radiation released at Hiroshima. Release of even a fraction of this cargo would spell unprecedented radiological disaster."[54]

Have you ever seen how long it takes for a train to stop when it's moving at a good rate of speed? It's like an oil tanker or an ocean freighter. Hit that undetected submerged rock, or squiggle off the parallel tracks, and it might take a few miles to come to a halt. Maybe in the middle of your town, or in your reservoir that supplies your city with drinking water. If a diesel fire starts aboard, feeding on the cargo, it often burns for days. Yet our Nuclear Regulatory Commission (NRC) has only required that these monster-sized casks be

tested to burn at 1475 degrees Fahrenheit for half
an hour. Alas, diesel fires burn at 1800 degrees
Fahrenheit![55]

As the NIRS/WISE analysis quoted above
states further: "Shipping is probably the weakest
link in the entire chain of irradiated nuclear fuel
management."[56] Although we are talking about
up to 70,000 fuel shipments - - and that is <u>with-
out</u> increasing our total number of nuclear plants
- - the truth is that there have only been 2,500 to
3,000 irradiated fuel shipments "in the U.S. since

the dawn of the Atomic Age 63 years ago. Even
the limited experience of such shipments...has
seen numerous incidents and accidents, including radioactive leaks beyond the vehicle, as well
as over 50 incidents of shipments radioactively
contaminated on the exterior of the shipping container, endangering not only workers, but also the
general public."[57]

Yet the federal government is currently allowing shipping casks "to give off 200 millirem per
HOUR at their surfaces."[58] This exceeds the total
amount of 'background radiation' most citizens
are permitted to receive in a YEAR, not including
radon exposure. (More on this in Chapter Five.)
"Nuclear workers, truck drivers, locomotive engineers, railroad workers, inspectors, toll booth at-
tendants, gas station employees and customers,
innocent bystanders at rest areas, residents living along transportation routes, and unsuspecting passersby on the highways all face radiation
doses [like this] if they come too close to such
shipments."[59]

In the late 1990's "activists and investigative
reporters revealed that 20-37 percent of all shipments into France's reprocessing facility were
externally contaminated above regulatory limits

- many emitting 500 times the permissible dose, and one emitting 3,300 times the permissible dose!"[60] That is what looms ahead for us Americans if we ignore what the Bush administration wants to do with its renewed push to intensify our nuclear 'advantage.'

Another question for you, Dear Reader: Did you know that the <u>original</u> plan for September 11th 2001 was to crash airplanes directly into nuclear power plants? This was reported in the Nuclear Monitor on September 13, 2002. However, Al-Qaeda leaders decided against this option for fear it would "get out of hand." Back in 2001, the quote was: "for now." Today, or someday soon, their "Department of Martyrs" may decide to send suicide bombers to detonate themselves in or near a few of our nuclear power plants, new ones or old ones, where security is ridiculously insufficient, as has been widely reported via our television networks.[61] Or shoot some form of missile or bazooka at these blatantly vulnerable, incredibly toxic targets.

Then the resulting released plutonium particles, and those of the other 500-plus radionuclides, can fly wherever they may...Causing you and/or your children to suffer for us just *having* to have that technology that produces its extremely

long-lived lethal legacy....while only generating perhaps 50-100 years' worth of electricity.

Chernobyl

On April 26, 1986 there occurred the worst man-caused accident in the history of the human race. Testing was going on at reactor number four at the Chernobyl Atomic Energy Station in the northern Ukraine when power was dropped to 7%, but suddenly surged to 100 times 100% of full power in <u>less than one minute</u>!!! A catastrophic steam ex-

37

plosion occurred that "flipped the reactor's massive cap like a coin and left it wedged and hanging askew inside the ruined reactor. The reactor's core caught fire, leading to the largest single non-military radiation release in history."[62]

Estimates vary, but nuclear physicist Dr. Vladimir Chernousenko, who supervised the clean-up (and subsequently died from cancer) "for a 10-kilometer zone around the exploded reactor, [stated] that 80 per cent of the reactor's radioactivity escaped - - something like 7 BILLION curies" out of a possible 9 billion curies. That is an unbelieveable quantity of radiation. A food irradiation plant theoretically holds up to 10 MILLION curies of radiation.

Of course, the "Russians and the International Atomic Energy Agency (IAEA) claimed in a 1986 report that 50 million curies of radioactive debris, plus another 50 million curies of rare and inert gases were discharged."[63] Baloney for the nuclear soul, that report was later "condemned as a cover-up."[64] Sadly, Soviet authorities cared so much for their people that they "neither officially acknowledged the explosion, nor warned their citizens until May 2, 1986."[65]

Meanwhile, "the fire in the reactor core burned for ten days," continuing to release radioactiv-

The Chernobyl nuclear reactor accident has killed at least 300,000 people so far, due to the toxic radiation released far and wide after the steam explosion blew the reactor's core apart.

ity for months afterward.[66] Yet (from Svetlana Alexievich's tragic collection of 'Voices From Chernobyl'):

"They suddenly started having these segments on television, like: an old lady milks her cow, pours the milk into a can, the reporter comes over with a military dosimeter, measures it. And the commentator says, See, everything's fine, and the reactor is just ten kilometers away. They show the Pripyat River, there are people swimming in it, tanning themselves. In the distance you see the reactor and plumes of smoke above it. The commentator says: The West is trying to spread panic, telling lies about the accident."[67]

Soviet authorities took advantage of their people's ignorance concerning radioactivity. That one cannot see, taste or feel radioactivity contributes to it being kind of unbelievable that it can kill you. Might I ask: Are Americans any better with their knowledge concerning radioactivity?

And what about the nuclear French, with more than 75% percent of their electricity produced by 59 nuclear reactors?[68] In the immediate wake of the Chernobyl explosion, "France, instead of taking precautions like other European countries, had its state television stations issue weather reports indicating that the cloud of radioactivity from Chernobyl had miraculously stopped short at the Franco-German border!"[69] Amazing how a society or culture, distorted by nuclear power, can have its people sacrificed to the radioactive gods.

Going slightly back in time now, to the scene of the disaster, from Sergei Vasilyevich Sobolev, Deputy Head of the Executive Committee of the Shield of Chernobyl Association:

"There was a moment when there was the danger of a nuclear explosion, and they had to get the water out from under the reactor, so that a mixture of uranium and graphite wouldn't get into it - with the water, they would have formed a critical mass. The explosion would have been between three and five megatons. This would have meant that not only Kiev and Minsk, but a large part of Europe would have been uninhabitable. Can you imagine it? A European catastrophe.

So here was the task: who would dive in there and open the bolt on the safety valve? They promised them a car, an apartment, a dacha, aid for their families until the end of time. They searched for volunteers. And they found them! The boys dived, many times, and they opened that bolt, and the unit was given 7,000 roubles. They forgot about the cars and apartments they promised - that's not why they dived. These are people who came from a certain culture, the culture of the great achievement. They were a sacrifice.

And what about the soldiers who worked on the roof of the reactor? Two hundred and ten military units were thrown at the liquidation of the fallout of the catastrophe, which equals about 340,000 military personnel. The ones cleaning the roof got it the worst. They had lead vests, but the radiation was coming from below, and they weren't protected there. They were wearing ordinary, cheap imitation-leather boots. They spent about a minute and a half, two minutes on the roof each day, and then they were discharged, given a certificate and an award - 100 roubles. And then they disappeared to the vast peripheries of our motherland. On the roof they gathered fuel and graphite from the reactor, shards of concrete and metal.

It took about 20-30 seconds to fill a wheelbarrow, and then another 30 seconds to throw the "garbage" off the roof. These special wheelbarrows weighed 40 kilos just by themselves. So you can picture it: a lead vest, masks, the wheelbarrows, and insane speed."[70]

41

There is a video of the clean-up showing this madness on the roof, each individual soldier's run actually lasting up to about 4 to 5 minutes worth of very high level radioactive exposure, from getting onto the roof, loading the wheelbarrow, or just a shovel, and then running it to the edge, where it could be tipped off and dumped over the side, then rapidly as possible exiting the roof.[71] Many of these men died, or their reproductive organs were severely compromised. Soviet wives, naturally, were averse to have sex with these men for fear that their babies would be congenitally damaged.

From historian Aleksandr Revalskiy: "A while ago in the papers it said that in Byelorus alone, in 1993 there were 200,000 abortions. Because of Chernobyl. We all live with that fear now."[72] Of malformed babies, or stillbirths, or children that will tragically develop cancer. Like the boy that was born with "a mouth that stretches to his ears and no eyes."[73] Or the girl born, that "wasn't a baby, she was a little sack... not a single opening, just the eyes....more simply: no pee-pee, no butt, one kidney."[74]

What about this, relative to getting volunteers for the clean-up, and what happened to one father and son, again from 'Voices From Chernobyl':

Soldiers on the roof of Reactor No. 4 (there are a total of four reactors located at the Chernobyl plant) pick up deadly pieces of radioactive graphite from the explosion and toss them down into the cauldron of the demolished reactor core.

"...they appealed to our sense of masculinity. Manly men were going off to do this important thing. And everyone else? They can hide under women's skirts, if they want. There were guys with pregnant wives, others had little babies, a third had burns. They all cursed to themselves and came anyway.

We came home. I took off all my clothes that I'd worn there and threw them down the trash chute. I gave my cap to my little son. He really wanted it. And he wore it all all the time. Two years later they gave him a diagnosis: a tumor in his brain....You can write the rest of this yourself. I don't want to talk anymore."[75]

Tragic. You *could* be cynical, and say, oh, that's not proven, there's no cause and effect. But, this poor child and his father were not the only ones to experience cancer in their families very probably related to the Chernobyl disaster.

There are at least 4000 cases of thyroid cancer, that "a limited United Nations study" verified, mostly in children.[76]

The radionuclides of iodine, including iodine-129 with its mind-blowing half-life of 15.7 MILLION YEARS, are basically responsible for these thyroid cancers. However, some doctors have been thrown in jail, or into psychiatric institutions in various parts of what once was the Soviet Union [the Chernobyl accident occurred on April 26, 1986] for doing their duty, trying to report radiation-related illnesses and deaths. New cases of thyroid cancer continue to turn up as the next generations of exposed children (and fetuses), living on contaminated land, ingesting contaminated nourishment, drinking contaminated water, become sick.

Dr. Vladimir Chernousenko, who was also the former head of the Ukrainian Academy of Science, stated that although a 30 kilometer radius surrounding the Chernobyl plant was eventually evacuated because of contamination, it should have been a 600 kilometer (375 mile) radius. But that would have then included the major cities of Minsk and Kiev, which probably would have made it difficult to accomplish, for political reasons.[77]

Remember that Byelorus, which is the <u>country</u> now, north of Ukraine, received the most radioactive fallout from Chernobyl, due to the winds blowing toward the north and northwest at the time of the steam explosion. One quarter of all the land there is contaminated as a result of the disaster[78] for at least 300-600 years. Mostly with cesium, which has a half life of 30 years. Though Dr. Chernousenko reckoned the contamination actually will last 100,000 years[79] (don't forget about the half-lives of plutonium-239 and iodine-129 being 24,000 years and 15.7 million years respectively, and these having to be multiplied by 10-20 times to get their 'hazardous lives').

As far as how many deaths occurred secondary to the Chernobyl accident, it has to be in the hundreds of thousands. Unfortunately, as you may see from the quote above about the "liquidators," no scientific tracking was arranged to follow their states of health. Estimates of their numbers alone commonly range around 700,000 individuals. Then there are all the other humans (and animals and plants) affected in contaminated areas,

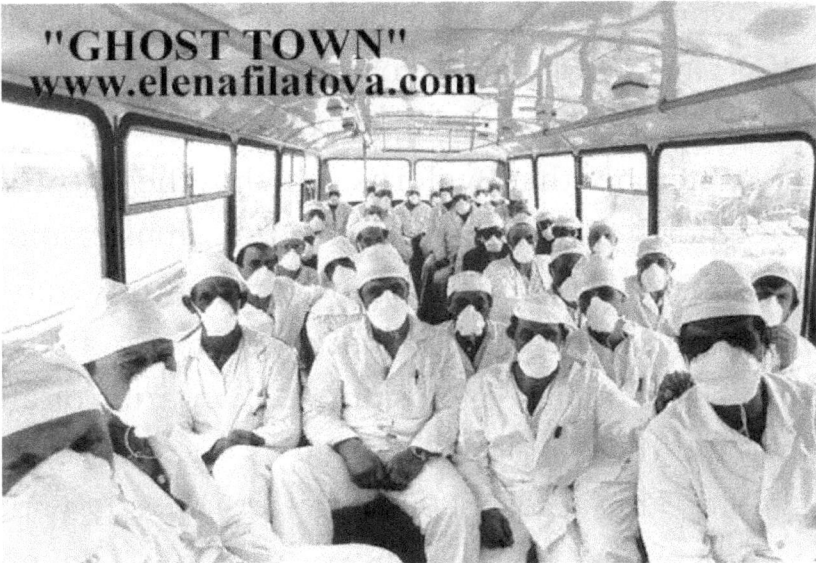

The "Liquidators" were recruited or forced to assist in the cleanup, or the liquidation of, the consequences of the accident.

As a totalitarian government, the Soviet Union provided many young soldiers to assist with the cleanup of the Chernobyl accident, but did not provide many of them with adequate protective clothing...or with any explanation of the dangers involved.

Over 650,000 liquidators were involved in the Chernobyl disaster cleanup during that first year. This group included those who built the containment building called the "SARCOPHAGUS" over destroyed Reactor No. 4.

Above from: http://www.hlswilliwaw.com/GhostTown/html/chapter6.htm
http://www.elenafilatova.com

and beyond, who may have unknowingly inhaled some plutonium fallout, for example, in Wales or even in the USA. Then you have the 300,000 death figure from Dr. Alexey Yablokov, president of the Center for Russian Environmental Policy, as quoted in the Introduction from his 2007 book[80].

Also, be aware that the number of <u>cancers</u> in such accidents of radioactive exposure usually is DOUBLE the number of <u>deaths</u> that occur.

Does this jive with nuclear power being "safe and clean?" Or "green?" Or the misleading falsehood that some brazen proponents of nuclear power continue to regurgitate, that only 31 people died at/from Chernobyl? What do these people think? That the effects from all those curies of radiation released have produced no cancers or deaths, nor will they in the future? Or are they just foolish liars, pushing their nuclear power agenda ideologically, at the expense of all living things on this Earth?

"Radiation health experts working for the National Academy of Sciences [state that] most cancers that result from radiation exposure do not develop until 10-20 years after exposure. The highest incidence of cancer is expected to occur over the next 5-10 years [from 2006], and therefore no accurate assessment of Chernobyl's overall impact can be made until this period has expired."[81]

Kofi Annan, the former Secretary General of the United Nations added: "At least 2 million children in Belarus, Ukraine and the Russian Federation require physical treatment (due to the Chernobyl accident). Not until 2016, at the earliest, will we know the full number of those likely to develop serious medical conditions."[82]

When Dr. Chernousenko was speaking in Austin, Texas back in 1994, amongst other things he revealed were the following. He was asked about the Chernobyl reactor's containment structure. Many nuclear power cheerleaders will repeat the mantra that Chernobyl was an inferiorly designed type of nuclear reactor, and had no containment.

The Soviet reactors at Chernobyl did not have an inferior design, and they did have a containment structure, Dr. Chernousenko stated. However, "the force of the explosion at Chernobyl exceeded the protective capabilities of this containment by at least ten-fold."[83]

Also, he told his audience that "Dr. Rosalie Bertell, who participated in the investigation of the [1979] accident at Three Mile Island, [in Pennsylvania,] can tell you, if a miracle hadn't occurred, and the hydrogen bubble within that containment hadn't dissipated, the accident within the United States would be comparable to the accident at Chernobyl. And the containment wouldn't have been able to protect from these dangers."[84]

Are we Americans ready to hear that? Dr. Chernousenko warned us all that "one more nuclear accident could destroy human civilization as we know it."[85] There are approximately 500 nuclear reactors in the world today[86], and the Bush administration has moved the goalposts toward planting more of them in civilization's backyards. Paying subsidies to an otherwise unsustainable mature industry, that can then use their $20.5 billion gift from the 2005 Energy Act, for example, to dole out money for advertising, propaganda, and political contributions to our governmental representatives to promote nuclear power, and all things nuclear. Skewing our essentially one-sided national "debate" that the media refuses to balance fairly with information like you are reading here. In effect, we are financing the nuclear establishment's deathwalk on the bones and souls of us and our innocent children with our own hard-earned tax money.

Oh, we hear that there could or will be a new generation of "inherently safe"[87] nuclear reactors. But listen to the words of the late Dr. Chernousenko, spoken to a Texas audience:

Nuclear physicist Dr. Vladimir Chernousenko, in charge of Chernobyl nuclear clean-up. Died from cancer, probably contracted during this service. Subsequent to his clean-up duties, he vehemently opposed nuclear power because of its inherent uncontrollable frighteningly powerful reactions "which occur within millionths of a second."

"To construct a safe reactor is practically impossible either here or in Russia... we simply cannot get energy from such enterprises. Because we are dealing with nuclear processes, with uncontrolled reactions, which occur within millionths of a second, and no matter what kind of protection mechanism you design, sooner or later the object must explode and they will. Why were they created at all? When they were created, constructed, it was understood that they were extremely dangerous, but at that point the physicists were told that they must save the world from Hitler at any cost and as soon as possible. And unfortunately the physicists accomplished this, which they regret to this day."[88]

One last statement from Dr. Chernousenko about Ukraine nuclear plants and the data concerning disease and cancer in their surrounding environs, that you may ponder lingeringly - - for you seldom hear about U.S. studies stated so simply and clearly:

> "We have conducted studies of the regions around 20 different nuclear plants in my country. In all of these territories we noticed an increase in the breast cancer rate— sometimes an increase of 15% over the normal level. We noticed a growth of anemia amongst children who lived in those areas, cardiovascular diseases, and cataracts. So from this you can conclude that even without the explosion of nuclear weapons there is quite a bit of danger to human lives."[89]

And just in case you think everything is under control in Moscow, twenty years after the accident, how about this report:

> "Nearly 20 years after Chernobyl, large amounts of radioactive goods are still reaching markets in Moscow from the west of the country and Byelorus. In 2005, some 830 kilograms of radioactive produce were seized by officials at markets in Russia's capital...Much of this produce consists of mushrooms and berries...all market places have a laboratory that checks goods before sale...[after] removing and treating the goods...[these] are classified as radioactive waste."[90]

Clap your hands if you think <u>ALL</u> the radioactive produce flowing into Moscow is detected as above....Then, when you realize the story is not over, you might as well know about the end of the line, or what should be the conclusion for nuclear power plants. Something very very expensive, called "decommissioning." When the utilities who own the nuclear plants have to own up to their responsibility to properly dispose of these plants that inevitably become too radioactive to continue operating.

Decommissioning or Disposing Of Over-Aged Nuclear Plants

Remember that when the question of <u>financial cost</u> of nuclear power is considered, this should include decommissioning. Plus, all the peripheral effects, including medical and community expenses of radiation pollution from the nuclear plant itself, and from mining and milling uranium; long-lived waste storage and safety; the exorbitant cost of enriching uranium with power most often supplied by coal combustion; transportation of waste, and security for waste shipments, in the time of terror paranoia; and now perhaps much much more of all of the above from Bush's proposed move toward reprocessing, as discussed earlier.

What is supposed to happen is that when their time is up, when nuclear power plants become unsafe to operate further, due to the effects of radiation bombardment and extremely intense heat on pipes, containment, metal, concrete, etc. When the erosion and radioactivity become too severe - - when their licenses *should* expire - - after about 20-30 years of operation - - there are supposed to be adequate funds, trust funds, available to pay for interring or entombing or chopping up the plants and their components, and sending them to a fool-proof nirvana where they can de-radioactivate without spoiling our land, water, or air.

Unfortunately, enforcing decommissioning has become a debacle due to the shabby integrity of our Nuclear Regulatory Commission (NRC) and the nuclear industry itself. Instead of being able to watch over, and correct any deficiencies in the corporate actions of clean-up, the NRC has effectively excluded the public's "right to review and intervene in utility processes that can amount to economic short-cuts and sloppy radiation controls resulting in excessive contamination to workers, the site, and uncontrolled releases into the environment."[91]

Besides that, projected costs themselves may be way off. You make your mess, you must put sufficient monies away to acceptably mop it up. Yankee

53

Rowe's 179 megawatt plant was supposed to cost $120 million to decommission, but actual costs of $500 million resulted. That was for a 179 megawatt plant, and the average plant produces 1000 megawatts. So, if we assume, probably incorrectly, that costs would be directly proportional for the bigger average-sized plant, we would expect close to three billion dollars to pay decommissioning costs for our common nuclear power plant. Of course, that is without complications, contaminations, unforeseen liabilities that could occur years into the hazardous lives of the radioactively tainted plant left-overs. Besides other factors that could produce inaccurate (higher) projected costs. . . .

Hear here that even for their small 179 megawatt reactor, Yankee Atomic Electric Corporation "acknowledged for the first time that they expect to raise electric rates in New England to help pay the cost of closing the reactor." This was reported in the New York Times on November 4th, 1994.[92] Imagine what this could mean for all the other reactors that have to be decommissioned, and how YOU could be suckered into paying THEIR bills, one way or another.

For example, "current bankruptcy law does not prioritize decommissioning costs above other creditors' [claims]."[93] "As much as 50% of the remaining projected final costs [for closed nuclear power plants] are left to future ratepayers or taxpayers not receiving one watt from the retired nuclear power stations."[94]

That's an edited snapshot of the money end of things. Then there is the actual level of radioactivity that would be deemed acceptable for us and the operator/owner of the nuclear power plant to cease monitoring the area and dream of its locale/remnant(s) as a "green field" where we can plant our corn and canola, and frolic in nearby waters. Again, the NRC has whirling dervishly danced away from its responsibility to set up sensible standards.

In the words of NIRS in their year 2000 decommissioning report: "The NRC has reclassified decommissioning as not constituting a major federal activity and an activity that can be conducted under the original operational license without the availability of a public hearing on any potential safety issues raised by a particular decommissioning process. Utilities are now allowed to submit vague plans without any public scrutiny of the actual chosen process."[95]

Thank you, NRC. Our watchdog of the nuclear industry. Recall that the NRC supposedly was created to replace the Atomic Energy Commission (AEC) because that agency had been behaving too much as a <u>proponent</u> of nuclear power, rather than serving the public as an impartial watchdog of the nuclear industry. Similarly, the International Atomic Energy Agency (IAEA) still acts internationally like the AEC did in the USA, pre-NRC. Remember that, as you read the news every day. And hear about the IAEA and its Director General Mohamed ElBaradei winning the Nobel <u>Peace</u> Prize, and censuring Iran, but still facilitating the multiplication of more nuclear power plants and thus, the threat of nuclear terrorism and weapons-spread worldwide.

These days we are concerning ourselves especially with Iran and North Korea, having the nerve to want to enrich their own uranium or develop a nuclear bomb. Does anyone ask why, if one country can have a nuclear bomb, or a nuclear power plant that can be used to produce <u>plutonium</u>, which in turn can be extracted from the plant's wastes, why can't another country do the same? Or, even better, why haven't more countries developed a nuclear bomb, besides the nine currently enrolled in the nuclear weapons brotherhood?

Maybe it is because of the Nuclear Non-Proliferation Treaty (NPT)? That 188 nations have signed onto. That went into force in 1970, and as a signed treaty is part of U.S. law. The main stipulation in the treaty is Article 6, which "commits the nuclear weapon states to good faith negotiations on nuclear disarmament in exchange for the promise by the nonnuclear weapon states not to acquire weapons."[96]

North Korea had signed onto the treaty, but withdrew in 2003. <u>Five</u> of the other nine nuclear weapons states remain signed onto the NPT, but are not following their commitment to reduce their arsenals. Not aided by the Bush administration's obstinate lead to rev up another nuclear arms race, like we had during the Cold War, before the USSR disintegrated in 1989.

Those five nations, you should know, are the USA, Russia, China, the United Kingdom, and France. Then there are the three nations that have never signed onto the NPT: India, Pakistan and Israel. These three states do not even have to submit to periodic inspections of nuclear facilities, as do the other 188 countries[97], which indeed includes Iran.

Who knows how many nuclear weapons states

there would be if not for the NPT? 25? 30? 40? Imagine what a nuclearly-rampant world we would have panicking us then! And recall that as blustery and foolish as the Bush administration - - OUR USA GOVERNMENT - - is behaving today, both George Bush and John Kerry agreed during their 2004 campaigning and debates that the biggest problem the world faces today is.......nuclear proliferation.

Of great concern for our country and the world is something that most Americans do not know enough about: the recent influx of profiteering in funding our nuclear ambitions. In addition to asking "Congress for $27 million to help jumpstart the country's first new nuclear weapons program in two decades...the money...[to]...be used to fund a competition between the Los Alamos and the Lawrence Livermore laboratories to fund and design a new generation of nuclear bombs to replace the country's entire nuclear arsenal,"[98] in blatant violation of the NPT, be aware that the Los Alamos National Laboratory in New Mexico is now being run, starting in 2006, on a 20-year no-bid contract netting about $2.2 BILLION per year. Plus the awarded consortium of companies involved, including Bechtel and Washington Group International, along with the University of California, also is able to garner extra management awards of up to $1.5 BILLION per year via Los Alamos.

Greg Mello of the Los Alamos Study Group of New Mexico. Concerned with the unprecedented increase in the profit motive in USA nuclear weapons policy. If New Mexico seceded from the USA, it would be the world's third greatest nuclear-armed nation with approximately 2,500 atomic bombs.

"We've never seen this kind of profit motive in the nuclear weapons business up to now,"[99] Greg Mello, of the Los Alamos Study Group, tells us. He adds that the consortium members also "get an entree or a leg up in the nuclear power business, which they expect to be growing."[100]

Do you think this amplified profiteering policy will lead our country in the right direction? Have we forgotten about the destructive power of nuclear weapons, and the deliberate, dedicated process taken over the past sixty years to prevent any nuclear holocaust from happening? Does anyone consider the word PEACE for our world and our children anymore?

Is it really as Greg Mello says, that we want our nuclear weapons to "continue to evolve and remain a central part of the goal of so-called <u>full spectrum dominance</u>,"[101] as discussed in the Star Wars chapter (Three) ahead?[102] "Full spectrum dominance" meaning dominating the full spectrum of the war fields of air, water, ground, and now <u>space</u>, utilizing nuclear weapons, or the threat of using nuclear weapons, instead of adhering to the NPT and similar treaties to reduce the chance of nuclear destruction? ? (See images from Vision For 2020 on pages v, and xviii.)

Do we see that the nuclear industry is indeed one big happy family, producing lifetimes of toxicity, threatening apocalypse, because of how it has evolved? First the weapons, then the bright idea of nuclear power, now and forever the ultimate radioactive wastes to deal with, that our entrepreneuring minds are channelling into smoke detectors (americium), airplane ballast ('depleted' uranium), tank 'penetrators' and shielding ('depleted' uranium - - that turns out to be excessively flammable, so that <u>both</u> our tanks' <u>and</u> the enemies' tanks' shielding can catch fire, incinerating our sacrificial soldiers and theirs - - having a half-life of 4 BILLION YEARS! too much of it strewn across the battlefields of Iraq, the Balkans, and recently Lebanon, to cancerize and kill present and future citizens of those areas) - - watches (tritium), food irradiation plants (using cobalt and cesium, generated in nuclear reactors - - the latter, cesium, being the major USA byproduct of nuclear fission/nuclear power, by volume; see Chapter Two), medicine (very short-lived radio-isotopes, which the nuclear lobbyists deceivingly present prominently as a portion of nuclear waste, though they are a minute percentage, and after rapidly decaying soon become the least toxic portion, most only being radioactively hazardous for a few days, compared to reactor and weapons' extremely long-lived radionuclides).

Where there is profit, there may not be conscience.

Front and center now, the G-8 countries and ex-KGB chief, Russian Demagogue Putin! Just in case you missed it, during the summer of 2006, the G-8 countries[103] decided in St. Petersburg, under Mr. Putin's leadership, that pushing nuclear power worldwide would be a great idea. Yes. But non-G8 countries "would not be allowed to enrich uranium fuel, or to reprocess spent fuel to extract plutonium."[104] The excuse was security concerns.

Be aware that part of the NPT deal for countries abstaining from developing nuclear weapons is the guaranteed "access to nuclear technology for peaceful purposes."[105] Sorry suckers, though you may wish to be treated equally on this unfair planet, while you "will be permitted to run [nuclear] reactors to generate electricity... [your second-rate country] will have to buy fuel enrichment and reprocessing services from G8 countries."[106] That's the way it's going to work.

And us haughty arrogant G-8 nations declare that we shall base our self-sustaining nuclear power dream/nightmare on reprocessing nuclear waste within our respective countries, our citizens be damned re the pollution and cancer, etc., that results; plus we shalt shine up and polish the defamed image of 'fast breeder reactors' to burn and produce plutonium! HARK AND BOW TO OUR BENEVOLENT POWER!

Mom, weren't we worried about nuclear proliferation, madmen (and madwomen) and terrorists getting their hands on plutonium?...Yet we want to produce as much of it as possible with one of the most dangerous technologies ever conceived, that has all but been abandoned for at least a decade? Sodium fires, the spectre of an explosion in a nuclear plant FUELED with plutonium?!! [which is how a fast breeder reactor might work] Are our leaders nuts?

Remember henceforth that Dr. Vladimir Chernousenko warned us that "business people and the military are behind the building of nuclear plants."[107] Regardless of what crazy kind they decide will meet their profit or destructive goals.

But, to top it all, Vladimir Putin, George Bush's pal, and winner of Time Magazine's 'Man Of The Year' for 2007, "has a plan for mass producing reactors, installing them on barges and selling them around the world as floating nuclear power plants!"[108]

BRILLIANT!!

Please let me have a Guinness to soften my brain further, and make this nightmare unreal, somehow. . . might I just drift away into the slithery mist?. . . and accidentally bonk into one of these Putinated mobile Chernobyls!? . . . . .

Ex-Russian President and ex-KGB chief Vladimir Putin, who wants to mass produce floating nuclear reactors to be deployed all about the world.

. . . I thought that was the worst foolish arrogantly concocted idea that could possibly actually occur, until I just read about the Toshiba mini-nuclear plants that this company wants to sell everywhere and anywhere. That will be fully automated, and need no monitoring, and maybe even no guarding. . . ? . . . Room sized, the contraptions could be buried in the ground, and electrify 'an apartment complex for the rich' with 'steady power' "for up to 40 years."[109]

For the theoretical bargain price of perhaps $3.5 million you could have a 200 kilowatt reactor right beside your house or business. That's equal to about 1/5000th the power of a typical 1000 megawatt commercial nuclear reactor. If Toshiba can have its way, these devices will crop up all over the world, for all sorts of dream profits.

Alas, they will still be powered by the fissioning of uranium, produce the same ultimately toxic, greater-than-500-in-number radionuclides, that our supposedly closely guarded, monitored 104 commercial nuclear plants produce. Could these weenier nuclear reactors leak? Or explode? And contaminate a city like yours forever?

Of course!

Toshiba initially optimistically fantasized that their mini-nukes would be first sold to a buyer in Japan in 2008. However, latest news is that the things are not quite ready. But by 2010, WATCH OUT![110] Some rich unconscious individual or corporation might try to site Toshiba's mini-nuclear reactor too close to you or someone you love.

You might want to know that Toshiba bought up what is now *its* Westinghouse nuclear-plant-making subsidiary from British Nuclear Fuels, Ltd., in October 2006 'for about 5.4 billion dollars.'[111] Westinghouse and General Electric have historically been the two main USA producers of nuclear power plants.

According to Ace Hoffman, outside of its laptop computers, camcorders, telephone systems, DVD players, etc., 'nuclear reactors and equipment for those reactors (and for other reactors) accounts for about 25% of Toshiba's business.'[112] He recommends boycotting buying all Toshiba products to send this company a bottom-line survival-oriented message.

As ridiculous as this all probably sounds to many of you, there also has been some talk festering about 110 megawatt Pebble Bed nuclear reactors, which, naturally, are claimed to be 'inher-

Hamm Pebble Bed Reactor, in Germany. The four year most prominent experiment at deploying this type of nuclear plant ended in disgrace and contamination after it was discovered that a radiation release blamed on the Chernobyl accident actually came from the Hamm Pebble Bed. Top inset of fuel balls or 'pebbles,' each ~ size of a lemon - 320,000 bobbing around reactor at any 1 time.

ently safe,' and need no evacuation zone around them - - just in case there *could* be an accident. . .

These gum-ball-machine type of reactors have been around for a few decades, and have not proven to be either safe or economically viable. There's a seven page section on them in the Appendix. The most prominent attempt at running these reactors occurred over a four year failure-of-a-period in Hamm, Germany that was terminated in 1988.

As with so many nuclear calamities, when the plant accidentally [?] released a serious amount of radiation at the same time that the Chernobyl accident occurred over in the Ukraine in 1986, the owners of

the plant lied about what happened. Instead of own-
ing up about the magnitude of their own release, they
claimed whatever was being measured by various
agencies and individuals all came from Chernobyl's
radioactive drift and fallout. When it was discovered
that 70% of the radioactivity in question about the
area indeed <u>did</u> come from the Hamm Pebble Bed,
the local citizenry became incensed. Two years later, the reactor was closed forever.[113]

By the way, weren't we frightened about the
danger of 'dirty bombs?' Which are conventional
explosive devices having nuclear materials imbedded in them. That can then contaminate our
cities for hundreds and thousands of years, with
just one bomb's ignition. Build those Toshiba mini-
reactors and floating Mobile Chernobyls! Those
killing reprocessing plants, and fast breeders IM-
MEDIATELY! More nuclear material to poison
us, more plutonium and cesium and strontium
to be made available, that shifty shysters can sell
on the black market in G-8 nations, like Russia,
or maybe even in the old USA....

Just BRILLIANT!!

And why not nuclear bunker busters that we
can dare to USE finally against non-nuclear nations!? Initially we were bold enough to concoct
them to be 1/5th the strength of the atomic bombs

dropped on Hiroshima and Nagasaki. But why stop there?! We were about to make them any strength we wished!

Who cared that the fallout from them can drift all around the planet, coming to rest in our food and our bodies, killing innocent people, including even us Americans? [So far, funding for these bunker busters is being denied.]

And we can save the world with nuclear power because it will reduce global warming. That is the latest faux-pas reason to re-ignite the nuclear industry. The Bush administration has been shying away from accepting that global warming exists. But just in case the media finally airs a fair debate of the possibility. . . Al Gore DID win an Oscar from Hollywood for his movie about it with all those facts and graphs - - we have the same media helping us omit the adverse killing health effects of nuclear power and nuclear waste. The frightening un-greenness of it is not being reported.

Could Americans know that our uranium "enrichment facility at Paducah, Kentucky requires the electrical output of two 1000-megawatt coal-fired plants, which emit large quantities of carbon dioxide, the gas [purportedly] responsible for 50 per cent of global warming."

"Also, this enrichment facility and another at Portsmouth, Ohio, release from leaky pipes 93 per cent of the chlorofluorocarbon gas (CFC's) emitted yearly in the U.S. The production and release of CFC gas is now banned internationally by the Montreal Protocol because it is the main culprit responsible for stratospheric ozone depletion. But CFC is also a global warmer, 10,000 to 20,000 times more potent than carbon dioxide."[114]

"In fact, the nuclear fuel cycle utilises large quantities of fossil fuel at all of its stages, including the mining and milling of uranium."[115] And what about all the carbon dioxide in the hot exhaust of those thousands upon thousands of trucks, boats, and trains that will be transporting those massive amounts of nuclear waste cross country to maybe Yucca Mountain, in Nevada, or perhaps some other inad-

U.S. uranium enrichment facility at Paducah, Kentucky. Requires two 1000 megawatt coal plants to supply its electricity, producing large quantities of carbon dioxide, our most popularized global warming gas. Paducah and Portsmouth, Ohio plant together also release 93% of USA CFC gases via leakage – CFC's 10 to 20 thousand times more potent as global warmer than carbon dioxide!

equate storage center, for 20,000 to 70,000 shipments, if we DON'T increase our number of nuclear reactors? The number of shipments, and amount of global warming exhaust, being even greater if we build more nuclear plants.

And what about the energy needed to store the radioactive waste for hundreds of THOUSANDS of years, wherever it may be placed?

About 35% of our nuclear plants pull more than a BILLION gallons of water from some river, ocean or sound into their 'once-through cooling systems' every day, and send the same amount of water back out. But this discharged water often is up to 25 degrees hotter than when it came into the reactor.[116] How does this affect the lives of the fish, plants, and people that depend on this body of water? What if this water somehow is contaminated with tritium or any of the 500-plus radionuclides that might leak into the plant's vessels or pipes? What about the contamination from venting plant vapors into the air? Why aren't these outgoing waters and ventings monitored precisely the way they should be, for each radionuclide possible, every single day, in a proper accurate reproducible manner?

Reportedly, there is an operator-dependent

non-standardized computerized detection-print-out produced at many of our nuclear plants on varied schedules to show what radionuclide concentrations should be flowing out into the nuclear plants' surrounding environments. Although it is known that this is done in a debateably reliable manner all too often, or not done properly all the time, the NRC allows such computer 'monitoring' to continue as if it were entirely acceptable.[117]

Is it just too 'impractical' to actually accurately measure the level of each and every radionuclide being discharged from a nuclear plant?

But isn't that what should be done if we want

to do things right? Just because it's nuclear power, and it's so expensive and complicated because of the vast possibilities of contamination, is that an excuse to turn a blind eye, and get a cancered lung or a stillborn baby?

As a physician, I cannot accept ignoring this technology's toxic effects. There are much more radioactive wastes in the poorly secured cooling ponds beside, and ABOVE, the containment structures of our nuclear reactors, than are inside their containment structures. 35% of our reactors are of the boiling water variety, with only sheet metal actually covering the pools situated six to ten stories above ground level (see image on opposite page) - while the pools' walls have a 'blow-out' design in case of increased pressure. What if a terrorist attempts to crash a Cessna into the reactor roof, or an accident occurs that could contaminate vast areas of our nation for incomprehensible periods of time?

Yet there are radioactive spills and leaks like those that poisoned Godley, Illinois' waters occurring at way too many of our other nuclear facilities, all across our great country. Just as it happened in Godley though, how many of those are ever reported? or reported honestly and faithfully to the surrounding community? Godley's parent nuclear corporation, Exelon, did not admit publicly to spills occurring as far back as 1996 until 2005! Perhaps that is why our most nuclear state, Illinois, with its eleven nuclear power plants - - equal to a little more than half the nuclear power capacity of the entire United Kingdom [with 20 nuclear plants] - - passed a Nuclear Release Notice Act in 2006. The very first legis-

Cut-away diagram of a boiling water nuclear reactor (35% of USA reactors). Note containment around functional core/reactor essentials > > center/lower left; and barge-like structure which is the cooling pond, top right, under only sheet metal roof, very vulnerable to plane crash, bazooka/missile attack, or steam explosion/accident of core itself.

lation of its kind in the USA, the act mandates reporting to the Illinois Environmental Protection Agency (IEPA) and Emergency Management Agency (IEMA) "the detection and reporting of unpermitted releases of "radionuclides" (instead of "contaminants" including radionuclides) into groundwater, surface water, or soil at nuclear power plants...Requires the owners of a nuclear power plant to notify the (IEPA) and the (IEMA) within 24 hours of an unpermitted release. Provides that the quarterly inspections shall be by both IEPA and IEMA."[118]

According to various sources in Illinois, Exelon may have reported a few "unplanned releases" of "contaminants" to the NRC in a very untimely manner over the past several years, and certainly not within 24 hours. Yet the NRC did not do its duty to promptly notify the citizenry immediately thereafter. Thus, the state of Illinois took matters into its own hands and produced this very important landmark legislation.

To take the impetus for better protection against radioactive contamination one step further, Will County (where Godley is located) Board Chairman Jim Moustis requested "a review of all laws governing the Nuclear Regulatory Commission."[119]

Presidential candidate Barack Obama, who happens to be the junior Senator from Illinois, co-sponsored a similar Nuclear Release Notice bill in the U.S. Senate to require immediate reporting of "unplanned releases of fission products and radioactive substances...to the NRC and the State and county in which the facility is located."[120] However, politics seeped into the attempts at passing the Senate bill. Politics and money.

Exelon Corporation, the USA's 'largest nuclear plant operator...based in Illinois,' indeed the operator of the Braidwood reactor in Godley, via its executives and employees, has given at least $227,000 to the Obama presidential campaign. 'Two top Exelon officials, Frank M. Clark, executive vice president, and John W. Rogers Jr., a director, are among his [Barack Obama's] largest fund raisers.' Exelon's chairman, John W. Rowe, 'another Obama donor...also is chairman of the Nuclear Energy Institute (NEI), the nuclear power industry's lobbying group, based in Washington.' Plus, David Axelrod, 'Obama's chief political strategist...has worked as a consultant to Exelon.'

Would it be any wonder then that the Senate bill eventually became watered down to only require voluntary reporting of leaks? '"Senator Obama's staff was sending us copies of the bill to review, and we could see it weakening with each successive draft," said Joe Cosgrove, a park district director in Will County, Ill.' reported the New York Times on its front page on February 3, 2008.[121]

So, for 2008, the Senate bill for immediate reporting of nuclear 'releases' is dead. Both Hillary Clinton and Barack Obama say nuclear power has to be part of the overall energy solution that our country has to create for the future. And John McCain, the Republican presidential candidate with the _zero_ environmental score from the League of Conservation Voters, is totally gung-ho for nuclear power. He thinks the USA should go 80% nuclear power, like the French[122], to generate our electricity. Oh, he also favors reprocessing, like the French.

But does he, or the Democratic candidates, know that 'less than 31 percent of the French public favor nuclear energy as a response to to-day's energy crisis. 54 percent are now opposed to investing 3 billion euros in the construction of a new reactor, while 84 percent favor the development of renewable energy'...?[123]

Besides all the usual worries about still being unable to safely dispose of radioactive wastes, and reprocessing wastes having 'a greater radio-activity per unit mass'[124] as compared to so-called 'depleted uranium' that results from uranium enrichment, perhaps the French are very aware that 'the only nuclear plant being built in the West that is well along in its construction' is being done so with 'a turnkey contract with Finland by AREVA.' AREVA is a nuclear company 85% owned by the French government which has agreed 'to absorb all costs more than 3.2 billion euros'[125] accumulating for the Finnish project.

In plain French (translated into English for

you) that means that the current cost of the Finnish reactor that 'has now escalated to 4.5 billion euros'[126] will be at least 1.3 billion euros more than the turnkey contract limit, and the French taxpayers will have to pick up most of the bill for the excess cost overrun, which may continue escalating before the reactor might be completed.

Even if the USA makes the foolish move to Go Nuclear! building as many new nuclear plants as possible, the first one won't go online until 2015. By then it has been estimated solar energy could be available at 5-10 cents per kilowatt hour. And the real cost of nuclear energy, borrowed or subsidized, would still be between 14 and 19.75 cents per kilowatt hour.[127]

For the suddenly fawning Francophile who wants to power the world with nuclear, despite its health and anti-economic effects, please read what renowned engineer and nuclear expert Arjun Makhijani states in his 'France's Nuclear Fix?' article published in January 2008:

'The French model of imposing added costs on its ratepayers and taxpayers, of polluting the oceans in the face of protests from neighboring governments, and of accumulating vast amounts of domestic and foreign surplus plutonium hardly seems like a model for the United States or anyone else to follow.'[128]

Though we may be lost in our own nightmare on this side of the Atlantic, thanks to our past three decades of energy misguidance, Americans should know that the European Union (EU) expects 20 percent of its electricity to be generated by renewable energy by 2020. This does not include nuclear power. Nuclear power is <u>not</u> a source of renewable energy. Not in the EU's eyes, nor in the eyes of anyone who realizes uranium is not a renewable resource.

Spain has advanced so far as to hope 'to generate almost 30 percent' of its electricity from renewable sources by 2010. Although they are number two in Europe (behind Germany) with over 15,000 megawatts of windpower on line[129], Spain is 'also pushing solar energy, removing obstacles to connecting renewables to the electricity grid [created] three years ago and recently requiring all new and renovated buildings to use solar power for part of their energy.'[130]

From various angles then, we know that solar and wind technologies especially can indeed supplant nuclear power in our energy blueprint for the very immediate future. Especially if we smartly generously spend our government funds to assist their rapid development, rather than wasting it on the nuclear option.

We should be watchful that we are not hood-winked to become the world's dumping ground for nuclear waste. Thanks to the Bush, and perhaps McCain, administration(s)'s nuclear manipulations, we could undergo a 'deja-vu all over again' Yogi Berra experience that almost happened in the early 1990's. That was when 'our' NRC facilitated designating so-called 'low level radioactive wastes' as '<u>B</u>elow <u>R</u>egulatory <u>C</u>oncern' (BRC) so they could be discharged into our community dumps <u>unmonitored</u>. This would have included radioactive pipings, resins, liquids, etc., from nuclear power plants contaminated with untabulated radionuclides – which could have meant any and all of those 500-plus radioactive elements from plutonium to strontium in whatever amounts we would have never known that were generated in the fission of uranium.

However, citizen outrage, thanks to national media debate on this issue, soundly defeated the measure back then. Alas, as Yogi Berra also says: "It's never over 'til it's over." See Chapter Five to discover the latest attempted travesty to discard radioactive wastes improperly, unmonitored, once again. The most outrageous volley apparently to be 20,000 tons of Italy's radioactive waste

- - to arrive by ship from a country that closed off its nuclear power program in 1990, following the Chernobyl accident. (Plus perhaps additional nuclear waste may be coming from England, negotiated by the EnergySolutions corporation.)

Before we fly into the skies to unmask the deception of 'missile defense,' which really is a Trojan Horse for missile <u>offense</u>, and the introduction of nuclear weaponry into space - - which violates many treaties near and dear to those who wish mankind to survive a bit longer on our humble planet - - let's now go to the biggest chapter in the book. On your food, what is good, what is bad, how our corporations want to adulterate it and irradiate it, plus how you can safely feed your family in this time of corporate control of our diet and our consciousness.

Genes, Chromosomes and DNA - - a brief explanation

'DNA' is short for a substance called de-oxy-ribonucleic-acid. 'DNA is wrapped together to form structures called chromosomes.' 'Genes are sections or segments of DNA that are carried on our chromosomes and determine specific human characteristics' like hair color or height. Each of us has 25,000 to 35,000 genes which can combine to form all sorts of possible characteristics.

Most cells in our human bodies have 23 pairs of chromosomes, with a total of 46 per cell. But our 'individual sperm and egg cells have just 23 unpaired chromosomes.' We receive half of our chromosomes from our mother's egg cell and the other half from our father's sperm cell. Information quoted above, and more, available at:

http://kidshealth.org/teen/your_body/ health_basics/genes_genetic_disorders.html

Chapter Two:
Your Food: Mutated, Irradiated, or Pragmatically Pure?

'Here comes synthetic food
And their big time money
And they want to control
Our body and soul
        –Ziggy Marley, from 'Tumblin' Down' [1]

A mericans can do anything! With our aptitude to harness and invent supermodern technologies, we can charge across the Earth and its oceans and skies to conquer any foe -- whether it be any country we deem to perceive as an 'enemy,' or even Mother Nature herself (or so many of us seem to think). Yet we still have to feed ourselves without losing the war on cancer, or poisoning ourselves and our fellow inhabitants of this planet in the process.

Because we want to laugh in the midst of our media-amplified paranoia post September 11, 2001, while currently being bogged down in the Vietnam-like quagmire of Iraq, many of us prefer watching 'American Idol' or 'Survivor – in Fiji,' instead of hearing about the very serious dangers we are building into our aggressive food policies under Washington's guidance (?misguidance?).

Although, Yes, most of us <u>do</u> want to have pure untainted food to eat, to give to our children, supplied and blessed by us supposedly wiser adults, that is not riddled with mad cow prions or pesticides or unforeseen complications that we could have discovered before we rushed that adulterated manna through the supermarket checkout line.....

Incidentally, our safety net of organic foods is growing at a 15-20 percent rate every year. Dairy foods and milk lead the way, skyrocketing upwards at a 132% rate of sale annually, on average. By the year 2020, our prophets foresee <u>half</u> of our food in the ole US of A could be produced organically. New Zealand is aiming to have all 100 percent of its food produced organically by that same year of 2020!

But people, and corporations, have to make money. To drive the economy. Better living through chemistry. Meat for the multitudes. Perfectly shaped french fries for the fast food fiscalities.

Of course, we got more milk! We also got recombinant Bovine Growth Hormone (rBGH) being shot up into 18 percent of our dairy cows every two weeks[2]. This produces up to ten times more insulin-like growth factor ('IGF-1') in those poor cows' milk. That our kids drink, and we use in our coffee, and whipped cream, and American-made cheeses, and heart healthy oatmeal, and other processed foods.

No, IGF-1 should not appear in such concentrations in <u>organic</u> milk and other <u>organic</u> dairy products - - which should not allow the use of rBGH.

How come our researchers have found that this increased amount of IGF-1 is linked to breast, colon, and prostate cancer, yet most of us Americans still don't know about it?

How come, of all the industrialized countries on Earth, only three: the <u>USA</u>, Mexico and Brazil, do <u>NOT</u> illegalize the inhumane use of this genetically altered hormone?

We belittle the French and the Europeans, but most Americans are not fully aware of the risks we are running as we try to force a dangerous genetic roulette upon the protoplasm of life. In fact, studies by Rutgers University and the Pew Institute in latter <u>2003</u> found that only <u>25%</u> of those

Americans interviewed thought they had <u>ever</u> eaten any genetically altered foods or additives! And that's with (today, in 2008) more than 90% of our soybeans being genetically altered[a], and 60% of our canola, 73% of our corn, and 83% of our cotton (that you eat as part of your inorganic potato chips, packaged nuts, bakery goods, etc., in the form of 'cottonseed oil') being genetically altered[3]. Plus, about 80% of our processed foods have genetically altered ingredients in them[4], but you'd never know because these do not have to be labelled as such.

Then there is the next threat that is supposed to be inevitable: half of all USA non-organic sugar could come from genetically altered sugar beets by the end of 2008. This will happen if genetically altered sugar beets can be decertified so they can become deregulated by our USDA (United States Department of Agriculture), and our food corporations do not reject the sugar coming from these novel patented plants. Unfortunately, you will probably not know which of your cookies or sauces contain this form of sugar, because it will not have to be labelled as genetically altered or

a 'Poll: Many Won't buy Genetically Modified Food,' U.S. & World News, May 11, 2008, http://cbs4.com/national/CBS.News.New.2.721469.html

GMO either.[b]   Not yet. (See two one-page letters from Kellogg's and the Organic Consumers Association in the Appendix, and the one-page pamphlet in the Introduction.)

What's the danger? What's the big worry? A study by Pusztai and Ewen, published in England's prestigious <u>Lancet</u> medical magazine in 1999 found that rats fed genetically altered potatoes developed increased size and proliferation of their intestinal cells, which could lead to cancerous growth. In addition, there appeared unexpected malfunction of the immune system, and changes in sizes of vital organs, including smaller brains, testicles, and livers, which did not occur in 'control' rats fed non-genetically altered potatoes.[5]

Another study found that the amplifying *Cauliflower Mosaic Virus* (CaMV) promoter- - used in the above genetically altered potatoes to promote the expression of desired implanted trait(s), and the most commonly used 'promoter' in the gene-alteration business:

> "has the potential to reactivate dormant viruses or
> [create] new viruses in all species to which it is trans-

[b] ALERT: Genetically Engineered Sugar to Hit U.S. in 2008,' http://salsa. democracyinaction.org/o/642/campaign.jsp?campaign_KEY=12700 See the full page pamphlet to take action against this imminent travesty in the Introduction.

ferred. In addition, because the CaMV promoter is promiscuous in function [efficiently functioning/infect-
ing plants and bacteria - - like yeasts, green algae and
E.Coli], it has the possibility of promoting inappropriate
over-expression of genes in all species to which it happens to be transferred. One consequence of such inappropriate over-expression of genes may be cancer."[6]

Should this alarm you? Why didn't you hear
about this in America, while this fueled a furor
in the United Kingdom? It is part of the reason
a book like this had to be written by a medical
doctor.

By the way, our government does have a role
to play here, via our agencies that are supposed
to be watching over our national health, rather
than advocating serendipity and overenthusiastic
approval of things like genetically altered fruits
and vegetables in the face of corporate moxie.

Monsanto, the company that gave you Agent
Orange and PCB's, and wanted to give you the
plutonium-powered coffeepot that wouldn't have
to be refueled for 100 years,[7] tells us that they
"don't have to vouchsafe the safety of biotech
food. Our interest is selling as much of it as possible."[8]

Based in St. Louis, Missouri, Monsanto produces 90 percent of the genetically altered crops sold worldwide.[9]

This company has massive influence over our government's policies toward food, and those who stand against these policies. Here's one interesting tidbit to ponder and masticate: whom do you think received more money than any Congressional candidate in the year 2000 round of elections from Monsanto? and also was "a leading advocate of policies to force Europe to accept genetically engineered foods?"[10]

Answer: your former Attorney General, John Ashcroft (when he ran for Senator of Missouri, and lost, - - "to a dead man," as Michael Moore reminded us). Remember this when protestors are batoned and teargassed at Free Trade Area of the America meetings, where genetically altered foods, and 'terminator' seeds that do not regenerate, and hormone-affected milk and beef, are promoted for trade acceptance by faceless ministers behind closed doors.

Also remember that with the developing miracle cornucopia of synthetically patentable food products being developed for 'the market,' the push has been for 'no safety testing, no labelling.' Because that would be like stamping a skull and crossbones on an entire industry, if we dared to do that.

Just like clearly labelling food that is irradiated, with cobalt or cesium or e-beams or x-rays.

This is America! Who is more important: the individual citizen/customer/potential canceree or the campaign contributing corporation? (Well, James Randall, ex-President of Archer-Daniels-Midland, or ADM, did say: "Our competitors are our friends. Our customers are the enemy."[1112])

Don't you want to eat food that is 'cold pasteurized?'

Sounds eerie, doesn't it? Maybe like how that smoking Smirnoff alcohol product could be processed, that you have to pierce your tires for, so you won't drive drunk in your GTO later....?... when the midnight Limp Bizkit concert starts....

Except 'cold pasteurized' actually means 'irradiated,' thanks to Iowa Senator Tom Harkin's overseeing the introduction of a few paragraphs into the Farm Bill of 2002[13]. Irradiated food isn't so good for you, especially when it's hit with 300,000 rads of ionizing radiation when a mere 500 rads can be fatal to a human being. Maybe the E. Coli 0157:H7 might die, but then there are all those unstudied radiolytic byproducts of irradiation that remain in the food that no one wants to talk about, or responsibly study. Especially here in America. The cyclobutanones, and the benzenes, and the formaldehydes that preserve Frankenstein's brain in that slimy bottle.......

Oops, those cyclobutanones have been discovered to 'promote cancer growth' by German scientists.[14] In fact, we now know that they are 'markers' for foods that have been irradiated because they last so long in them, and do not occur naturally anywhere on Earth.[15] Isn't it terrific that our own schoolchildren will now be the guinea pigs for 'cold pasteurized' hamburgers served up unlabelled, requiring no announcement in school cafeterias all over America?[16]

(But not in the Los Angeles Unified School District, serving 700,000 students in its 677 schools. A unanimous vote of the board in mid-2003 forbid the buying of 'irradiated meat from the U.S. Department of Agriculture's (USDA) National School Lunch Program...Calling it "ludicrous"'[17] to subject children to 'irradiated food when the long-term health effects are unknown.'[18] The Program countrywide involves 27 million schoolchildren. 93 percent of comments invited by the USDA opposed the inclusion of irradiated meat in children's lunches. Yet the Bush administration ignored these comments.[19])

Well, why bother to eat meat anyway. Those mad cow prions that even food irradiation can't kill, keep popping up in cattle herds south of the Canadian border lately. But, at least fish are safe. Especially those that live up in those idyllic isolated mountain fish farm ponds. And fish are supposed to be good for your heart, with all their cardio-protective omega-3 fatty acids!

Wonderful! But once again, man marches forward too fast. Although 'aquaculture,' with its fish farms, is the wave of the future, and is the quickest growing sector of our agricultural industry, say it ain't Sosa, Sammy, but the cork is: those fish farms aren't all perfectly pure and pristine far beyond the pollution of the world we have to live in way below, over here.

For example, the Atlantic salmon that we actually farm mostly in overcrowded pens along our coastlines, too often receive antibiotics, fungicides, and colorants to keep their meat pink rather than gray. Not very organically produced then, unfortunately. Does it have to be that way? Perhaps, when you pack 50,000 fish into an unnaturally confined environment, and their excrement and undigested chemicals drop into the ocean below, it does. This led the ever-business-first Bush administration federal bureaucrats to blueprint the SELLING/ PRIVATIZATION OF THE OCEAN!! to circumvent local and state objections/legislation concerning the alarming damage these fish farms are wreaking upon our waters and coastline.[20]

Fortunately, this privatization/sale did not happen, because there came the Offshore Aquaculture Bill of 2004. Now aquaculture entrepreneurs can <u>lease</u> a piece of ocean somewhere beyond state and local scrutiny, three to 200 miles from the shoreline. And it will be easier to introduce the genetically altered salmon to American waters, far beyond adequate oversight of those who would protest against such an action. (See more about this bill in the section in this chapter about Aquaculture.)

Where is the wisdom and logic we need to craft a world that we can proudly hand over to those who follow us as the Earth's next generations? Americans are ambitious and ingenious, especially when it comes to making money, and getting things done. (Well, OK, many of us *are* satisfied and passive, ready to live in peaceful communion with our lives and our surroundings, though you might not know it from TV and movies.)

Why can't we also be patiently brilliant in choosing our alternatives, especially when ridiculous scams that could heftily wound our precarious only-planet, could be methodically considered, and rejected, if necessary?

Do we need more Starlink corn getting into our food chain, causing possibly fatal allergic reactions amongst our own people, while threatening our economic well-being when countries like Japan shockingly receive contaminated shipments that they angrily reject, when we could be naturally raising organic cobs of maize instead?

Is it smart or hare-brained, granting the Monsantos and Syngentas and Aventises international domain over seeds and life-forms that they want to patent exclusively so they can then devise soybeans or wheat that will match their <u>pesticide/herbicide</u> polymers for increased sales of <u>BOTH</u> entities, crop and pesticide/herbicide, as opposed to their mantra that they want to feed the hungry and/or improve crop yields?

Look out for the Aquabounty 'transgenic' salmon coming into your dining room. Will it be better than the wild variety? Will its altered protoplasm, or the chemicals used to treat it, cause untoward reactions in us consumers? Will it be labelled as 'genetically altered?' Will inevitable escapes from supposedly 'secure' pens lead to extinction of wild species that have taken 10,000 years to develop into what they are today?

Be thankful that one of your basic grains amped to appear on your table in genetically altered form: your bread, from your wheat, managed to avoid the Gene Giant corporations' assault on our marketplace. 'Monsanto, the world's biggest seller of genetically modified (GM) seeds, has announced that it is stopping all further efforts to commercialise its controversial GM wheat,'[21] announced Friends of the Earth on May 11, 2004. They said this only happened because 'virtually every major wheat-user in the world rejected this product before it was even allowed on the market.'[22] Yet our biotech-friendly regulatory agencies had already approved the altered wheat for USA nationwide sowing, rearing and sale.

Ann Veneman, our ex-Secretary of Agriculture, formerly was a board member of Calgene, maker of the ill-fated Flavr Savr tomato, the first biotech food to be marketed. Even though 7 of the 40 test rats force-fed the altered tomatoes died in two weeks[23], Ms. Veneman enthusiastically still told us: "We simply will not be able to feed the world without biotechnology."[24] Our new Biotech Advisory Panel which she appointed in April 2003 to tell her what she and the industry wanted to hear, was noted to be disproportionately weighted with agribusiness/Gene Giant companies.[25]

Meanwhile, be aware of the Organization for Economic Co-operation and Development (OECD) report *Agricultural Outlook 2004 to 2013* informing us that "The growth rate in the production of most agricultural products is expected to be <u>larger</u> than that for consumption, leading to a continuation of the long-term decline in real prices [adjusted for inflation]." In other words, for the next half-decade we'll have <u>more</u> food than we need on Earth. <u>Distribution</u> of this excess food to the hungry 800 million people who can't seem to get their hands on it is the actual problem.

Ann Veneman, ex-Secretary of Agriculture, Biotech True Believer, as are many members of the Bush administration.

Might all this be enough to give you a nightmare? Or

a daymare? just while you're nodding off, feeding your fragile little baby her bottle of organic rBGH-free milk.......A snowstorm blowing across the valley outside your windows. The color of the flakes *yellow*. Bees and birds swerving suddenly into view, veering in all directions, seeming confused, disappearing, swallowed by the raging blizzard...

Strange sorts of monster plants loom up through the haze, branches all craggy and misshapen, with funny iridescent green and pink and neon blue kernels visible between the gusts of whistling winds...

You wake up and eat your mix of corn flakes and what is supposed to be organic crispy rice, and feel a deathly gag in your throat. You see stars, your throat is swelling, you struggle toward the phone - - feeling desperate as the storm of cross-pollination rages outside your dusty dimmed windows - - intent on dialing 911. You hear the radio voice promising the future feeding of Africa with genetically modified foods. Remember the Monsanto company suing hundreds of farmers for 'patent infringement' for having genetically engineered crops growing on their property without paying them 'royalty' payments, even though the farmers didn't want the durn plants within windshot of their fields, but the currents of breeze inadvertently landed the mutant pollen where it wasn't supposed to be anyway.

You try to talk, but you sound like a frog. Your wife is not there to kiss you or save you. You sprawl out on the pollen linted carpet, looking through the skylight at the yellowed clouds, knowing you are about to die. Recall the little girl who began wheezing until she became unconscious in the restaurant, after eating a corn taco when she never was allergic to such a thing before. An asthmatic, she was, like you. The ambulance drivers and EMT's rushing in, sticking tubes and needles into her, blood spurting, her family screaming and crying. And now there you are, taking a grunting breath, seeing your life flash past, your arms and legs spasming, your vision filled with sparks, an Aventis FDA[26] spokesman in a maize-blue blazer and tie proclaiming StarLink corn showed "no evidence" of causing allergic reactions in humans, feeling the lights go out in the auditorium/your living room, Death about to claim you by anaphylaxis and asphyxiation, your tongue gorging up like a purpling anvil blotting out the skylight, your daughters never to see you again; your father in the nursing home waving his hand at you, dismissing your worries about global warming, ovarian cancer and the dumbing down of the world -- the last image that follows you through the pollen, the blizzard, Death's gray cobweb-cape.....

.....To awaken in a sweat, the sun just rising, your alarm clock buzzing as your hand knocks it off the night table along with your bottle of spring water and your albuterol inhaler.

You try to re-orient yourself. That it is morning. That the baby somehow is in her crib across the room – your wife must have put her there. That there is no pollen storm raging outside. But you think of the tree farms that Monsanto and their experimenting think-a-likes want to foster, with their supertrees and massive amounts of pollen. Like the stuff that blankets your car annoyingly in the spring so you have to clean it off every morning before you go to work. But the mutant pollen toxic with *bacillus thuringiensis* ('Bt') can blow for hundreds of miles, in so-called 'genetic drift,' wreaking havoc on the environment, interbreeding with other trees, while unselectively killing off lacewings, ladybugs and other organisms that help keep the crops of organic farmers healthy.

Your twin daughters run into the bedroom,

jump on top of you and your wife. You hold them and kiss them, glad that your nightmare (daymare) was just that.

"Daddy, you're all wet!" Cindy says.

"You're stinnnnky." Ruth chuckles, holding her nose, trying to slide away from touching your sweat-soaked pajamas.

"Daddy had a bad dream." you tell them, acknowledging their childhood sensitivities. "But now I have to go to work."

They try to follow you into the shower, but you smilingly close the bathroom door, thinking of the mad millennial world blooming outside their naiveté. And that the yellow pollen in your nightmare most likely would have come from corn, the plant around which the greatest controversy, research, and genetic modification has been centered as the debate on accelerated food alteration simmers just below the boiling point. Glad that you are not allergic to corn - - yet. But the geneticists and biotechies that produced Star-Link and other unknown variations in their laboratories could cause your nightmare to become reality any day soon when *you* bite into a taco shell or a golden piece of corn bread. Because even in December 2003, it was reported that 1%

of the nation's corn was still contaminated by the StarLink genetically altered variety, even though this patented unfit-for-human-consumption allergenic strain 'had not been grown by farmers for several years.'[27]

As Greenpeace scientist Doreen Stabinsky warns us: "The StarLink lesson is that contamination is to some extent irreversible."[28] For once a genetically altered 'plant is released on the market, it {may be} impossible to keep its pollen from spreading, thereby leading to its existence in the food supply forever. According to the European Union, this is one of the reasons for its decision to resist growing and importing genetically altered foods.'[29] Or getting genetically altered foods off of supermarket shelves <u>almost totally</u> in the United Kingdom today, after 60% of these goods had genetically altered ingredients back in 1999.[30]

Corn

In America, we picture corn as the delicious yellow kernels that grow on a cob, and sometimes are eaten right off of it, when our teeth are fit for the feast. Otherwise, we have learned how to harvest and can the kernels to the benefit of our society.

However, if you speak to a European, 'corn' might mean something else. In England, what *they* call corn is what *we* call wheat. And what the Scotch and Irish call corn, *we* call oats. So, 'corn' usually means the most common cereal/grain of the region we are looking at, specific to that culture.

When the Puritans came to America they planted their 'corn' to sustain themselves in the new hemisphere. Alas, all their crop died from plant diseases, etc., on its maiden excursion into our environment. What was really wheat had not adapted to radically different elements of nature here on this continent.

Fortunately for our ancestors, the natives of North America benevolently shared 'maize' with them, greatly assisting their survival. Maize had come from Central and South America, nurtured and bred in its different colors to become the food staple of the Western Hemisphere. 85,000 sub-varieties exist today in Mexico's Oaxaca region alone, such diversity making it near impossible to wipe out this essential crop into tragic extinction.

Mexico's Oaxaca region has 85,000 subvarieties of maize, threatened by contamination with patented genetically-altered monocultured kind of corn.

Contrast that with our monoculturing the few genetically altered varieties of corn, soybeans, cotton, canola, etc., today to match the herbicide or pesticide the Gene Giant corporation makes.

With time and science, maize, or our 'corn,' became a favorite medium for our geneticists to manipulate and develop as the twentieth century rolled out its carpet. Hybrid corn seed was a ridiculed product of genetic theory that revolutionized practical agriculture, eventually. Resulting in higher crop yields, increased sugar, and lowered starch content, uniform plants could even be bred for mechanical harvesting.

The hybrids became the norm, representing most commercially grown corn types. (But, according to biologist Arpad Pusztai, generally "hybrids lose their vigor after the first year"[31] of their existence.)

Although tamales, tortillas, hominy, grits and such, feed much of the humans of the Americas to this very day, in the United States of America, <u>half</u> of our corn crop is used to feed our animals. In Europe, this is just about exclusively what 'our' corn is used for (or <u>was</u> used for) there.[32]

What has been happening over the last two decades however is of a different magnitude of change than what has occurred previously. The 'life industry' has stepped into the labs and campuses of America to implant pieces of viruses and bacteria, and genes even from fish and other non-plants, into the DNA chains of various crops and animals for 'progress' and profit.

While previous cross-breeding techniques used similar strains from similar plants, now the sky is the limit. Many scientists - - and sensible citizens - - worry that untold complications may result from this insufficiently restrained tinkering with nature. That new viruses may be formed from the imprecise gene-gunning of a promotional complex into a chromosome, that doesn't often land where it is supposed to. That these viruses may multiply to threaten other life forms with unanticipatable epidemics that could threaten our civilization, so corrupted by commercialization, that measured guidelines were trampled before they could be rationally evaluated. That a flood of unlabelled altered novel foods could expose the human population to unforeseen allergic reactions as the volume and variety of these products *tsunami* into our delicatessens and supermarkets.

In fact, the pioneers of genetic engineering foresaw such dangers being very possible, leading them to call for a moratorium in 1975 in the Asilomar Declaration. But pressures were applied to cut the moratorium short, landing us in the maelstrom multiplying around us today.[33]

The StarLink Corn scandal highlights the overall problem as the new millennium begins. StarLink corn had been allowed for animal consumption, but banned for human consumption because of concerns about allergic reactions in people (what about allergic reactions in animals? one might ask). There were also serious concerns about Aventis' (StarLink's producer) 'secret formula' employing virus and bacterial components, along with antibiotic resistance genes, to construct StarLink's patented DNA sequence.

Somehow, StarLink corn became mixed with the U.S. corn supply. This was first disclosed in food company data submitted to the Food and Drug Administration (FDA) in September of the year 2000. Thereafter, incidents of allergic reactions attributed to corn related products were reported. In fact, 210 such reports were made in the following two months.[34]

Typical reactions like the one in the nightmare preceding this section of the chapter, with symptoms of difficulty breathing, throat closing, tongue swelling, were documented. When this happens, you, the person affected, must seek immediate treatment, or you may die if the symptoms progress beyond the point of reversal. Intravenous steroids and antihistamines, and sometimes adrenalin itself, often must be emergently administered either in 'the field' by an ambulance crew, or in an Emergency Medicine hospital setting, to save your life.

People allergic to peanuts, for example, know to avoid foods containing peanuts. Similar things can happen to them, and have killed many of us. But, peanut-containing foods are so-labelled, helping us to avoid this tragedy in modern America. Frighteningly, in 1995, it was discovered that although <u>animal tests</u> had proved <u>negative</u>, a Brazil nut gene introduced into a nearly marketed soybean strain could indeed cause fatal allergic reactions in <u>humans</u>. Research on the product was stopped.[35]

Yet the StarLink corn fiasco had proceeded so far that the tests by the Department of Agriculture showed 22 percent, or nearly one in four grain samples, were StarLink positive. 'This has led to the voluntary recall of nearly 300 products, including more than 150 brands of corn chips and taco shells. It has recently also turned up in corn dogs, corn bread, polenta and hush puppies.'[36] Hundreds of millions of dollars were lost by Aventis, StarLink's producer. (In fact, Aventis 'sold off its crop seed subsidiary'[37] due to the StarLink disaster.) Countries all around the world continue to be reluctant to buy USA corn. But the worst thing is that foods that are genetically altered, like StarLink corn, or other genetically altered foods, still do not have to be labelled as such, as do foods containing peanuts, for example. Not in the United States of America. So the next catastrophe is just waiting to happen, unhindered by our pro-biotech government and its agencies like the FDA and USDA.

Of course, the biotech industry minimizes any claims against them, saying people are just taking advantage to sue them and make lots of money. They also point to an FDA study that apparently was a whitewash on the issue. In that study only 20 people were tested for allergic reactions, and not with the suspected Cry9C protein directly extracted from StarLink corn. Instead, the Cry9C came from a bacterial surrogate protein grown in E. coli bacteria. These two Cry9C proteins were different, had different molecular weights and structures.

When you have an allergic reaction to something, let us call it the 'allergen,' that can cause an 'antibody' to form in a reaction to it, first, there must be the initial exposure to this 'allergen,' so the 'antibody' can be formed. The next time, or perhaps after several exposures to the 'allergen' -- like whatever it may be in the StarLink corn, probably its very own Cry9C -- then your 'antibody' can suddenly react to it. Voila, the possibly life-threatening allergic reaction that can shock through your system at lightning speed, and kill you in a matter of minutes.

'Antibodies to StarLink corn Cry9C may not recognize the bacterial surrogate used in {the FDA} testing, resulting in "false negatives."', a critique on the FDA study proclaimed.[38] A 'false negative' means your <u>test</u> comes out 'negative,' or here, no allergy; but that result is incorrect or 'false' because you did not use the actual allergen directly extracted from the StarLink corn. It seems to be 'negative,' but it is not, as the testing was <u>not</u> done appropriately using the true allergen in question.

Plus, extra sugar molecules (not present in the substituted bacterial E. coli protein) 'increase the likelihood that StarLink Cry9C is allergenic.'[39]

Probably due to the StarLink contamination, and other problems with relatively high profile genetically altered corn, the percentage of our U.S. corn crop of the genetically altered variety shrunk to 38% by 2003[40]. However, with practically no big media revelation of the dangers of genetical alteration of our food to the American public, this percentage rose to 73% by 2007, according to the U.S. Department of Agriculture (USDA).

Switching over to soybeans now, people like me, and those of vegetarian ilk, are very concerned that more than 90 percent of U.S. soybeans are genetically altered.

Tofu, non-dairy yogurt, baby foods, soy-formula, breads, soy lecithin, all sorts of vegetarian unchicken, non-turkey concoctions, contain soybeans in some form. Soybean related foods are a major staple of diets averse to meat and dairy, also including those devised recently by people newly afraid of getting mad-cow disease. And for those non-vegetarians out there, be aware that approximately <u>80 percent</u> of all our processed foods, in fact, contain soy or soy derivatives in some form[42]. That goes for your hot dogs to your pizza to your margarine to your soy sauce to your knishes. And any fast food probably fried in soybean (though it very likely also could be corn) oil.

Again, the scary problem of increasing allergies, this time to soybeans, has arisen. However, although it was reported that the York Nutritional Laboratory in England found a 50 percent increase amongst the citizens they tested in 1999 for soybean allergies, this could <u>not</u> be shown to be due to genetically altered varieties of soybeans.[43] Though this led to a furor, erroneously based or not, the USA to this day, continues to try to avoid safety testing and labelling foods as genetically altered.

Instead, our food policy declares out of hand that any genetically altered tomato or NewLeaf potato is 'substantially equivalent' to any unpatented cousin tomato or potato, even though the vegetable is claimed to be different, a new, novel life-form never before existing on this Earth, as a basis for its being patented in the first place.

The logic is shaky, at best, but the disregard for the welfare of our citizens and those of other countries, by our own government and companies like Monsanto, disavowing responsibility for any adverse reactions to their novel products, is astounding to me.

Think of our medicinal drugs, to keep your sanity on this issue. We test them to insure that they are safe, and 'do no harm' — one of the primary principles of medicine. If a drug will kill you, or injure you, drug companies are held responsible by our government.

There have been occasions where drugs have been tested, deemed 'safe,' introduced and hyped up for maxi-marketing, but produced unforeseen complications including death to the patients that consumed them. Examples: the antibiotic Oroflex killed about 11 people in Europe before it was unceremoniously pulled from commercial existence. Then there was the miracle antibiotic Trovan that started to hit its stride, when a young woman, a doctor's wife no less, died secondary to its toxic affects on her liver after receiving it intravenously for a few days in 2000. Don't expect to receive that drug too soon, my fellow Americans. And don't think Pfizer, Trovan's producer, didn't fork out hundreds of thousands of dollars for its liability/responsibility in this matter.

Then there was Duract, one of the non-steroidal anti-inflammatory drugs in the motrin, Advil, Aleve, Vioxx family. Whose precise mechanism of action is not really understood. Ah, medicine is an 'empirical' science. That means we learn by trial and error most of the time, in developing our techniques of treatment. That's the reality of 'modern medicine.'

The Duract drug representative used to bring us samples of his wondrous red capsules and tell us it was recommended for twice a day use, but <u>he</u> used it four times a day for his back at double the recommended level dosage and he felt great now from it! Studies were handed to us on its efficacy and safety. We gave the elaborately packaged samples to patients who had no insurance, or no money, or after the pharmacies had closed for the night. FDA testing had proclaimed it safe, and allowed its release for use in treatment.

But then, it was found that Duract caused liver toxicity. *Whooooooooop*! Duract was suddenly gone, and I never saw that drug rep again either, now that I think about it.

Basically, in medicine, and with any product, I believe, that can cause harm, we should operate on what is universally called 'The Precautionary Principle.' Make <u>sure</u> it is safe <u>before</u> you put it out, and test it appropriately for enough time, so that Trovan-induced or Oroflex-induced deaths should not occur. Most countries, including the USA, <u>do</u> function utilizing the Precautionary Principle in relation to most drugs, goods, etc.

If people are desperate, like AIDS patients, who are willing to take drugs to impede their disease process that are not sufficiently tested yet, they could be allowed to access them if an experimental arrangement can be devised. Doesn't that make sense? They would realize the dangers, take the responsibility, in conjunction with their prescribing doctors, if any untoward complications occur.

Yet with the life industry and genetically altered foods and whatever else Monsanto, DuPont,

Syngenta, Aventis, or Dow AgroSciences concoct wherever, the powers-that-be have the audacity to <u>defy</u> applying the Precautionary Principle, to the detriment of their citizenry. And yes, also potentially to the detriment of plants, non-human animals, and the environment in general.

People should not have to be subjected to the anxiety of eating some food that could unexpectedly adversely affect their health, whether it be by an allergic reaction or some long term toxic or cancerous or infectious result of consuming unlabelled gene-altered products, designed to profit unrepentant corporations. Not if it can be prevented.

Perhaps you were not aware that this feeling is close to unanimous right here in our country, though the picture seems to be skewed for us by our media in its overall presentation of the issue. A June 2001 ABC News poll found that <u>93 percent of Americans</u> polled want the federal government to require labels on foods saying whether they are genetically altered. Plus, 52 percent believed such foods are unsafe, with another 13 percent unsure about them. 62 percent of <u>women</u> polled believed genetically engineered foods were unsafe.[44] A similar CBS News/New

York Times poll in 2008 found that 53 percent of pollees said they wouldn't buy genetically modified foods, and 87 percent wanted these altered foods to be labelled.[c]

Archer-Daniels-Midland or ADM, your 'supermarket to the world,' has been segregating their soybeans, besides extensively marketing Genetically Engineered-Free 'identity preserved' soybeans to meet market demands around the world. In addition, they have aired ads on midwest radio stations telling farmers their mills will only accept crops 'that have full feed and food approval world wide.'[45] That does not sound like genetically altered crops are what ADM wants to mill right now. "We don't want another StarLink," quoth Larry Cunningham, an ADM spokesperson.[46]

This is all because, while the USA and its citizens seem to be living in an informational vacuum, countries and citizens around the world are enacting mandatory labelling laws for genetically altered foods. The European Union nations, New Zealand, Australia, Saudi Arabia, Israel, Brazil, Japan, South Korea, Indonesia, and even China, are some of the most recent countries to do this[47]. Plus

---

c Op. Cit., 'Poll: Many Won't buy Genetically Modified Food.' http://cbs4. com/national/CBS.News.New.2.721469.html

China has banned the cultivation of GE rice, corn, soy and wheat (but they are growing genetically altered cotton).[48] In fact, today, approximately 40% of the world's population has laws and regulations requiring labelling of genetically altered foods.[49]

Spain has been the only country in Europe commercially growing any genetically altered crop, but this was temporarily stopped in April 2004 due to concerns about their Syngenta Bt 176 maize possibly causing resistance to antibiotics. And on April 29, 2004 the Independent Science Panel in England's House of Commons stated that 'the accumulation of evidence...casts doubt on the safety of genetically modified (GM) food... demand{ing} an inquiry into GM food safety.'[50]

Of course, Europeans have been most adamant about what they call 'Frankenfoods' (named after Professor Frankenstein and his monster from Transylvania, from the book authored by Mary Shelley), attempting to ban them, importing their soy and corn increasingly from Brazil and other countries where crops are being identified as <u>not</u> genetically altered. This is because Europeans have been getting the news about studies showing genetically-altered-food-tested animals dying prematurely, or developing tumors and hyperproliferation of intestinal lining cells; DNA from genetically altered foods finding its way into human cells and organs, and the bacteria in our intestines, though it was proclaimed that this could not happen. (Amongst many other adverse effects, including genetic pollution of the environment and time-tested crops, etc.)

Cancer, you should know, is actually a proliferation or continuing growth of misshapen anaplastic cells that multiply and become tumors or leukemia -- the latter being a cancer of the blood cells. Those of us who have, or develop, irritable bowel syndrome, Crohn's disease, ulcers, and other gastro-intestinal disorders, will certainly worry about what genetically altered foods might do to us. Especially if we hear that consuming such products 'may have unforeseen consequences, and some of these may be irreversible,' as world renowned biologist Arpad Pusztai warns.[51]

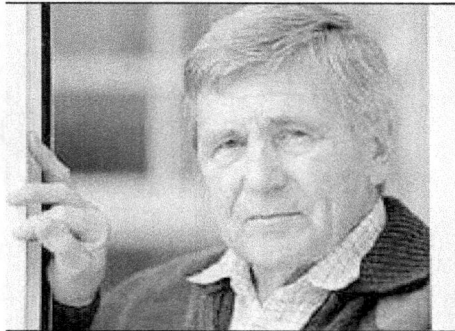

Arpad Pusztai, ex-Senior Biologist at Scotland's Rowett Institute was suspended, retired and gag-ordered after he voiced concerns about genetically altered foods.

The editors of the bimonthly on-line Bio-

Democracy News estimate that '70-80 percent of consumers {in Europe and Asia} remain firmly opposed to "Frankenfoods.'"[52]

This does not mean that <u>nobody</u> wants genetically altered foods now, or in the future, when the technology can be better refined. People just don't want the unlabelled, potentially dangerous foods and products forced down their throats, and into their backyards, forests, and waters by immoral profiteers, without having a choice, or being able to identify them, or apply the practical brakes on an industry gone hyperwhacko.

Not to be totally negative, let us even take a further step back, and contemplate the words of Prince Charles of England, relative to genetically altered food:

"At the moment, as is so often the case with technology, we seem to spend most of our time establishing what is technically possible, and then a little time trying to establish whether or not it is something we should be doing in the first place."[53]

Do we really need genetically altered crops with their implanted virus and bacterial gene sequences, their pollen's genetic drift, their allergic and intestinal dangers? What about all the other crazy constructs like industrial 'Super Fish,' gene altered trees and grass, 'terminator seeds' that Monsanto and fellow corporations can sell that will not regerminate, so farmers must purchase another round of seed from them each year and be ever beholden to these almighty Gene Giants? Do we have enough food on Earth pre-gene-alteration industry to supply the planet's Earthlings or not? Should life be patentable, as 'free-traders' espouse? Is this debatable or re-bottleable after the genie appears to be going wild outside of its container? Who should be making these decisions? Why is this happening when most Americans want purer, safer food, the majority of us now saying we would prefer food that is <u>organic</u>?

Rather than serving our children artificial products with herbicides or pesticides or strange genes or viruses engineered into them that could ruin their health, most Americans realize we could be giving our kids fortifying toxin-free foods that we are learning more and more about every day.

Yet our federal government seems to be corrupted by commercial forces that care little for purifying our food supply, or reliably refining our codes to protect the American people from unwanted manipulation of Nature's simple bounty.

Let me give you a terrible example to shake your faith in what our government could and should do, when mandated by its own citizens to bureaucrat rightly.

Sabotaging The Concept of 'Organic' Foods

Organic farmers and the organic industry had been working with the federal government, and the United States Department of Agriculture specifically, to hone down the definition of 'organic' foods. However, with the market for organic foods increasing by 15-20 percent per year, and an increasing majority of shoppers desiring organic food, in latter 1997, Dan Glickman, then the Secretary of Agriculture under Bill Clinton, came up with the following double-cross on 'organic' standards: all the following would henceforth be deemed 'organic' - - irradiated foods; sewage sludge treated foods (which could contain any imaginable contaminant from dioxin to disease causing germs); genetically engineered foods; factory farmed products; and animals cannibalistically fed with diseased or waste parts of animals, including brains and spinal cords, plus offal and blood—the practice which gave us Mad Cow Disease – cows normally do not eat meat anyway (nor do they need antibiotics in every bag of their feed).

These newly proposed standards including the five above techniques/practices, shocked the developing organic industry. None of the 33 private certifying agencies and 17 states involved in the negotiations, allowed any of these five techniques/practices, which were amongst those that everyone expected would not even be <u>thought</u> to be considered 'organic.'

Yet Glickman had done the deed, to be "balanced." That is, the big agribusiness companies wanted to get in on the organic craze, but wished to continue their cruel and careless practices that Glickman fawningly proposed to henceforth be 'organic.'

Not stopping there, it was prohibited for any disgusted organic entity or organization to set up independent organic standards outside of Glickman's standards, or design a new label that showed foods were organic beyond the US Dept. of Agriculture's [USDA's] declared auspices.[54]

This was totally unbelievable.

The battle has continued into today, with a very under-reported rider to a 2005 Agriculture Appropriations bill. Basically, what this Congress-approved rider has done is shift more control of what could be 'organic' into the hands of the USDA, rather than the de-fanged National Organic Standards Board (NOSB). More synthetic ingredients like ethylene, mono and di-glycerides, chemicals that act like preservatives like sulphur dioxide and sodium citrate, plus upwards of 500 synthetics may soon be part of your 'organic' food. And animals that may end up in your 'organic' section of your meat freezers may be eating more and more genetically altered corn and soybean and cotton in their pre-slaughter feed.[55] Watch this story best via www.organicconsumers. org for it is not over yet.

Meanwhile, on the international front, the maxi-aggressive pro-agribusiness Bush administration is fighting for its corporate backers thusly:

Going back to February of 2001, President Bush threatened to invoke punitive tariffs of over 300 million dollars against the European Union for their rejection of hormone treated U.S. beef (94% of our 'feed lot cattle' have hormone pellets imbedded in their ear[56]) and other unsavory

U.S. food products, while pushing to end their 'de facto moratorium' on genetically altered foods. Following the invasion of Iraq, Bush renewed the USA challenge against European Union rejection of genetically altered foods by making a formal challenge in World Trade Organization [WTO] court. Although six countries immediately decided to individually maintain their moratoriums regarding genetically altered foods (France, Italy, Luxembourg, Austria, Greece and Denmark) in 2001,[57] the conclusion is yet to be written concerning President Bush's latest assault on the sensibility of many wary nations and their concerned people. (See more on this in Chapter 4 and the Appendix.)

For while threats may be heaved across WTO dispute panel backrooms, Europe does not want our genetically altered crops. Taking corn as an example: in 1996, the USA exported $305 million worth of corn to Europe, but due to our genetic alteration of this crop, by 2001 this number had crashed down to but $1.6 million, a greater than 99% decrease.[58]

As the Bush biotech bullying extends south into Africa, utililizing the 'World Food Programme' in which our country is "the majority stakeholder,"[59] another embarrassing chapter is unfolding that most of you are not hearing about. Did you know that several African nations are attempting to reject our ultimatum to either accept food donations that are laden with genetically altered, basically unsellable surplus crops and unmilled seeds, or starve? "Beggars can't be choosers" was a phrase actually used by one USAID spokesperson. But such shameless coercive behavior in the name of food aid by our own country and our leaders is being fought angrily, not only in Africa, but also by countries all over the world. From Sri Lanka to Bosnia to the Philippines to Guatemala and Ecuador and Colombia, exasperated governments and citizens are trying to stem President Bush's tide of politico-economically motivated dispersion of as-yet[?] unwanted genetically altered foods.[60]

Zambia, for example, was "offered $50 million on strict condition that it only be spent on genetically modified maize from the USA."[61] Although that African country may have been eyeing "vast surplus stocks of {Indian} rice – 65 times as much as Africa needs,"[62] available at half the price of the genetically altered USA maize, the deal prohibited such a much more practical purchase. Zambia's President Levy Mwanawasa rejected the offer. Mozambique, Lesotho, Zimbabwe and Malawi were offered similar deals.[63]

Be aware, my fellow Americans, that <u>every African representative</u> in the Food and Agriculture Organization (FAO) (except South Africa's) signed the following joint statement, called 'Let Nature's Harvest Continue' in 1998[64]:

> "[We] strongly object that the image of the poor and hungry from our countries is being used by giant multinational corporations to push a technology that is neither safe, environmentally friendly, nor economically friendly to us....we think it will destroy the diversity, the local knowledge and the sustainable agricultural systems that our farmers have developed for millennia, and that it will thus undermine our capacity to feed ourselves."

We are engaging in a policy of diplomatic illwill, but most of us Americans who are not reading this book are being blanked from the news about it by our negligent corporate media.

And it gets much more serious.

Did you know that Great Britain's Ministry of Agriculture, Fisheries and Food (MAFF) officially warned our own FDA:

1) Transgenic DNA can spread to farm workers and food processors via dust and pollen.

2) Antibiotic resistance marker genes may spread to

bacteria in the environment, which then can serve as a reservoir for antibiotic resistance genes.

3)   DNA is not readily degraded during food processing nor in silage, so transgenic DNA can spread to animals in animal feed.

4)   Foreign DNA can be delivered into cells of mammals {like us humans}, by bacteria that can enter into the cells.

5)   The ampicillin resistant gene in transgenic maize undergoing 'farm-scale' field trials in the United Kingdom and elsewhere is very mutatable, and may compromise treatment for meningitis and other bacterial infections, should the gene be transferred horizontally into the bacteria. The potential hazards of horizontal gene transfer are unlike those we have ever experienced.[65]

'Horizontal gene transfer' here means going sideways across species lines, like fish to people, or virus to plant, as compared to <u>vertical</u> transfer down from father to son, for example, in the same species (e.g., man). Viruses and bacteria are always changing and multiplying anyway, developing resistance to things that could kill or impair their life functionings. But by breeding and implanting antibiotic resistance genes into life forms, things could go haywire.

It works like this: the gene blaster wants to see

which cells successfully received the genes he or she tried to transfer into the cell nucleus' DNA. Since the process is so imprecise, and so few cells of thousands on a plate indeed successfully received the foreign gene, why not attach a marker, an antibiotic resistance marker (ARM), to the foreign gene? Then when the specific antibiotic, say ampicillin, is applied to the plate, only those with the ARM will survive. And those will thus be the cells that also successfully incorporated the desired foreign gene into their DNA!

BRILLIANT! right? Absolutely! The biotech touters say no danger here, no way this cell and its DNA could pass through our digestive system into any part of our body. You aren't what you eat.

Unfortunately, studies were done finding that ingested DNA <u>did</u> find its way into the liver, spleen, intestinal wall and blood of experimental subjects via eaten food. Specifically, in 2002, one study found that horizontal gene transfer of the herbicide-resistant gene used in soybeans was found in the bacteria of colostomized patients , and that 'a relatively large proportion' of this genetically altered DNA from said soy had survived the passage through the intestine, which was not supposed to happen[66]. "Everyone used to deny that this was possible," said molecular pathologist Michael Antoniou, of London, England. "It suggests that you can get antibiotic marker genes spreading around the stomach which could compromise antibiotic resistance."[67]

Also, this has to make you worry about the long term complications of taking in supposedly safe, stomach acid/enzyme destroyable, novelly-altered DNA. Monsanto had told us our stomach's digestive chemicals would wipe out that gene altered DNA so it would never make it into the small intestine.[68] The above 2002 study took place in the United Kingdom, where exposure to genetically altered soy and such is relatively minimal, compared to what we are exposed to every day here in America. What do our intestinal bacteria look like? and what about our intestinal villi that could be overgrowing, possibly on the road to forming cancer and those other intestinal diseases I mentioned before?

Horizontal gene transfer in action!

And your corn may be unlabelled Bt corn that has an ARM for ampicillin.[69] As this corn spreads around the prairies and fields of the USA, Mexico, Argentina, and wherever else the pollen blows, people can unknowingly chow down their corn-food along with that ampicillin resistant marker into their bodies, possibly making them resistant to this very commonly used antibiotic when they get their streptococcal infection in their throat. Plus, those affected bacteria in their intestines can pass out into the environment and spread this resistance to whatever other humans and life forms you can imagine...............

So, now you can read that such important organizations as the World Health Organization (WHO), Britain's House of Lords, even our own American Medical Association (AMA), and the Royal Society of the United Kingdom, have made the call to phase out any further use of antibiotic resistance marker genes.[70]

However, with our biotech laboratories multi-plying across the country, and oversight political-ly hampered, the process of 'evolution' is being experimentally dangerously accelerated.

Yes, throughout our more natural previous eons of evolution, certain mutations would oc-cur. Many would prove to be unreproduceable dead-ends that would perish, never to be seen again. But today's geneticists have discovered that certain promoters and novel gene sequenc-es that are gene-gunned into a cell can possibly produce remarkable results. Although it is also known that the technique of gene-gunning itself is imprecise: that experimental gene sequences very often land randomly in a cell's DNA during the process, and might not be expressed as anticipated, or may disrupt the function of the cell.

Yet there still exists the concept that with the right vibrant promoter nearby the desired gene sequence, a viable magnificent improvement in a plant might occur that could allow it to survive in cold weather when normally it could not. For example, perhaps thanks to a successfully implanted protective gene from some Arctic fish's DNA, a fair-weather plant might live through ice and snow and then reproduce a long line of hardy descendants in Wisconsin or North Dakota. All sorts of possibilities for plant or animal survival could become reality. Maybe one day tropical mangoes could be grown in Alaska, or super-cod could be bred to grow in your backyard pond down in Dallas!?

But don't worry, no, without that promoter placed nearby it, a lonely misplaced novel desired gene would pine away like a lovesick maiden in the middle of the Mojave Desert, never to be expressed. Horizontal gene transfer is not that likely to occur. Though, yes, it could and does, but, usually, it does not, and will not in most cases.

"The wisest man that ever was, when they asked him what he knew, answered that he knew this much, that he knew nothing. He was verifying what they say, that the greatest part of what we know is the least of those parts that we do not know; that is to say that the very thing we think we know is a part, and a very small part, of our ignorance."[70b]

- -Montaigne

That is what the pro-genetic alteration proponents tend to say to comfort us. Yet, the British Medical Association was concerned enough to call for a moratorium on genetically engineered foods and crops until more 'independent' research could be done. They cited "gene inter-action of unexpected kinds which might take place." Dr. Michael Antoniou warned about "unexpected production of toxic substances."

Your next example: scientists trying to genetically alter yeast to increase its fermentation were shocked to discover that their maneuvering also somehow increased the 'levels of a naturally occurring toxin by 40 to 200 times'[71]! In the paper they published, they admitted that 'their results..."give some credence to the many consumers who are not yet prepared to accept food using gene engineering techniques."' They also pointed out that 'their' yeast {did not implant} foreign genes, but instead used 'multiple copies of the yeast's own gene.'[72] Their paper told us that their product 'was not "substantially equivalent" to normal yeast,'[73] though it would have been deemed so by many pro-biotech countries' policies, including ours in the USA. Especially with no safety testing being required.

What could happen microscopically here goes something like this: a predominant amount of our genetically altered foods use a gene complex incorporating the Cauliflower Mosaic virus (CaMV) as the promoter that turns on the desired nearby gene spliced into the DNA sequence. Though cells normally maintain a balance between genes, turning some 'off' and some 'on' at various times relative to their expression, while protecting cell DNA from 'foreign invaders,' the CaMV promoter "is designed to overcome a plant cell's defensive devices to prevent foreign DNA from being expressed." 'The CaMV's light switch or promoter' complex is permanently turned 'on,' 'set to high intensity...working 24/7, non-stop, in all cells of the plant...[and] is a key element enabling the virus to "hijack a plant cell's genetic machinery and make many copies of itself," as Michael Hansen of Consumers Union puts it.

Besides turning on the desired gene to which it was attached before gene-gunning, the CaMV promoter also can turn on some native genes (to the cell) as well, including those that have remained dormant during the cell's evolutionary existence, and some on distant different chromosomes. This can cause 'a flood of proteins that are totally inappropriate,' or wake sleeping viruses that have been present along the DNA for eons, or "create highly virulent new viruses." The CaMV promoter also may turn 'off' or silence other native genes including one that may 'prevent the expression of some toxin, the net result of the insertion [then being] an increase in the level of that toxin.'[74]

In addition, studies show that a promoter like the CaMV type can 'create a "hotspot" in the DNA. This means that the whole DNA section, or chromosome, can become unstable. This can cause breaks in the strand or exchanges of genes with other chromosomes. According to [Joe] Cummins, a promoter can have "the same impact as a heavy dose of gamma radiation."'

Jeff Smith notes that 'scientists from all over the world have expressed concern about CaMV, calling for an immediate ban.'[75]

Dr. Michael Hansen compares the possible situation in the genetically altered plant to that of a computer. You put one new thing into the computer that was working just fine, and somehow this causes the computer to mysteriously crash. With the Cauliflower Mosaic Virus gene complex, which Dr. Hansen calls a 'hyper-promoter' because it works like a highly amplified complex turned up 10 to 10,000 times louder than other genes around it, better leading a desired gene nearby it to be expressed successfully, it also may drown out the symphony and harmony of the plant's functioning. Unanticipated changes in the plant's now-different DNA sequencing may alter the essence of the plant. It may not be the same plant in too many ways that it once was, before the CaMV promoter and its attached gene sequence were gunned into it. It may not be 'substantially equivalent,' in reality, to the plant that it used to be. . . [†]

Let's now go back to a study wherein potatoes altered via the utilization of the Cauliflower Mosaic Virus promoter were fed to lab rats. The rats' stomach linings were afflicted by a severe viral infection, their immune systems were damaged, some organs became abnormally enlarged, while brains, testicles and livers were smaller than normal. The control rats in the study eating the un-altered parent potatoes developed none of these symptoms.[76] Something was wrong with those genetically modified po-

---

[†] As discussed in telephone conversation with Michael Hansen of Consumers Union, 7 25 2001

tatoes....that you could have been eating soon....
maybe at McDonald's.

Yet, believe it or not, McDonald's and other
big buyers were the ones responsible for Monsanto withdrawing its NewLeaf Potatoes from the
global market in 2001![77]

Unfortunately, to this very day, to move
things along the biotech approval conveyer belt,
our government continues to call too many new
constructs 'substantially equivalent;' the new
(gene-altered) tomato or potato is 'substantially
equivalent' to the old reliable tomato or potato
that we have come to love and trust and devour
with relish, not worrying about allergenicity or
toxic affects. Therefore, concludes the bureaucrat approver, minimal or no testing needs to be
done on the new gene-altered tomato or potato
for it to be released into the market. That's the
current way of the USA.

According to Brewster Kneen of the *Ram's
Horn* up in Canada, U.S. representatives to the
Codex Alimentarius - - the organization that has
inherited the responsibility to set up worldwide
food standards now, with the World Trade Organization's blessing - - are twisting arms to have
U.S. food policy become Codex's norm. To 'har-

monize' the standards of all the countries of the world to be what the Bush administration wants. But that seems to have backfired, with Codex telling Bush biotech pushers that our current 'no pre-market safety testing' and 'substantial equivalence' doctrines on genetically altered foods, are not acceptable.[78]

Instead, Codex has adopted the stance that 'substantial equivalence' might only be deemed a 'starting point' for further examination of novel foods, rather than an 'end point.'

More testing, and safety testing, has to be done to find out if a genetically altered crop or food can be accepted to grace the dinner tables or finger bowls of our planet's populace.

Theoretically, it appears our government has come on board with the rest of the Codex community. However, as Michael Hansen states: "the devil is in the details..." "There's too much room for some funny business." Only the future will reveal how we comply with protecting our own people and environment, as the representatives to Codex Alimentarius hope we will do.[79]

Americans must realize that today, in 2008, although the number of hectares planted with genetically altered crops is slowly increasing, recently crossing the 100 million hectare mark, people are not really accepting the crops and technology with open arms. 80% of the stuff is being funnelled into animal feed. No GMO [Genetically Modified Organism] whole food has ever made any success with us human beings, except maybe Hawaiian papayas. But these do not have to be labelled properly in most situations.[80]

According to Ronnie Cummins, National Director of the Organic Consumers Association, "the only reason for sales [of GMO products] in countries like the USA is the lack of labelling. If GMO ingredients end up on the label of some products, the big processors and distributors will just stop carrying them...and it is unlikely, if consumers had the right to know, and these foods were labelled, if the use/sale of gene spliced foods could continue.[81]" Mr. Cummins also told me that due to a "backlash by nature," all sorts of resistance is developing to the one crop/one pesticide combo in the fields of the world where the Monsantos and Syngentas are marketing their products. Older pesticides like 2-4 D and those in the Agent Orange family are being resorted to in various places on the planet to combat this.

With the increasing cost of petroleum projected for this dwindling fossil fuel/resource, expenses for GMO's production shall increase in the future for transport and fertilizers.

These are several reasons why farmers in nations like Poland and especially other European Union countries are opting to enter the organic market, rather than raise GMO crops. And don't think even Monsanto and the Gene Giants are acting stupid here. Latest news is that Monsanto is investing 50% of its research and development [R & D] monies into 'marker assisted breeding' and gene mapping. Syngenta is reportedly investing 80% of its R & D dough into this much more practical technology. Where genes are being mapped all over the world, at all sorts of institutions, linked to various expressed characteristics, catalogued for worldwide sharing in many instances, so that more conventional plant-to-plant hybrid breeding can then be arranged to produce newer, hardier, more delicious crops. Without the dangers of imprecise gene-gunning and virus implantation.[82]

At this point, seven nations have planted 99% of GMO crops, mostly corn, canola, soybeans and cotton. Here is the hectarage in millions, as supplied by the pro-GMO ISAAA: USA 54.6, Argentina 18.0, Brazil 11.5, Canada 6.1, India 3.8 [cotton only], China 3.5 [cotton only], Paraguay 2.0 [soybeans only][83]. Outside of these seven countries, there is little conscious acceptance of gene-spliced or gene altered or GMO foods and crops.

With all the information available in Britain concerning the adverse affects of GMO foods, which were once present in 60 percent of food products back there in 1999, British supermarket shelves are now practically empty of all GMO-containing products. Labelling has been a primary reason for this, besides the awareness of very concerned shoppers - - though most of us Americans probably know nothing of this.[84]

For the broad picture is not something the television networks, or U.S. mass media will report to the American public. But we should know where we sit with the rest of the world and their reaction to our inadequately regulated food technology. A major food in the picture that you may lack perspective on is milk and the injection of recombinant Bovine Growth Hormone (rBGH) into our cows.

Milk and recombinant Bovine Growth Hormone

Every industrialized country but the USA, Mexico and Brazil <u>bans</u> recombinant BGH use.[85] There are concerns about increased incidence of breast, prostate, and colon cancer because of elevated levels of 'Insulin-like Growth Factor-1 (IGF-1).'[86] [87] Eli Lilly and Company, a corporation that never actually manufactured rBGH commercially but just purchased Monsanto's Posilac (brand name for Monsanto rBGH) business[†], reported a ten-fold increase in IGF-1 in the milk of cows receiving the hormone, back in their 20th century studies done on what *used* to be their *potential* product... [88]

IGF-1 is produced by the pituitary gland, our sort of supervisor of all endocrine glands, and circulates via the blood to each cell of our bodies. It helps to regulate cell growth, division and differentiation, especially in children.

The University of Illinois Medical Center's School of Public Health Cancer Prevention Coalition states that cows injected with rBGH show heavy localization of IGF-1 in their udders' or

---

† email from Michael Hansen, August 1 2008.

breasts' epithelial cells. This does not occur in untreated cows. And <u>industry data</u> shows up to an 80 percent increase in incidence of 'mastitis,' or udder/breast inflammation/infection in hormone treated cattle, with resulting contamination of milk with significant levels of pus.[89] Who would want to report all that to their television viewers or radio listeners?

Wait: so with the likelihood of mastitis being increased, antibiotics have to be used for the cows, which leave antibiotic residues that pass into the milk[90] you may be drinking, Mr. and Mrs. American Human Being, and Dear Children. Isn't that exactly what you want when you feed your family that wholesome white liquid?

Of course, the more relevant question here is why are we sanctifying our money-mad agribusiness factory farmers injecting growth hormone of any sort into innocent milkcows? Think about it: growth hormone is basically that. It promotes <u>growth</u>: of our and a cow's bones, and vital organs. Normally, when the animal, or each of us, grows to adult size, feedback systems turn off growth hormone production within our bodies. Otherwise weird things happen. Like in human beings, something called acromegaly develops. Bones that shouldn't grow, continue to do so. The jawbone overgrows and thickens, as do other bones in the face. Hands and feet get huge. The skin thickens and its secreting glands enlarge and produce more *sebum* - - the stuff that teenagers hate because of the acne that results. Acromegalics also suffer more commonly from diabetes and arthritis than do normal non growth-hormone-over-stimulated individuals.

It has been said that the way a civilization treats its animals is a measure of its greatness. If that is true, boy are we failing here. Injecting cows every two weeks with growth hormone so they will produce another round of milk more quickly while they stand in their constricting stalls, to die much sooner than they would normally, is such an inhumane way to nurture our milk makers.

Well, we could label our growth hormone amped milk, couldn't we? you might suggest. Recall our government's antipathy to labelling genetically altered foods, as delineated above. Then learn that our FDA has <u>prohibited</u> dairy producers and retailers from labelling their milk 'hormone-free.' Because this could be 'false or misleading,' under federal law.[91] OUCH! So, who is protecting who?

Monsanto has been suing several milk producers for using such labelling over the years.

The intimidation that has resulted still restricts you, dear consumer and parent and person, from having the right to choose non-hormone treated cowmilk <u>or</u> rBGH treated milk.

Conscientious farmers are afraid to label their milk for the most part, though the majority (82%) of U.S. dairy cows are <u>not</u> being injected with the drug. Yet the two forms of milk are being co-mingled, without any labelling, meanwhile, for the U.S. dairy industry's captive patrons. (This may change with the very aggressive Monsanto corporation dropping out of the rBGH picture to focus on their seeds, GMO and marker assisted breeding crop sales, and herbicide production[d].)

Food Irradiation

Another deplorable chapter in our government's policy toward food and our growing desire to have less, rather than more, alteration of its natural goodness, comes with the renewed, but failing, push for food irradiation.

You may have heard terms like 'cold pasteuri-

---

d 'Monsanto Looks to Sell Dairy Hormone Business,' By Andrew Martin and Andrew Pollack, The New York Times, August 6, 2008 http://www.nytimes.com/2008/08/07/business/07bovine.html?...

zation' and 'electronic pasteurization,' wondering what exactly this or that was...?...? Well, basically, it is irradiation of what you might eat, not with some '?cold heat' or '?electronic heat' actually, but massive doses of bombarding ionizing rays.

90 percent of our current functioning food irradiation plants could use either cesium or cobalt from nuclear waste that the nuclear power and nuclear weapons industries would love to find a use for. The other 10 percent of our approximately 50 food-irradiation-capable plants use x-rays, the current latest trend, or 'e-beam' or linear-accelerated electronic beams shot out at the speed of light at some piece of meat or food you may never know was processed in this manner.

And the question of labelling again has the same answer: avoid it, or use something instead

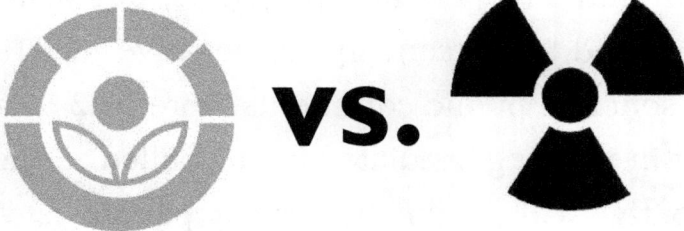

The radura flower image on the left does <u>NOT</u> mean 'Approved by the Peaceful Florists of America.' But if you were trying to sell irradiated food, would you like to use the more obvious, though less appealing, traditional radioactive symbol on the right?

of the common danger radioactive waste circle with its alternating yellow and black triangles inside. What about a teeny stamp with an innocuous looking *radura* flower pasted on packages of gamma rayed chicken? (that's the insignia we're using today, thanks to the nuclear industry).

According to Mark Worth of Public Citizen, if a food is irradiated in any of the above manners, it must either be labelled 'treated by irradiation' or 'treated with radiation' and have the little *radura* symbol somewhere on the packaging. Various forces are working to eliminate this safeguard today, since with food irradiation, the onus seems to be as alienating as being posted as a pediophile for the whole neighborhood to know.

A CBS 1997 News poll found that 77 percent of people would not eat irradiated food, including our major food shoppers and preparers, our women, who polled 91 percent in the negative. In 1999, FDA-solicited public comments were 98.2 % adamant in wanting irradiated food labelled as such.[92]

So the will of the American people is very clear here.

Yet Senator Tom Harkin of farm-state Iowa, as the manager of the Farm Bill of 2002, (approved in May, 2002) shepherded a few paragraphs into the verbiage that allow applicant companies to use such a term as 'cold pasteurization' to label irradiated food, if the Secretary of Health and Human Services decides it's OK[93]. If Ralph Nader was the Secretary concerned, would we worry? But this is the Bush/Cheney cabinet Secretary Mike Leavitt.

People working on this issue for many years anticipate that this could be the beginning of the end for forthright labelling of irradiated food.

Concerned parents should also know that another little-publicized addition to the same Farm Bill of 2002 included allowing irradiated food into the National School Lunch Program and the Child Nutrition Lunch Program – both of these under the auspices of the US Department of Agriculture (USDA), as mentioned before. This inclusion had been attempted administratively in 2001 by Agriculture Secretary Ann Veneman, but public outcry against this was so great, it had to be rescinded.[94]

However, persistence wins the battle – at least it seemed to, for most places in America. Irradiated ground beef could appear in your child's cafeteria any day soon without any labelling or notice being required or given. America's kids could be the guinea pigs for the irradiation industry's biggest distribution program[95], and many of us parents will be very angry – once, and if, we find out that it is actually happening to our precious children. Of course, we can stand up and fight back like the citizens in the Los Angeles Unified School District did! And good news: school districts in Washington, D.C., Berkeley, California, even Iowa City, Iowa, in the corn and beef heartland, plus several other locales , have also risen to follow suit[96].

Plus, more good news: as of June 25 2004 both the House and the Senate have passed the Child Nutrition Act reauthorization which does the following:

1) requires that irradiated food only be made available at the request of state and local school systems -- it cannotbe mandated by the USDA;

2) irradiated food cannot be subsidized by the federal government, meaning that the USDA cannot offset the *increased costs* of irradiated food to encourage their use;

3) factual information must be supplied to the state and school food authorities on food irradiation, including the notice that irradiation is not a substitute for safe food handling;

4) irradiated foods distributed to federal meal programs must be labelled as irradiated. However,

the school is NOT required to pass the labelling on to the students eating the food;

5) co-mingling of irradiated and non-irradiated food is prohibited;

6) schools using irradiated foods are encouraged to offer a non-irradiated alternative.[97]

Overall, this is terrific, a stunning act of protection for our Oh So vulnerable children in an era when business always seems to come first. Why, soon we might even be like the Emilia Romagna Region (near Bologna) of Italy, where it has been legislated in fortunate school districts, that 100% of foods served in schools to children up to ten years of age must be organic by 2005.[98] By that same year, 35% of cafeteria foods in advanced schools, universities and hospitals must be organic.'[99] 100% organic foods for all Italy's school children has been promised for the not too distant future[100] in Europe's number one provider of organic fare.

But there are people starving around the planet; three million North, South, and Central Americans die of food borne illnesses every year; foods are spoiling that could possibly be 'saved' by irradiation.[101] Up to eighty percent of our chickens carry Salmonella[102]; Campylobacter or E. Coli germs were found on 88 percent of broiler hens by food inspectors in 1996.[103]

According to John Robbins, in his book <u>The Food Revolution</u>, Campylobacter is our number one agent causing food poisoning, with more than 5000 people per <u>DAY</u> falling ill because of this germ. Chicken flesh seems to be the most common source. And up to 70 percent, or 7 out of every 10 chickens, Robbins states, have enough of this bacteria present to cause clinical sickness. Bloody diarrhea, fever, abdominal pain, often take a week to occur after eating the offending fowl, and there is a 20 percent relapse rate. More than 750 deaths per year in America are attributed to Campylobacter.

And up to 40 percent of Guillain-Barre cases are connected to infection by Campylobacter.[104] This is a frightening disease that paralyzes you in a progressive ascending fashion, going up from your legs to sometimes even paralyze your ability to breathe, and/or cause coma. It is a very serious illness that, yes, can kill you. Even those that recover do not always get back to the way they once were.

As bad as chickens are though, turkeys are worse, chiming in at a <u>90 percent</u> illness-capable contamination load of Campylobacter, as reported by the Center for Science in the Public Interest.[105]

The Clinton administration's answer to the problem was to put out a rule that as few as 20 birds out of 67,000 would be OK to check for fecal contamination.[106] And to push for food irradiation. And then to let the meat companies inspect their own meat[107] as the new millennium spreads its poultry wings across our shrinking planet.

Central in your mind should be the statement from the National Broiler Council declaring that 'normal cooking kills the same harmful microorganisms as irradiation would.'[108]

As to 'cold' and 'electronic' pasteurization, Webster's Dictionary defines plain old pasteurization, or pasteurizing as:

'to heat (as milk) to a point where harmful germs are killed.[109]

Dorland's Medical Dictionary defines it as: 'heating {e.g., milk} to a moderate temperature, often to 60 degrees Centigrade [140 degrees Fahrenheit] for 30 minutes. By this heating, pathogenic bacteria are killed and other bacterial development is considerably delayed.'[110]

Questions about vitamins being destroyed, and a myriad of sometimes toxic 'radiolytic products' resulting from whopping doses of irradiation splitting up natural chemicals and bonds in foods have many scientists and citizens worried about food irradiation. Just think: a chest x-ray exposes us to 0.01 to 0.02 <u>rads</u> or one-two hundredths of a <u>rad</u> of radiation; 400-500 rads will kill us via radiation sickness over the next two weeks post-exposure as our hair falls out, and our immune system implodes; but food irradiation zaps potatoes and chickens and whatever else the nuclear industry can fodder up to their cesium or cobalt wastes, or x-rays, or e-beams, to <u>HUNDREDS OF THOUSANDS</u> of rads of destructive bond-breaking ionizing power.

Sounds ridiculous, doesn't it? How wholesome could those obscenely be-blasted foods be? But 9 ½ percent of our spices are currently irradiated right now. And this irradiation has been going on at an increasing level for the last two decades, basically unbeknownst to our all too trusting eyeballs. Turns out if the spices are in processed foods, or were sold to restaurants, hospitals, schools, or nursing homes, or are not designated as a 'whole food,' nay, they need not be labelled. Nor does the consumer/eater have to be told she or he is eating something that has been irradiated.

For example, if you would buy curry powder, which is a _blend_ of various spices, and the turmeric in it was irradiated, nope, that bottle you picked up does <u>NOT</u> have to be labelled 'treated by irradiation' or 'treated with radiation,' and you will miss the sight of that cute faint _radura_ flower. However, if the product was the 'whole food,' <u>just</u> the irradiated turmeric, then, yes, it would have to be appropriately labelled.

With perceptions being what they are, the irradiation business is working very cautiously to erect the foundations for legalizing various foods to be zapped. It is highly doubtful you will see any chicken or hamburger at your supermarket labelled as irradiated. Not quite yet. Who would buy it?

What is being done is selling irradiated chicken to a hospital in Alabama[iii], where the patient would never know what she or he is getting on that long awaited dinner plate. Trying to get food processors to buy irradiated products to silently incorporate into their frozen dinners. Serve up irradiated hamburger meat to our schoolchildren. Set up the rest of the world for the irradiation industry's growth and prosperity, sow the seeds, butter up the golden calf.

Wal-Mart, our nation's leading discount retailer, may be ready to sell boxes of frozen, irradiated beef patties. They could join forces with IBP (Iowa Beef Processors), the USA's largest beef processor, to utilize 'electronic pasteurization' and test market the partially sterilized product via ionizing radiation, in a limited number of stores.[112] Interesting is a statement an IBP official made to the FDA imploring them to allow "accurately describing the process to which a meat-food is subjected to *low level treatments*" of "less than 7.5 kilograys" of radiation or 'Electron Pasteurization' or even better, 'Cold Pasteurization' or 'Gamma Pasteurization.'[113] (Italics mine.)

"Low level treatments." Learn that, and remember: a gray is 100 rads, and a kilogray is a thousand times that, or 100,000 rads. "Low level" at less than 7.5 kilograys or 750,000 rads when just 500 rads can kill you does not seem very low level to me.

Dr. Samuel Epstein, a molecular biologist and Professor of Environmental Medicine at the University of Illinois School of Public Health, who also was a recipient of the Right Livelihood Award

(Alternative Nobel Prize) in 1998, is concerned that all forms of ionizing radiation zapping our food, especially at the penultimate levels mentioned above, can produce "unique radiolytic chemical products," some of which have been "implicated as carcinogens or carcinogenic under certain conditions,"[114] as he quoted from U.S. Army 1977 analyses.

For example, the family of chemicals known as 'cyclobutanones' were first identified in 1971. Because they remain in foods for such a long period of time, more than a decade, they have been used as 'markers' to tell scientists that foods have been irradiated. But it was not until 1998 that German scientists started discovering genetic and cellular damage in both rat and human tissues caused by variants of cyclobutanones.[115] Translate that to mean such variants are possibly cancer-causing. Yet our Food and Drug Administration (FDA), aware of these cyclobutanone findings, persists in extending legalization of more and more foods to the irradiation process.[116] Rather than doing the prudent things: await the conclusions of further cyclobutanone studies, and conduct extensive studies of our own under public scrutiny.

Dr. Epstein and other concerned scientists,

wanting to fight a sane war against cancer, have urged that radiolytic products be isolated and tested specifically. But the FDA admitted that "it is nearly impossible to detect {and test radiolytic products} with current techniques."[117] The FDA's "abdication" on this testing persists to this day, according to Dr. Epstein.

Furthermore, the chairperson of the FDA's Irradiated Food Task Committee, which reviewed the mere five studies used out of a possible 441 published (prior to the early 1980's) to decide that irradiated foods were 'safe,' "insisted that none were adequate by 1982 standards, and even less so {by standards} of the 1990's."[118] In addition, "detailed analyses revealed that all were grossly flawed and non-exculpatory,"[119] (that means: not clear from alleged fault or guilt.)[120]

Elevated levels of cancer-causing benzene at ten times higher than cooked levels in beef;[121] poorly documented radiolytic products and their long term effects; dangers from radioactively contaminated foods which cannot be detected by sight, smell or taste; reduction of levels of vitamins A, B2, B3, B6, B12, C, E, K, thiamine, folic acid and amino acids;[122] possible production and multiplication of resistant bacteria, viruses, and other germs; are amongst the plausible reasons why irradiated food can't seem to find itself accepted by American consumers.

And very importantly, if a germ produces a toxin <u>before</u> the food is irradiated, this toxin is <u>not</u> destroyed by the irradiation process. However, the bacteria, for example, that could start the food smelling to warn us that it is bad, will be killed for the most part. Result: instead of us knowing not to eat the rotten food because of its foul odor, with its toxin that can injure us still present, we could become very ill from swallowing down that irradiated food.

Now let's get specific on the major forms of food irradiation being promoted in the USA as the new millennium begins. The oldest, scariest, and most dangerous, yet still not eliminated from consideration for future use here or overseas is irradiation via nuclear wastes like cesium and cobalt. We have hundreds of <u>millions</u> of curies of cesium that we have to dispose of one day. Cesium and strontium are the most <u>voluminous</u> wastes resulting from our experimenting with nuclear power and nuclear weapons.[123] Each of this type of irradiation plant would need half a million to 10 million curies of cesium to operate,

and be replenished annually. That means, besides the danger of an accident at the food irradiation plant, and nuclear contamination of the
neighborhood and surrounding countryside, we
also have to worry about transporting and trans-
ferring millions of curies of cesium all about our
nation's byways and highways. Especially if 1000
plants that are projected as necessary would be
constructed.[124]

The Chernobyl nuclear plant explosion (not
'meltdown') in 1986 released hundreds of millions
of curies of cesium. 300,000 deaths and counting
have resulted from the Chernobyl catastrophe
over the following decades (as discussed in Chapter One) from various radiation-caused cancers
about the Ukraine and Byelorus and points north
and northwest where most of the microscopic
bits of radioactivity were carried.

This is a terrible technology. Cesium 137 has
a 'half life' of 30 years. That means that half of
the cesium will still be radioactive in 30 years.
Most experts agree that its 'hazardous life' is 10-
20 'half lives,' or, cesium 137 is hazardous to your
health, and those of your descendants and future generations, for 300-600 years once it is let
loose into the environment from whatever cause

--- Chernobyl nuclear plant explosion, or your local cesium food irradiation plant accident. Not a good splat of garbage to land on your lawn either, if a transportation truck manages to overturn on your street.

Cesium comes from the Latin word for 'sky blue.' It was discovered back in 1860 by Gustav Kirchov and Robert Bunsen -- who perfected what we know today as the Bunsen Burner, which was actually invented originally by the father of electricity, Michael Faraday[125]. Cesium-137 acts like potassium, can accumulate in muscle, and was found to cause profound kidney damage in many victims of Chernobyl[126]. Like any other radioactive isotope absorbed in excessive quantities, the danger of cancer, and spontaneous abortions must be anticipated.

However, especially because of an accident that occurred in Decatur, Georgia in 1988 with cesium capsules that the Department of Energy deemed leak-proof, that leaked, much of our radionuclide type irradiation plants now use cobalt 60 instead of cesium. Most of the cobalt comes from Canada, and is not water soluble, as is cesium. But, same problems, same perceived problems, in that radioactivity still could be transferred into the irradiated food and you would not know it unless you shop with a Geiger counter – though this contamination seems to be very unlikely, but still possible. Plus the radiolytic by-products, unbeknownst germ toxins, resistant germs, vitamin loss, etc.

Another shameful problem is that the liability of these nuclear waste type of irradiated plants is inadequately mandated by the NRC. For example, the GRAY*STAR Genesis Irradiator that was to be sited in Milford, Pennsylvania, only was required to have $75,000 coverage. However, this had actually been deemed required for "decommissioning and decontamination," according to Judy Szela, who lived one mile from that plant. She has spearheaded the fight against GRAY*STAR, and food irradiation in general, documented on the website www.nocobalt.org<. In an email of June 26, 2004, she informed me that a second GRAY*STAR Genesis under-water irradiating unit, 'being kept secret from this community,' would require an additional $113,000 coverage.[127] Cost to clean up and decontaminate the Decatur, Georgia plant and surrounding environs after the accident there: $47 million. And that was back in 1988 dollars; and you and I, we fellow taxpayers, footed the bill.

Cost for the Chernobyl disaster: in the <u>hun-dreds</u> of <u>BILLIONS</u> of dollars. Just to give you an idea of financial magnitude. And Chernobyl did not involve an urban area, like New York City and its metropolitan surroundings, or a city like Los Angeles.....

If there had been an accident at the Milford, Pennsylvania plant that exceeded the above min-imum coverages, which did not include liability for injury, deaths, etc., who do you think would have had to pay the bill? Highly unlikely the GRAY*STAR folks with their 'inherently safe'[128] technology would hang around town to com-passionately fork out the dough, especially if it totalled up to millions or billions of dollars. [In fact, GRAY*STAR left Milford by 2005 and is now working on setting up a new site for business on the island of Oahu, in Hawaii. Irradiation of mangoes and papaya seems to be next on their current agenda.]

You would assume that our government would disassociate itself from this form of food irradia-tion with the newer x-ray and e-beam technologies not having the contamination dangers of cesium or cobalt type plants, yet those who are supposed to know best continue to support research on cesium food irradiation with GRAY*STAR.[129] Though we may be street-smart in America, maybe in Cambodia or Kenya they might accept that cesium we have to dump somewhere. Might as well irradiate some food with it, as we graciously sell another waste product of our nuclear industry to the ignorant or desperate, struggling to survive the political inequality of this world.

According to Mark Worth of Public Citizen, x-ray irradiation is the most utilized form of food and medical supply radiation currently employed in the USA. The same concept of blasting meat or chicken or potatoes or spices or medical syringes with hundreds of thousands of rads of ionizing radiation is used here. But you don't eat the syringes.

Radiolytic by-products like formaldehyde (that's what we use in embalming fluid at our funeral homes) and the rest are still to be worried about. Plus all the vitamin loss and the resistant bacteria and viruses. Contamination of the food with radioactivity should not be a concern. But it does get hot in these plants and lots of ozone forms, which can adversely affect the lungs, mostly of the workers at the plant. This ozone does have to be vented periodically.[130]

X-ray food irradiation is what could be the immediate future, if this overall technology can take off above popular rejection. And it may stand a much better chance of acceptance on a food processing line in some foreign country too. However, due to the higher expense of x-ray irradiating food, as opposed to medical supplies, this form of food irradiation is currently stuttering on the starting line.....

Then, from Star Wars, and television technology, we now have the 'ebeam' or 'electronic beam' type of food irradiation. A beam of electrons is accelerated at near to the speed of light and shot at a target to partially sterilize hamburgers, garlic powder, vegetables, fruit, whatever is amenable to its capacity to penetrate almost four inches of water density. That translates to 8 hamburgers scannable in one second, or 40,000 pounds of chopmeat zappable per hour. An ebeam machine can be miniaturized to 300 square feet and incorporated into a processing assembly line.[131] Wal-Mart and ebeam could be a match made in chainstore heaven.

Alas, good as it seems, no worries about undetectable radioactive contamination, but, still the same concerns with ionizing radiation producing those radiolytic by-products, the resistant bacteria and viruses, the vitamin losses, not removing the pre-zap germ toxins, plus the technology has only been minimally tested with food, without long term studies of untoward effects perhaps not even conceived of yet.

SureBeam Corp., the owner of a several ebeam plants in the USA, projected expansion to Brazil.[132] They were dreaming of branching out into the worldwide food market. Until they had to file for liquidation bankruptcy in January 2004. Titan, the people who make the rockets, is their biggest creditor, according to Patty Lovera, now working with Food and Water Watch. Titan could somehow charge up the ex-SureBeam facilities, unless Lockheed takes them over....It's a big mess.[133]

Yet just think: places like Brazil, where 50% of the food spoils before it ever gets to the consumer, could see an improvement to only 25% spoilage if ebeaming can help save the day. Shelf life could be prolonged. Strawberries could be blitzed with electrons and last 15 days, instead of just 5 days. Foods could be packed, and zapped in their containers, and slow boated up to the Caribbean or the USA to feed starving children and adults. Diseases could be killed before they kill you and yours.[134]

The world could be a better place.

But you still have to cook the food anyway, even after it is irradiated. And you still have to be clean and careful with the human end of the food handling, where the hepatitis and E. Coli and Campylobacter and Norwalk virus germs lurk in microscopic skin folds.

And you who fear Mad Cow Disease, be aware of this: food irradiation does NOT kill Mad Cow 'prions,' which cause this terrible disease. More research has to be done here, of course.

Who really wants to eat irradiated foods, in this era of increased food awareness and desire for organic purity? 80% of the vitamin A in eggs is destroyed by ionizing radiation, for example.[135] Most likely the poor and the choiceless will end up at the bottom of the totem pole, selected out to receive the irradiated relatively empty nourishment that could exact an unhealthy toll if one had to eat irradiated food every day.

On the other hand, in this post-yuppie-I-want-everything-now-Here! era should we really have irradiated mangoes in February in Aspen, Colorado? What about eating foods in season where they normally grow? How much 'progress' is mandatory with us modern folk? This is a debate for the new millennium.

I hope that Dr. Epstein is wrong when he says: "The day will soon come when international opposition to American exports of irradiated food will surpass that of genetically engineered food. If Congress fails to ban irradiated food, American farmers, ranchers and the economy could pay a devastating price."[136]

The American people have expressed their distaste for coming near irradiated food. They want it labelled so they can reject it, or, if they are in dire straits, buy it knowing what they're getting, thinking optimistically of the next un-irradiated meal.

We expect the government to protect us: the eaters and growers of the food. We have children and grandchildren to raise in a healthful environment. Unfortunately, now that globalization has erected the World Trade Organization (WTO) atop the ruling forces of the Earth when it comes to anyone's laws, it hath been declared that there shalt be NO LIMIT anymore to the dose of irradiation a food can be subjected to, in the world of corporate-managed trade! This WTO decision guts 'international food irradiation laws, in place since 1979.'[137] Public Citizen decried this action and "The UN and WHO (World Health Organization) {who} have abandoned their mission to protect the health and welfare of the world's population,"[138] for standing by haplessly, letting this occur.

Some nations in the world, yes, have allowed food irradiation to make inroads into their food chain. Of course, not the most enlightened or free nations. China now has 50 facilities irradiating food, both ebeam and x-ray. China is also the nation that allowed that chemical called <u>melamine</u> made from coal, to be added to wheat gluten and pet food and who knows what else, as filler, leading to many deaths of dogs and cats in the U.S. and the recall of 60 MILLION packages of pet food. And do you remember when some Chinese company was using India Ink to color its soft drinks? How about that Chinese soy sauce made from human hair?[139]

Also, watch Thailand-irradiated mango, pineapple, litchi, thanks to our first reciprocal trade agreement of its irradiated kind made with the Bush administration. Yes, Thailand is a military dictatorship these days. And our visionary current president made it a priority in March 2006, when he visited India, to proclaim "the United States is looking forward to eating Indian mangoes." After reaching a trade agreement promoting 'irradiation as a postharvest treatment for fruits and vegetables...India is the largest producer of mangoes in the world, accounting for half of all production.'[140]

But most of the world has not fallen to the ruse. However, if you follow the money, you will discover that much of the research at our universities leads to skewed proportions of projects experimenting with food, including what might be genetically altered and irradiated. When you earn your bread with compromise or misplaced passion, sometimes your common sense might get inordinately twisted, no matter how smart you think you are.

In that vein, I would especially like to see what happens to the family of Christine Bruhn, the enthusiastic food irradiation proponent who directs consumer research at the University of California at Davis. 'At home, she buys only irradiated ground beef that is shipped frozen from Omaha Steaks in the midwest.' She is willing to pay the extra cents for "that piece of mind...that safety" that irradiation brings her, when her brood can chaw down their hamburgers and meatloaf and maybe even vegetable produce.[141] Let us hope that her kids or her don't end up with cancer. Be interesting to find out what the future bodes for the Bruhns.... unfortunately.

Also, the Corbins and Clemmons, whose fathers pull publicity stunts like eating E. Coli contaminated spinach that has been irradiated that they dump salad dressing on for the cameras. They champion their strangley named Sadex company's Sioux City, Iowa plant that they took over from defunct SureBeam.[142]

Enthusiastic food irradiation proponent Christine Bruhn, director of consumer research at Univ of California at Davis. "At home, she buys only irradiated ground beef... shipped frozen from Omaha Steaks in the midwest." An experiment on a college family that should be followed through the future, hoping for no tragic cancerous results.

Food irradiation just is not taking off, no matter how nuclearly nutty the lobbyers are who try to convince us of its worth. Basically, we have two plants in the U.S. irradiating food really: the Sadex plant that uses ebeams, and one in Florida using cobalt. The politically corrupted FDA is currently so warped, it may even try to eliminate any labelling of certain irradiated foods, while further attempting to legitimize the use of the word 'pasteurized' when it actually means 'irradiated.'[143]

Factory Farming and Industrial Agriculture

Right now in America, factory farms and mechanization of the U.S. food industry may favor food irradiation, because it might make their end product seem cleaner, freer of contamination. Taking the pork industry as an example: where the family farmer 'husbanded' his or her hogs, treating them humanely, factory farms may 'shoehorn up to 850,000 hogs into cramped cages too tiny for the animals to even turn around.'[144] The animals tend to stand in their own feces, and according to Robert F. Kennedy Jr., President of the Waterkeeper Alliance:

'The hogs endure short, miserable lives, kept alive by constant doses of antibiotics. To stimulate growth, the pork factories force-feed them hormones and heavy metals...A single hog produces [4-6] times the waste of a human being. A pork factory with 100,000 hogs produces the same amount of fecal waste as a city of [600,000] people![145] But while cities must treat sewage before discharging it, these corporate criminals discharge their sewage untreated and use their corrupt political clout to <u>escape enforcement of environmental laws</u>....In 1993, for example, meat factory microbes sickened 400,000 people in Milwaukee (half the population!) and killed 114 individuals."[146]

Besides poisoning groundwater and sickening people, extensive fish kills of up to a billion fish have occurred from individual untreated discharges.[147]

Regretfully, on his very first day in office, President Bush II suspended new hog factory regulations that were designed to stop the above described pollution.[148]

Photo: Marlene Halverson, Animal Welfare Institute
Life in a cage at the Hog Factory 'Farm'

North Carolina, apparently because of 'its lax regulatory scheme, is rapidly overtaking the midwest as "hog butcher to the world."'[149] Now ranked number two in hog production amongst the 50 states (Iowa is #1), North Carolina had 20,000 independent hog farmers fifteen years ago producing 2.5 million hogs. 'Today, they are all gone, replaced by 2,200 hog factories, 1,600 of them owned by a single company – Smithfield (Foods).'[150] In fact, with a hog population of 10 million, there are 2 million more hogs than people in North Carolina today.[151][152]

According to RFK, Jr., as few as two workers may tend to a factory of 10,000 hogs.[153]

When we get to the 'state-of-the-art Smithfield Packing Co. in Tar Heel, North Carolina, 32,000 hogs per day are killed in the 973,000 square-foot plant. Each worker on a processing line is required to cut apart a pork shoulder every 17 seconds for the entire 8 ½ hour workday.'[154]

Going over to chicken processing plants, 70-90 birds per minute roll past each worker, with the worker handling every other one.

Yet, with all the possible contamination, 12 thousand food inspector positions have been eliminated over the past 15 years by our Department of Agriculture. So that today we have only 7,800 USDA food inspectors for our more than 6,200 meat, poultry and egg plants.[155] This is occurring even as corporate food safety directors like Dell Allen of Excel, the USA's number two beef processor, confess "Nobody can {ensure meat will be completely free of E. Coli bacteria}. It's like a roll of the dice or a game of Russian roulette." 'E. coli, or Escherichia coli 0157:H7, first emerged in beef herds in the late 1970's and is now present in 28 percent of cattle entering midwestern slaughter-houses, according to the U.S. Dept. of Agriculture (USDA).'[156] However, our USDA further 'estimates that 89 percent of U.S. beef ground into patties contains traces of the deadly E. coli strain."[157]

Just to refresh your memory, the 0157:H7 strain of E. coli was the Jack-in-the-Box bacteria that killed several children in the state of Washington and sickened hundreds of others back in 1993. E. coli used to normally inhabit human and animal intestinal tracts, helping with digestion. But with the evolutionary adaptation typical of micro-organisms, the 0157:H7 strain appeared somehow, and multiplied, to sicken hundreds of Americans today, and every day. It likes to attack the colon, or large intestine, and cause bloody diarrhea and cramp up your abdomen.

Sometimes it can cause a hemolytic uremic syndrome that can kill. Can you imagine your child dying from kidney failure (uremia)? because she or he ate a hamburger? That is supposed to be an old person's disease, going on dialysis machines, and looking all yellow. Red blood cells get broken up (hemolysis) in such large numbers that they clog up the kidneys, which are trying to clear out the circulation. Sometimes this gets to the point of no return, when too many cells of the kidney are damaged or die. Then dialysis or a new kidney is required, or clinical death intervenes.

Why would this happen in the first place? Well, the fact that animals like cattle and chicken and turkeys routinely get antibiotics in their everyday feed could be the main reason. It's so crazy right now in America, that 80 percent of antibiotics used are not used on people to treat disease, but are given to animals, mostly in their feed. And these animals are not being given these antibiotics because they are sick (yet). It's just that the typical overcrowded feedlot where

Photo: Marlene Halverson, Animal Welfare Institute
Chickens by the thousands packed into the factory
'farm' building, some of them 'de-beaked' inhumanely
so they cannot peck their neighbors to death.

cattle are fattened, or the warehouse or cage in which chickens are housed in the tens of thousands could and does breed disease.

All too often the animals have to stand in their own wastes, or have the animals above them in slatted metal cages, rain their wastes down onto the animals below. Chickens are routinely *debeaked* in wired-in spaces where they cannot even spread out their wings. Because they are so confined so unnaturally, they start behaving violently, pecking at their neighbor that invades their personal space, most of the time un-intentionally. What a way to live! But that is what we have too much of today: factory farming, industrial agriculture.

Animals collected for slaughter en masse, rather than feeding naturally outside on grass. So 0157:H7 E. coli and Campylobacter tend to become epidemic, growing in the feces and the urine and on the carcasses of the crammed-together creatures who also deserve their own measure of dignity in both life and death.

Evidence operations like the biggest one in Connecticut that houses (?housed?) 4,700,000 chickens, controlling 90% of the egg market in that state, having an outbreak of avian influenza.

The farm had to be quarantined, and because of the magna-size of the operation, the country of Japan temporarily banned all USA poultry imports in March 2003!

Not unlike what happened in Pennsylvania when another mutated strain of avian influenza hit the overcrowded super-chicken farms in that state and 17 million chickens had to be destroyed. Plus many other birds numbering in the 'hundreds of thousands' also died from the disease.[158]

Think about this, just in case you didn't know - - again from John Robbins, son of the Robbins of the formerly number one ice cream company Baskin-Robbins:

> The original inhabitants of Earth were bacteria and other micro-organisms, and for more than half of the time that life has existed on Earth they were the planet's only life forms. Even today, though too small to be visible to our eyes, they remain the *dominant organisms on the planet* [my italics].....
>
> ....Micro-organisms are everywhere in the environment, on and in everything that can support life, including our bodies. A square centimeter of human skin harbors 100,000 microbes. The number of bacteria in the gut of each of us is greater than the total number of people who have *ever* lived on Earth. Ten percent of human body weight is composed of bacteria. Each and every human being is, in fact, an aggregate of quadrillions of bacteria.

Most microbes are not harmful, and many play critical roles in human life functions. Without the aid of
certain micro-organisms, we could not digest our food
and absorb its nutrients. Other microbes help us fight
off pathogenic intruders.[159]

In other words, the mutant bacteria and the
pathogens that cause all sorts of illness for us humans have evolved out of the original life forms
that have long inhabited Earth and our own bodies. Those that normally would help us stay alive
and healthy during our days of profligate greed.
Those which our corporations would wipe out in
their ignorance, indiscriminantly trying to kill off
the offending 0157:H7 E. coli and Campylobacter
and such. Yet these corporations persist in their
unsanitary larger and larger operations to make
more and more money, providing increasing mis-
ery for animal and mankind alike.

The Jack-in-the-Box case was not just one rare
episode on the fast food normally safe-and-ster-
ile trail. Remember those 'recalls' of 25 MILLION
POUNDS of ground beef by Hudson Foods in
1997, and the 35 MILLION POUNDS of hot dogs
by Michigan's Thorn Apple Valley firm in 1999?[160]
What about the granddaddy of them all (so far),
the 2008 recall of 143 MILLION POUNDS of

ground beef because 'downer cows' that could be infested with Mad Cow Disease and cannot walk were videotaped being kicked and fork-lifted toward their termination/execution when they are supposed to be banned from the food supply? This is the Westland/Hallmark Meat Company recall, from a company based in Chino, California.[161]

Unfortunately, we are not talking about recalls of Hondas or Suzukis here, where the aggravated owner comes rushing in to get his vehicle fixed immediately and expects a loaner replacement to use in the meanwhile. USDA records tell us that most of this contaminated meat does not make it back onto the meathook.

Like the recall by a Minnesota company of 170,780 pounds of ground beef that only ended up recovering 2,818 pounds in 1999. Similarly, Dr. Richard Raymond, USDA under-secretary for food safety, admitted that the "the great majority [of the Westland/Hallmark meat] has already been consumed." And this is not getting better. Cutting the number of inspectors, corruption, public-relations-infused confusion on what would be ethical practices of our meat industry has actually led to more recalls of E. Coli tainted beef as this decade has gone by: 21 recalls to be exact in 2007, as compared to eight recalls in 2006, and five in 2005.[162] Way too much contaminated meat is progressing into America's kitchens and intestines to sicken unsuspecting consumers like you and me.[163]

Despite the mantra we keep hearing that our meat and poultry industry is 'the safest in the world,' with those CEO's assuring us any problems always get fixed, modern technology is putting germ-contaminating "outbreaks...behind us.....Nothing could be further from the truth.... Unless these cases are part of a widespread outbreak that cannot be ignored, the public never hears of them." writes journalist/researcher Nicols Fox in her book *Spoiled*.[164]

And just to highlight the meanspiritedness of the meat industry: on November 1st of that same year of the Jack-in-the-Box tragic outbreak, 1993, 'the American Meat Institute filed a lawsuit seeking a permanent injunction against the USDA to halt its testing of hamburger for E. coli 0157:H7'!!!!!![165]

So, Viva Food Irradiation! Besides the nuclear industry needing to dispose of its wastes, the factory farms and slaughterhouses of America, with their deplorable hygiene and overburdened workers, espouse the benefits of exposing their meat, pork, and chicken to ionizing rays. To kill off pesky germs and erase the concerns raised by aware citizens like RFK, Jr., and those in the farm community who still advocate smaller farms with their more attentive humane husbandry to produce safe healthy food for our dinner tables.

Aquaculture/Fish Farms

I would be remiss if I did not tell you something about our fastest growing sector of American agriculture today: Aquaculture. This can be defined as 'the cultivation of the natural products of water, such as shellfish and aquatic plants.'[166] About 20-25 percent of our fisheries production currently comes from so-called 'fish farms.' Industry prophets tell us that by 2025, the majority of our fish products will come from aquaculture, and may even surpass cattle-meat production.[167]

Think about this heinous image:

Don Tyson of Tyson Chickens purchases the Arctic-Alaska Fisheries Company and three other fishing companies, which operate $40 million football-field-length industrial super-trawlers that:

"...pull nylon nets thousands of feet long through the water, capturing everything in their path, typically taking in 800,000 POUNDS OF FISH IN A SINGLE NET-TING."[168]

What does that do to the sea-life ecosystem that has just about been raked clean? How will this affect the prospects of individual fisherfolk who do not use such massive, gouging machinery, radar and satellite technology, etc., to outsmart nature's aquatic bounty? As mankind revs up his ultrapowered engines to take all that he can from our oceans, we see our wild stocks of fish plummetting, causing us concern for our future on this planet. 'Only ten percent of big ocean fish remain' in our marine waters.[169]

This is not helped by the fact that per human Earthling, over 400 pounds of hauled-in fish were discarded without even being eaten! There are approximately 30 pounds per Earthling sold per each of us, that we do consume annually.[170]

And yes, per person fish supply has dropped from 32.1 pounds per year in 1987, to 29.3 pounds in 2001[171], perhaps also due to the increase in world population, and the more widespread knowledge about the health benefits of eating fish.

For many of us want to help our hearts, maintain normal blood pressures, fortify our supply of prostoglandins, and protect the integrity of our cell membranes. Fish and their predominance of <u>omega-3 fatty acids</u> (over omega-6 fatty acids) work at the cellular level to keep nutrients inside the cell, where they belong. And by being transformed into prostoglandins, the long-chain omega-3 fatty acids from fish infuse our bodies with a better ability to regulate pain, inflammation, kidney and digestive functions, hormone production, nerve transmissions, and response to allergies.[172]

<u>Wild</u> salmon, as an example, tends to have a 2.5-5 to 1 ratio of omega-3's to omega-6's. That is good! That is the type of ratio you want: more omega-3's than omega-6's.

Meat swings the other way: beef ranging only around 1 to 8, omega-3's the <u>lower</u> number of the ratio compared to the omega 6's. Chicken runs about 1 to 13; and pork ranges from 1 to 8 to 1 to 13, depending on the cut of the hog.

<u>Farmed</u> salmon, however, tend to lose the advantage of the better normal <u>wild</u> ratio of 2.5-5 to 1, commonly only being 1 to 1, depending on the species.[173] This reduces their beneficial health effects.

Alas, farmed salmon are not usually situated

in some mountain pond like many of us envision, feeding naturally on smaller fish and microscopic forms of life. Instead, they often live in pens in the tens of thousands and receive vegetable or grain based feed with preservatives and binders, plus coloring agents to ensure that muscle that could end up gray, ends up pink instead. Experts claim nibbling this sort of feed, instead of eating their normal wild smaller fish and fishlife naturally richer in omega-3 fatty acids, is the main factor in flattening out farmed salmon's omega-3 to omega-6 ratio. The you-are-what-you-eat phenomenon.

Yes, there are movements in the aquaculture industry to correct this by feeding the penned-in salmon, fish, or more fish, like herring and sardines, in their diet. But this type of feeding is more expensive, and though many aquaculturists may love their work, it's basically a business. More expense all too often means less profit. [Unless government subsidies are somehow provided.....]

As a result, most of the feed supplied to these farmed fish comes from soy and corn[174], two of our most genetically altered crops (90% and 73% respectively of these USA crops are genetically altered, as mentioned previously[175]). Plus, now our information-insulated scientists are discovering that canola oil could be added to the fish feed mix[176]. But canola is one of the most prevalently genetically altered crops: at least 60% of USA canola is, as is at least 90% of Canada's canola.[177] Can we be aware of this, and evade contaminating another crucial component of our food staples?

Consumers have to be the ones to push the envelope back under the ticket window here. Ask and inform your local fish salesperson/company/store/supplier about the components of the fish feed used on your fish, and tell them, if they want to supply farmed fish, only to get those that are fed regular old non-genetically altered corn and soy or whatever, or you won't purchase their product. That sends the message. And tell them to read Jeff Smith's book Seeds of Deception; or this book you are reading now, if you want to give them a broader view.

Then there are the basic same types of industrial monoculture agriculture factory-farm problems inherent in our mass production of so much of our foods: too many fish without enough room to move about, their wastes settling below their cages, smothering sea life, leading to the production of algae, the development of diseases and sea lice infestations multiplying in one-species-only constricted environments, the farmed fish meat not always being as 'pink' as wild salmon, etc.

That is why pesticides, fungicides and antibiotics have to be used in varying degrees in different salmon operations to temper the dysfunctions. In the USA, according to Sebastian Belle of Maine's aquaculture industry, the use of antibiotics has to be ordered and monitored by a veterinarian.

He also told me that the majority of the modest sized group of aquaculturists in Maine hardly ever use antibiotics. However, Maine's lobsterers have managed to evade this norm by obtaining 'exemptions' to the mandatory oversight by a veterinarian. As a result, Maine's 'embayed' lobsters can use 'medicated feed' (read that *feed medicated with antibiotics*, usually oxytetracycline) whenever the businessman sees fit.[178]

Again, as with our terrestrial factory farms, overuse of antibiotics still can result, and resistance develop by offending feared organisms. Plus, not all the antibiotic-laced feed ends up being consumed by its intended salmon, but instead drifts off into the ocean to taint the environment beyond the bounds of the salmon pens.

But the worries about sea lice, as an example,

fuel the fears that lead to such antibiotic/pesticide usage. These small crustaceans act like parasites on salmon and other species of fish. Just like potato bugs prospering and multiplying on endless fields of exclusively planted Russet Burbank potatoes to the detriment of the harvest, sea lice feast on and infest the crowded-together thousands upon thousands of penned-in salmon. The salmon get sick, the sea lice multiply, and end up getting flushed out into the environment.

One study showed that 38 MILLION sea lice larvae PER DAY were being flushed from one Atlantic salmon farm into a marine ecosystem where wild trout happened to be migrating.

Those wild trout found and tested close to salmon farms off Ireland's shores showed the highest infestations of sea lice, in an Irish study. And Norway's researchers found 10 times the sea lice infestation on <u>wild</u> salmon discovered near salmon farms, compared to wild salmon where no fish farming was going on. Evidence from studies like these led Ireland's Western Regional Fisheries Board to proclaim that 'an increase in sea lice from fish farms {to be} the sole cause of {that region's} sea trout collapse.'[179]

Similarly in America, citizens are fighting against rampant erection of insufficiently regulated fish farms along our shores. This may especially be flaring due to the push by companies like AquaBounty to introduce bigger better 'transgenic fish' into the USA's food pipeline. AquaBounty's genetically altered version of a North Atlantic salmon will supposedly grow seven times faster than the un-altered wild or usual fish-farmed variety(ies). For some reason it seems to also be more attractive to the opposite sex of the common salmon.

Detractions are: though it may grow faster and bigger, and attract the opposite sex better, it also is projected to live a shorter time. So while it may overtake wild salmon in its mating habits, great concern has been voiced over future inter-mated offspring multiplying and selectively wiping out varieties of normal wild species over future generations. 40 generations to extinction is the number Friends of the Earth quotes from a Purdue University study, should just 60 transgenic salmon escape into the wild.[180] In addition, it is not known how hardy the bigger badder transgenic salmon will prove to be. What if *they* don't make it through the trial of evolution? Then, we will have lost or damaged our stocks of wild salmon just because another huckster wants to make some more money.

Promises to the contrary, people worry that transgenic fish will escape from their pens. Just in British Columbia, a hotbed of Atlantic salmon fish farming on the west coast of Canada, it is estimated that OVER ONE MILLION FISH have eloped from their cages over the past twenty years.[181] Maryland has passed a law banning any transgenic fish operations that would function in lakes or ponds connected to state waterways.[182] California, Washington, and Oregon have passed laws prohibiting outright any production of trans-genic fish in state waters, which extend three miles from shore.[183]

As so many of us know, invasive species are second only to loss of habitat as a cause for the extinction of another species of life. Witness the frightening alien 'snakehead' fish invasion of that pond in Maryland in mid-2002 that led authori-ties to poison the entire pond ecosystem to 'save it' from the foreign Chinese-native marauder that could also walk on land. Authorities feared the larger fish and its offspring would devour so much of the life in the pond, and possibly then clomp on into the Little Patuxent River just 75 yards away[184], that this alien species could then be a threat to who-knew-how-much of Maryland's waterways and the life within them, and onwards and up-wards, plus what about those snakeheads climb-ing onto people's porches and getting into their houses with those big scary jaws and teeth and and...and....?...? (P.S. Even after all this, snake-heads were found in that same pond once again in 2004!)

The dreaded snakehead fish that can 'walk' on land, originally from China. An example of foreign invasive species, threatening the waters of the state of Maryland.

And here comes Monsanto, champion-ing the use of geneti-cally altered bovine growth hormone, as featured injected into 18% of U.S. dairy cows, to synthetically speed up the growth and increase the size of farmed fish like tilapia and shellfish![185]

No wonder the business-boorish Bush admin-istration seriously considered PRIVATIZING THE OCEAN!!!!!!! for the poor unfortunate aquacul-turists who might be pushed from the shores of America where people like to catch natural native fish, swim in clean water, and enjoy the healthful ocean environment they/we deserve! A big reason for this, according to Phil Lansing of the Institute for Agriculture and Trade Policy (IATP) was that "corporations don't want to risk investing offshore without ownership rights, which presumably will increase their chances of a federal buyout if things go sour."[186]

This could have been one of the most stunning travesties most people probably will never hear about, as our world increasingly promotes the commodification of practically anything in the beginning of this, so far, privatize-crazy millennium. Selling plots of ocean to corporations so they can continue their factory fish-farming kind of business out of sight and out of mind!

However, instead of outright privatization, with the Offshore Aquaculture Bill of 2005, Senate bill number 1195, our National Marine Fisheries Service (NMFS) intends to set up leasing plots of ocean to aquaculture corporations in what it calls "the U.S. Exclusive Economic Zone [EEZ]". Which translates to federal waters 3 to 200 miles offshore: in other words, beyond state and local control. This does not make Alaskans very happy, as their laws prohibit any fish farms, period! A fish farm sited a few yards beyond state waters could have very detrimental impacts on Alaska's treasured wild fish populations, with fish escapes and sea lice infestations, as you have been informed by reading this chapter.

Our Department of Commerce, of which the NMFS is a sub-agency, would be the prime mover to expand marine aquaculture. Talk about reversing or diminishing the seafood gap has been spun to the public, but commercial fisherfolk and environmentalists are very concerned that the 2005 Aquaculture bill is too broad and gives the Department of Commerce "total discretion on environmental regulations."[187]

Will the obviously pro-business Department of Commerce bureaucrats protect our waters and our fish properly? one might ask. Be aware that: "Remarkably, the agency {NMFS} admits that getting outside of state and local authorities and their regulations that protect the environment is a major reason for moving offshore."[188]

And think about these shocking fingings: a study done by Professor Emeritus Arthur H. Whiteley of the Department of Zoology at the University of Washington showed that four fish farms produced as much waste as 830,000 Seattle residents. Know that the Seattleans' waste has to be properly treated, while the fish farm waste just washes off to pollute the seas surrounding the pens.[189] Very similar this is to what happens near those industrial hog farms, with all their stinking wastes, resistant microbes, run-off chemicals, and toxic fumes produced, polluting the surrounding environment.

How will the Department of Commerce deal with this specific problem as its subordinate NMFS facilitates leasing out more and more fish farm sites to both national and international corporations?

What about what Phil Lansing has gleaned from NMFS documents (amongst others reported in his article 'Privatizing The Ocean' from iatp. org), in relation to state regulations mentioned above outlawing transgenic fish: "Prepare for commercial introduction of genetically engineered fish upon their approval by the Food and Drug Administration....Include plans for recapture of escaped genetically engineered fish and record keeping of genetic modifications to help monitor those"[190] escaped genetically modified fish that cannot be caught again.

Amongst other troubling considerations, here once again we have the problem of a government agency *supposedly* behaving as both an advocate and a regulator. This too often mutually exclusive schizophrenia tends to breed lopsided policies.

For example, when the NMFS drafted a "Code of Conduct for Responsible Aquaculture Development in the U.S. Exclusive Economic Zone," did it ask any commercial fish organizations for input? Nope. Just aquaculturists. Who might have wanted to buy themselves plots of ocean to breed transgenic fish, or keep their then less-regulatable operations way over the horizon, while commercial fishermen would have had to travel about our seas anxiously hoping to catch healthy wild fish not inadvertently contaminated by manipulated/genetically altered stock.

Phil Lansing suggests that instead of adventuring onto this foolish trail, this "special interest advocate"[191] agency should be contemplating sustainable sensible ways of developing safe salmon and shrimp fishing. The NMFS could help protect ecosystems and local economies instead of mashing and crashing them.[192] And stealthily commodifying our oceans, compromising the healthy lives of their precious inhabitants!

Meanwhile, our government worries about how necessary "genetic improvements" are, and will be, to keep up with that mystical 'competition' out there. Right now, the USA ranks number 5 on the global Aquaculture hit parade, with the overwhemingly largest portion of our fish production contributed by catfish. (280,000 metric tons of catfish are produced per year, compared to a distant second 15-18,000 metric tons of salmon.[193])   Our catfish are mostly produced in inland facilities in the southeast. As William R. Wolters, who heads the Agricultural Research Service Catfish Genetics Research Unit in Stoneville, Mississippi, contends: 'Genetic improvement of channel catfish [the most common kind] is essential for long-term viability of the U.S. catfish industry.'[194] Will this possibly include transgenic catfish? Stay tuned in to transgenicfish@iatp.org as one way to keep abreast of the unfolding developments here.

As a further unsettling piece of news re catfish, and the regulation of their aquacultural raising: fish farming companies want to re-register a "pisicide" called antimycin A to kill off undesired scaled fish in catfish ponds. "Pisicide" is not a very pretty word, to start with. Neither is the fact that the EPA has not done sufficient studies to determine the safety of this pesticide.

The EPA's own Health Effects Division [HED] states, "Based on its use pattern there are potential exposures for workers during application, children or adult recreational users of treated lakes/streams via swimming, anglers harvesting fish after treatment, and dietary exposures from the catfish farm use, and potential drinking water exposures from the treatment of lakes/streams/ reservoirs that could be used as a drinking water source."

In addition, the HED admits "there are insufficient data at this time to conduct a quantitative human health risk assessment for antimycin A." Plus, though the EPA may test for pesticide residues in food in their so-called "Total Diet Study," residues for antimycin A are not included.[195]

Food and Water Watch notes that antimycin A kills catfish at 20 parts per billion [ppb] concentration relative to its dilution in water. They added that trout and carp died at 5 and 10 ppb exposure, as found in a 1966 study. Yet the residues in the flesh of these two fish that YOU eat were found to average 76-201 parts per MILLION, which is a <u>higher</u> concentration. The carp, which survived the antimycin A longer than the trout, had two to three times more residue. Since catfish survive longer than carp, the antimycin A residue in the catfish flesh could be even higher.

Another study from the Bureau of Sport Fisheries and Wildlife "determined that the effects of antimycin A on fish are likely irreversible, even if the fish is placed in water that does not contain any of the pisicide."

So ideas like not harvesting the catfish for a year after exposure to the antimycin A are not good enough to accomplish the ensuring of "reasonable certainty of no harm" to people and the environment in general, as mandated in the Food Quality Protection Act of 1996.[196]

There needs to be more appropriate regulation of pesticide, pisicide, and whatever-other-cide use in fish farming. And the question must be asked: why can't we raise fish in a more humane and wise manner if we want to feed each other, our fellow human beings, without poisoning the planet and our own flesh and internal organs??

Stepping back, we should know that "the vast majority of global aquaculture, about 85 percent,[197] uses non-carnivorous fish species - like tilapia and catfish - produced in land-based ponds for domestic markets. Most ponds are ecologically integrated into the agricultural, industrial, and community fabric, meaning, for example, that wastes become fertilizers rather than pollutants."[198]

It all started thousands of years ago in China -- which still ranks number one in the world today in terms of volume of fish produced. China bred and grew freshwater carp in ponds, the practice beginning especially after waters lowered following river floods. Hawaiians constructed fish ponds like the one at Alekoko beginning around 1000 years ago.[199] Eventually aquaculture "spread to Europe, where tilapia, turbot, cod, sole, catfish, and sturgeon [were] raised in ponds and land-based tank systems."[200] These types of fish were herbivores, unlike salmon, which eats other fish, and theoretically causes "a 'net loss' of protein in the global food supply."[201] For "it takes two to five kilos of wild fish to produce one kilo of salmon,"[202] if fish rather than synthesized/non-animal-based feed is used on those salmon farms.

Then there are the oysters, mussels, clams, shrimp and scallops often farmed in mangrove swamps along tropical coastlines. This could be sustainable, but all too often, instead of integrating a shellfish farm in a non-polluting viable fashion, too much of the mangroves are chopped down, wastes build up too quickly, and often in three years, POOF! the operation has to move on, leaving ruination in its wake. The local people have lost their essential natural breeding grounds for sea-life, their beaches are polluted, their wild shrimp and shellfish have disappeared, as the plundering shellfish farmers move on to rape the next mangrove forest.

However, the economics seem compelling, in the coke-bottle short-term lenses of the rural rice farmer. Just imagine: say you earn the equivalent of $500 to harvest an acre's worth of rice, but you can earn $20,000 by shrimp farming the same acre? What do you choose?

Yes, the shrimp farm may typically only last three years, especially if you over-use your antibiotics, pesticides, and fungicides in overcrowded shrimp pens. And then you may have to uproot your family to move to the next mangrove forest. But, heck, you might have earned $60,000 worth of USA greenbacks!

Some shrimp/'prawn'[203] notes now for you shrimp/'prawn' lovers:

-- Today, about 90% of all shrimp consumed in the USA is farmed and imported.[204]

-- Most of the farmed shrimp we buy comes from Asia.[205]

-- 30 % of worldwide shrimp comes from shrimp farms.[206]

- - Wild shrimp are predominantly caught in trawl nets.
- - "Tropical shrimp trawling has the <u>highest</u> 'by-catch' of any commercial fishery"
  - 3-20 pounds[207] of "undesired animals are caught and discarded per one pound of shrimp." This includes sea turtles, but USA laws mandate special netting to let fish and sea turtles escape from sea trawling nets to provide a low by-catch alternative.[208]
- - Shrimp trawlers are responsible for 1/3 (33 %) of the world's by-catch, yet only contribute 2 % of the world's seafood.[209]
- - "Up to 25 % of seabed life can be removed in one trawl,"[210] wrecking the marine ecology and business of fishermen in countries like Guinea-Bissau, Mozambique, Venezuala and Greenland, via developed-world subsidizing of such trawling.[211]
- - Such "trawling is the greatest single threat to sea horses."[212]

However, to amend the above statement re seabed life, an Australian study found that <u>95 percent</u> of trawled deep water ocean bottom off Tasmania showed just bare rock post-trawl, as compared to 10 percent in untrawled areas.[213]

As far as the so-called 'seafood gap' goes amidst our multibillion dollar overall trade deficit, we find that $3.7 billion worth of <u>imported shrimp</u> is the biggest contributor to our total $8 billion dollar 'seafood gap' or 'seafood deficit.'[214] Much of it comes from destructive practices as listed above, from countries like the number one seafood exporter (in terms of dollars generated) on the planet: Thailand. Yes, <u>Thailand</u>, at a valued $4.4 billion annually for total fish exports. (China ranks second with actual fish <u>exported</u> valued at $3.7 billion.)[215]

Photo of a sustainable aquaculture farm in Honduras, Central America. Instead of chopping down mangroves and polluting the mudflats along the Pacific coastline, Grupo Granjas Marinas has been in business for over 30 years planting new trees, growing shrimp in balance with the estuaries and environment.

The USA aquaculture industry wants to become a player here, with government support. And why not? We consume about 400,000 metric tons of shrimp per year![216] Overall, we are the second largest seafood importer on the planet (after Japan[217]). We have a burgeoning willing band of aquaculturists. But can we do it sustainably, without just going for the money? As is being done at the model being held up for the world at the Grupo Granjas Marinas (GGM) in Honduras, Central America.

There on the country's pacific coast mudflats, GGM has built ponds, some of them still in service after 30 years. Instead of chopping down mangroves, they plant them, use low flushing techniques so as not to poison the estuaries surrounding the ponds, use low protein feeds, and keep the amount of growing shrimp stocked at relatively low population densities.[218]

Dead Zones, (GMO) Corn,
Fertilizers and Ethanol

Somewhere along our Gulf Coast, perhaps we could do the same sort of thing. Though first we may have to correct a few problems. Because, as BJ Narog, former Assistant Editor of _Aquaculture_ magazine extrapolated from the Environmental Protection Agency (EPA) 2002 National Coastal Condition Report: "almost all of East coast waters, and the Gulf of Mexico {are} deemed unfit to sustain life."[219] Perhaps due to the loss of up to "50 percent of the land to urban sprawl...that used to {filter and} buffer ocean waters,"[220] – at least, along the Gulf.[221]

Then there is that 'dead zone' in the Gulf that has an area of 20,000 square kilometers[e]. No fish. No shrimp. Richard Manning, author of <u>Against The Grain: How Agriculture Has Hijacked Civilization</u>, blames its appearance on runoff of nitrogen from fertilizer into the Mississippi River, and on out down south into the Gulf of Mexico. Referencing an Army Corps of Engineers study, he states that 'seventy percent of the Mississippi's nitrogen comes from a relatively small six-state area that is the heart of the nation's corn belt.'[223]

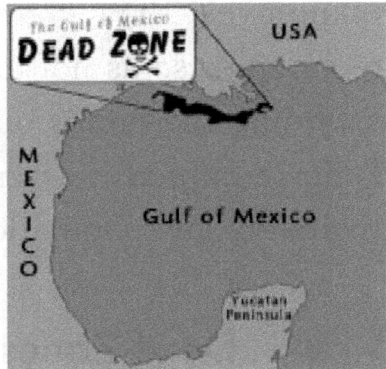

'Dead Zone' in the Gulf of Mexico, from outflow of Mississippi River. ~20,000 square kilometers = ~8,000 square miles. No fish. No shrimp. Bound to increase in size due to amping up mid-west ethanol production. [222]

A recent June 16, 2008 U.S. News & World Reports article notes that 'fertilizer runoff...is

---

e 20,000 square kilometers equals about 8000 square miles, which could also be visualized as a square area 90 miles on each side. Though, actually, the dead zone in the Gulf is spread mostly along the immediate coastline west to Texas, and east to Alabama.

the source of three quarters of the nitrogen and more than half of the phosphorus in the water... Unlike nitrogen, which eventually evaporates as a gas, phosphorus lingers in the water, contributing to dead zones of the future.' A worsening of this problem can be blamed on 'ethanol mandates,' encouraging farmers to plant 'record-size crops.' Especially corn, with about 15 million more acres of it planted between 2006 to 2007.[f] And what percentage of this corn was genetically modified? More genetic and chemical pollution has surely resulted from the current ethanol rush our government is dys-economically sponsoring.

Americans should know that recent research has shown that ethanol produced from 'corn requires 29% more fossil energy than the fuel produced.'[g] However, sugarcane, which Brazil is using for their ethanol, provides eight times more energy output than fossil energy put in to produce it.[h] Thus, Brazil's ethanol economically 'provides the fuel for 40% of [its] domestic transportation.'[i] However, be aware that sugar cane requires four times as much potash as fertilizer than do soybeans or wheat.[j] (Potassium carbonate has been the traditional form of potash, and the English word 'potassium' came from the term potash. Saltpeter is another form of potash, potassium nitrate.[k])

Cornell University Professor David Pimentel states: "The [U.S.] government spends more than $3 billion a year to subsidize ethanol production when it does not provide a net energy balance or gain, is not a renewable energy source or an economical fuel. Further, its production and use contribute to air, water and soil pollution and global warming," ...He points out that the vast majority of the subsidies do not go to farmers, but to large ethanol-producing corporations."[l]

Can we get on the sugarcane ethanol bandwagon instead of the corn ethanol subsidy express? Maybe in our warmer states like Florida and Hawaii? Can we or Brazil or whichother countries start producing sugarcane ethanol do so without the complications of industrial sugarcane production: excessive pesticides, herbicides, fertilizers?? Corn-based ethanol production is just another form of corporate welfare that Americans don't need to expend our good funds upon, nor pollute our environment with.

Realistic Prospects And Measures For Improvement For U.S. Aquaculture

Getting back to the upside of aquaculture, experts tell us "the deserts of Arizona...the clay ponds of Alabama and Mississippi, raceways on the Atlantic Coast...{places} 300 kilometers from the nearest ocean"[224] NOW have shrimp ponds!! So, sustainable shrimp farming within America's waters and territorial borders is already more than a possibility!!

And we can also make improvements with our salmon pens along our northern coastlines so that sewage and antibiotic/fungicide/pesticide discharges are minimized or excluded from occurring into our oceans. The David Suzuki Foundation suggests 'closed-loop containment systems' to prevent escapes and environmentally contaminating releases. Also suggested are openly monitoring drug use and especially drug-resistant diseases. Siting of aquaculture operations not imperialistically, but in communion with the involved communities by doing so equitably and democratically, also is a very practical idea. As is making the responsible companies develop and fund reclamation programs for their enterprises, while maintaining insurance for "full ecological restoration costs of disease epidemics, escapes, genetic pollution, and other catastrophes."[225] Clean up your mess, Johnny. Prevent Superfund sites before they fester needlessly.

As with all of our earthly enterprises, we can either create them realizing our shrinking planet is a place we need to preserve, be a part of, respect its remarkable as-yet-incompletely-understood interworkings; rather than stomp like radioactive monster-rapists wherever we decide to erect our next temporary money-camp, planning to dominate our soon-to-be urinal/dump before running amuck, clearcutting our way to the following paradise we will destroy shortly.

Aquaculture holds so much promise. Country-of-origin labelling was due in 2004, and has been postponed repeatedly, most recently to the year 2008 (yes, the regulations must be tweaked not to be counterproductive); Americans want to make sure all that fish they will be eating in the future will be safe; wild fish harvests are not projected to increase to meet the projected acceleration of worldwide demand for fish.

Yes, it would be terrific if we could aquaculturally produce organic fish, but there are some barriers to such a designation at the present time.

So, for now, find out where your fish is coming from. Ask your storekeeper or fish department manager or restaurateur about this, and tell her or him what <u>you</u> know. Ask if whomever supplies the fish use antibiotics, meat colorants, genetically altered corn, soy, or canola in their feed, etc.

Big tip on salmon for example: the farmed type at present <u>usually</u> is Atlantic salmon, but chinook (king) or coho (silver) salmon are also farmed.[226] If it is Pacific or Alaskan salmon, probably it is wild. Atlantic salmon raised in Pacific ocean pens are not really native to that ocean, but are there because Norway developed the technology. Rather than change everything around, Norwegian companies that operate from Canada to Chile[227], with Norway also being the number one aquaculture producer in Europe, keep Atlantic salmon in their pens wherever these are sited.

Sometimes the environment is not suitable for these non-native species, requiring more chemical and antibiotic nurturing than knowing consumers would like. But, for now, that's the way it is. So keep that grain of salt in your pocket when you decide to buy your salmon.

Plus, as a further reality check, not to be deterred, the burgeoning lobbying aquaculture industry has managed to get an Offshore Aquaculture bill introduced into the House of Representatives on April 24, 2007. Its number is H.R. 2010. This though the Senate bill never actually even made it out of committee to be debated in the full Senate.

However, with the House bill, states are promised that they can "opt out" of the bill's provisions for fish farms allowed in their waters for <u>twelve miles</u> out from their shores. Sounds good, but the Senate bill allowed the opting-out to go <u>200 miles</u> from non-approving state shores.

Then there may be exemption of the aquaculture sites from the Jones Act, which currently requires fishing and aquaculture activities to be subject to U.S. laws.

Another further complication could be the allowance of "guest workers" on these offshore enterprises: in other words, non-Americans, probably for less pay, without benefits, the whole "free-trade"/corporate-managed trade intimidatory arrangement unscrupulous businesses and corporations love to foist upon the planet only for *their* profit.

Big dreamers think big, if not right all the time.

Apparently many of the new facilitated aquaculture fish may be the carnivore kind, like halibut and black cod, that would be fed other smaller fish in their crowded offshore pens. Net protein loss resulting over the fish's lifetime would be about <u>20 pounds per pound of carnivore fish</u>, as compared to smaller omnivore fish where it commonly might be <u>3-5 pounds of protein</u> loss per pound of fish. Many of the irreverently labelled "trash" fish extracted from poorer countries' waters might be fish people of Peru and Chile and such nations normally eat as part of their diet. Besides being an intricate portion of the marine food chain. And the reaping of these fish likely will be done in an industrial indiscriminant manner as described at the beginning of this section by huge netting ships.

At the present time, we do not know if and when genetically altered fish like carnivorish salmon will arrive in offshore pens. The FDA is not telling us, because business is business, and since such fish are patented, information about them is "proprietory." Common citizens shall not be privy to find out what we should know in a transparent democratic manner, to maintain the integrity of the oceans and the life-forms naturally safely living in our waters. The Department of Commerce is a business advocacy department of our government unfortunately, supposedly regulating our oceans for the betterment of our environment and our human populace.

Anne Mosness, who has been following fishing issues for many years, agonizes that the aquaculturizing of big fish like black cod and halibut in state waters like those of her state of Washing-

Pebble Open Gold and Copper Mine and Bristol Bay Watershed Area

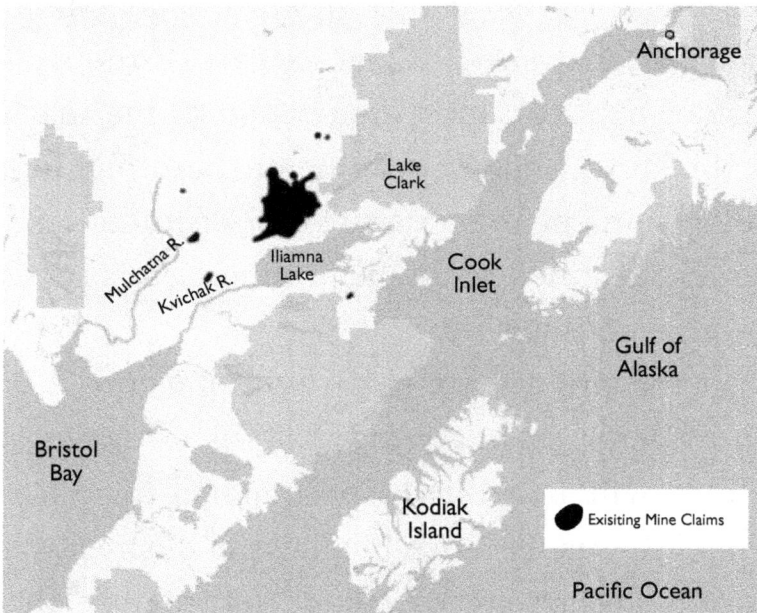

Proposed Pebble Open Mine Site (in black) would threaten Alaska's Bristol Bay watershed, "the source of the most productive commercial and sport salmon fisheries in the world." Mine site would include largest dam in the world, larger than China's 3 Gorges Dam, but be made of earth, not concrete – to hold back toxic mining waste that will include arsenic and cyanide.

ton could bankrupt the currently viable fishing industry and its coastal communities. This could lead to distraut displaced workers, who had prospered on fishing and the complementary economy it produced, getting desperate and seeking jobs detrimental to their overall benefit. For example, they might accept work in oil drilling and toxic mining of gold and molybdenum to employ and destroy them and their beloved environment, once the coastal fishing industry has been subjugated to humiliating relative worthlessness.

Specifically Ms. Mosness mentioned the proposed Pebble Open Gold and Copper Mine that would despoil the Bristol Bay watershed up in Alaska. The location for this unfortunate potential adventure would be at the headwaters of the "two most famous salmon producing river drainages in Alaska - - [those of] the Mulchatna/Nushagak and Newhalen/Kvichak [rivers], both of which feed into the renowned Bristol Bay. The proposed Pebble Mine, which would be the first of many, would include the largest dam in the world, larger than the Three Gorges Dam in China, and made of earth not concrete, to hold back the toxic waste created in the mining process."[228]

We're talking about arsenic and cyanide here, for the mining marauders to use in their process.

Poisoning the rivers, the Bristol Bay, the resplendent area where 43,000 sockeye salmon migrated in just one month's time to breed and temporarily exist in 2006.[229] The area that has been called "the source of the most productive commercial and sport salmon fisheries in the world."[230] Though the caribou and moose, bear and other animals cannot voice their displeasure, dependent as they are on these unpolluted waters, local native human opposition did register about 75% in a recent poll. The former Governor of Alaska, Jay Hammond, said "I can't imagine a worse location for a mine of this type unless it was in my kitchen."[231]

Providing the bigger picture for your viewfinder, Anne Mosness warns that what is going on is the Bush administration "is promoting the idea that we must rush to develop the offshore fish farm industry before other nations rear caged fish and flood our markets. [However,] the species chosen are not going to feed the world, but instead will accelerate the loss of fish at the bottom of the marine food chain. One third of ocean fish are now harvested for feed for animals and caged fish. And the amount plundered from oceans will increase if there is a substantial expansion of caged carnivores [like halibut and black cod].

When heavily subsidized, imported salmon began flooding markets in the U.S., the value of commercial salmon fishing licenses in Alaska decreased in value by 90-95% in some regions. [Meanwhile,] the region is facing oil and natural gas drilling, after President Bush lifted the moratorium [on these activities] in the winter of 2006. Consumers, after learning of the health benefits of wild salmon, and the serious environmental and economic impacts of fish farming, as well as the health risks of feedlot produced fish, can protect the watersheds and coastal regions of wild places by purchasing wild salmon. Only then can exploitation of unrenewable resources be halted, and sustainable wild fisheries continue as they have for thousands of years."[232]

Food, Seeds, Patents, Yesterday, And The Future

Now, let us step back a bit from the microscopic lens focused on today, the present, and take a broader view of what has happened, and probably will happen in the future with our food and the supply of it.

Allow me to share a few paragraphs from two of our greatest living thinkers on the big picture with you.

First, Ivan Klima, one of Czechoslavakia's finest writers, who remained in his country from the beginning of its Soviet occupation after the 'Prague Spring' of 1968, until the present day, enduring all sorts of humiliation

Ivan Klima, who remained and wrote in Czechoslovakia during the Soviet occupation.

and intimidation, but continuing to write all the while. From his January 1980 essay 'The Powerful and the Powerless,' please ponder this:

Sometime around the middle of this [20th] century people living in the Euro-American region of the world once again embraced the foolish illusion that they (genuinely, this time) were entering the promised land. This illusion was based on the false assumption that they had essentially managed to solve the most basic social problems: how to guarantee people a decent living, offer them the chance to realize their worldly dreams without too much effort, and enjoy the gifts of life: eating, drinking, travelling and living to the full. In fact they achieved nothing of the sort: an extravagant lifestyle was attained for one or two generations of a tiny percentage of mankind, but at the cost of the unimaginable devastation of the entire planet and the squandering of energy stored up over millions of years. We can scarcely imagine the price we have paid for such an illusion. I am not thinking of ecological crimes alone, which our grandchildren's grandchildren will still be wrestling with (if they survive at all), but of something far worse.

In its efforts to organize the greatest number of forces to 'overcome nature,' to 'suppress its enemies,' to promote 'further growth,' or to 'defend gains already made,' modern society has generated huge administrative, military and police structures. They were intended to serve society, its citizens and every individual, who was to be recognized as the source of their power, which was merely delegated to them. In the beginning they did no doubt recognize this. But then these structures began to behave like everyone to whom power is delegated: they began to usurp it for themselves, to the detriment of those from whom it originally derived. Certainly, there are societies where such structures are subject to some kind of control, but in most countries, they are not. They are no longer governed: they govern.'[233]

Klima wrote this whilst the Soviets were still ruling his land and the minds of his people, which they did until 1989, when the Berlin wall fell, and communism/The Soviet Union all but evaporated. Back then, if you reported that some one or some factory was polluting the Danube river, more likely than not you'd end up in prison, with at least a little bit of torture under your toenails, while the pollution proceeded unimpeded.

Nevertheless, this macroview of 'Euro-America' prosperity at the price of plundering the planet can be coupled with Vandana Shiva's perspective on 'Piracy Through Patents.' Ms. Shiva is a brilliant physicist and ecologist from India, who also was a winner of the 'Alternative Nobel Peace Prize' known as the Right Livelihood Award, in 1993. Here is a different historical portrait of the last 500 years than what most of you probably were taught during your schooldays:

Vandana Shiva, one of the Earth's most brilliant visionaries. Winner of Right Livelihood award 1993.

'On April 17, 1492, Queen Isabel and King Ferdinand granted Christopher Columbus the privileges of "discovery and conquest."
One year later, on May 4, 1493, Pope Alexander VI, through his "Bull of Donation," granted all islands and mainlands "discovered and to be discovered, one hundred leagues to the West and South of the Azores towards India," and not already occupied or held by any christian king or prince as of Christmas of 1492, to the Catholic monarchs Isabel of Castille and Ferdinand of Aragon. As Walter Ullman stated in *Medieval Papalism*:

The pope as vicar of God commanded the world, as if it were a tool in his hands; the pope, supported by the canonists, considered the world as his property to be disposed according to his will.

Charters and patents thus turned acts of piracy into divine will. The peoples and nations that were colonized did not belong to the pope who "donated" them, yet this canonical jurisprudence made the christian monarchs of Europe rulers of all nations, "wherever they might be found and whatever creed they might embrace." The principle of "effective occupation" by christian princes, the "vacancy" of the targeted lands, and the "duty" to incorporate the "savages" were components of charters and patents.

The Papal Bull, the Columbus charter, and patents granted by European monarchs laid the juridical and moral foundations for the colonization and extermination of non-European peoples. The Native American population declined from 72 million in 1492 to less than 4 million a few centuries later.

Five hundred years after Columbus, a more secular version of the same project of colonization continues through patents and intellectual property rights (IPRs). The Papal Bull has been replaced by the General Agreement on Tariffs and Trade (GATT) treaty {see Chapter Four}. The principle of effective occupation by christian princes has been replaced by effective occupation by the transnational corporations supported by modern-day rulers. The vacancy of targeted lands has been replaced by the vacancy of targeted life forms and species manipulated by the new biotechnologies. The duty to import savages into Christianity has been replaced by the duty to incorporate local and national economies into the global marketplace, and to incorporate non-Western systems of knowledge into the reductionism of commercialized Western science and technology.

The creation of property through the piracy of other's wealth remains the same as 500 years ago.

The freedom that transnational corporations are claiming through intellectual property rights protection in the GATT agreement on Trade Related Intellectual Property Rights (TRIPs) is the freedom that European colonizers have claimed since 1492. Columbus set a precedent when he treated the license to conquer non-European peoples as a natural right of European men. The land titles issued by the pope through European kings and queens were the first patents. The colonizer's freedom was built on the slavery and subjugation of the people with original rights to the land. This violent take-over was rendered "natural" by defining the colonized people as nature, thus denying them their humanity and freedom...

These Eurocentric notions of property and piracy are the bases on which IPR laws of the GATT and World Trade Organization (WTO) have been framed. When Europeans first colonized the non-European world, they felt it was their duty to "discover and conquer," to "subdue, occupy, and possess." It seems that Western powers are still driven by the colonizing impulse to discover, conquer, own, and possess everything, every society, every culture. The colonies have now been extended to interior spaces, the "genetic codes" of life-forms from microbes and plants to animals, including humans.'[234]

Relating this to food, we see Monsanto and the Gene Giants patenting seeds and plants that have been grown and tended for hundreds and thousands of years by different 'indigenous' civilizations. Some big. Some small. Because the world was not what it is today. Communication was not as miraculously advanced. People saved those good seeds, selected the best ones that produced the best plants, to grow those fortifying delicious natural foods. In the mountain forests of South America, the plains of Africa, the oases of Iraq, the midwest breadbasket of the USA.

'Improvements' like pesticides and fertilizers were introduced only at the tail end of the last 500 years to increase crop yields, but also increased pollution of ecosystems, while profit was a great motivation not to be discounted.

The Gene Giants want to patent and own 'genomes,' life forms, life itself actually, so their 'product,' their seed(s), their bioengineered potato or tomato or rutabaga can be protected under 'intellectual property rights' like a movie script, a new invention, or a Bob Dylan song. They are using the World Trade Organization (WTO) and its TRIPs agreement, as mentioned above by Vandana Shiva, to implement their agenda, which in reality, is a sort of biologic and economic imperalism.

The locals in India have their rice, which they have nurtured and grown and cross-bred over the centuries, in many different forms. Then some 'bioprospector' from our great country comes on in, talks with the locals, gains plenty of local knowledge on what works best, and in 1997 Texas-based RiceTec is granted a patent for "Basmati" rice.[235]

Under the TRIPs agreement, all 151 WTO member nations have to uphold the monopoly marketing rights for patented products sold in their country. That might make sense for pirated copies of the movie 'Titanic,' but with Basmati rice, which RiceTec admitted in their patent application Indians and Pakistanis have grown for generations, this caused quite an uproar.

'A coalition of eminent Indian civil society groups sent a letter to the U.S. ambassador to India...stating: "The truth is that the U.S. is pirating the intellectual property of the farmers, healers, tribals, fisherfolk of India and other developing countries."'[236]

Remember that India as a civilization has been in existence for over 5000 years, utilizing their natural resources and developing their own laws in their own unique ways. India patent laws <u>banned</u> patents for substances "intended for use, or capable of being used, as food or as medicine or drug."[237]

The upstart USA, in relative terms, is maybe 400 years old going back to Plymouth Rock, or 225-plus, going back to our Declaration of Independence in 1789. We're new, younger, faster, not as traditional as the cultures of India, China or the threatened U'wa civilization in northern Colombia. And with the WTO and all its manic corporate-plunder-facilitating policies approved in 1995, the world as an oyster to be shucked royally for its various jewels of potential wealth is paying a heavy price. As portrayed in the Klima quote above.

Here are some facts to garner your attention on this:

Corporations or individuals in industrialized countries hold 97% of all patents worldwide![238] In 1995, more than half of global royalties and licensing fees were paid to the corporations from the <u>US of A</u>.[239]

Even within developing countries, 80% of the patents granted belonged to residents of industrialized countries.[240]

The other side of the coin is that to attain the

patent and enforce it worldwide is too expensive for some Amazonian tribe or indigenous farmers in Namibia (Africa). For example, a Namibian community wanted to patent a local plant with medicinal properties 'to prevent bio-piracy by multinational pharmaceutical companies.'[241] That way, they perhaps might be able to profit from it monetarily, while also being able to continue using it. But when the costs were figured up, they ran to nearly $500,000 for ten patents covering a single invention in 52 countries, plus the additional costs of enforcing the patents in various courts.

The conclusion then was "there is no way a community in Namibia could possibly afford to jump on the patent bandwagon. The costs involved make patents the domain of the rich and powerful."[242]

With the TRIPs Agreement, patent protection of cloned or genetically altered cell lines is <u>required</u> by WTO signed-on countries. What the big corporations do is claim that they altered or produced an innovative new form of the Basmati rice or the Neem tree or the corn plant so that they now own it, and anyone that grows it or uses it illegally is subject to being sued for doing so. They patent their product, and the border gets dimmed between, say, RiceTec's Basmati rice, and all other forms of Basmati rice. Sometimes, because of marketing and distribution realities, farmers cannot purchase non-genetically altered varieties of the seeds they plan to sow.

Often, they are not even aware of the type of seeds they are buying. Poor suicide-prone Indian and third-world farmers - - including Iraqis suffering under Order 81 imposed by the USA-led Coalition Provisional Authority and administrator L. Paul Bremer III to outlaw Iraqis saving new seeds in favor of establishing a transnational corporate-controlled seed market in their country, prohibiting what farmers have done there for millennia[243] - - are not the only ones susceptible here; American farmers in the USA suffer the same fate.

Part of the reason is that with their dauntingly large distribution, litigation, and marketing networks, the Gene Giant corporations can control vast portions of markets for seeds, foods and drugs this way, with the WTO's backing, and threatened often-multi-million dollar trade sanctions.

All too frequently, these gene alteration claims are insignificant, but patent examiners lack the capabilities to test the alleged "new trait." As a result, the patent gets granted, and the 'validity of the claim is left to civil litigation, which is too costly for indigenous communities to undertake.'[244] What then happens is that, in the case of a crop of Basmati rice that is 'illegally' planted, the WTO member nation has the responsibility to enforce the seed companies' patent rights by either uprooting these 'illegal' plants, or collecting the fees from subsistence farmers.

Do you envision those protestors uprooting genetically altered testcrops when you read that last sentence? Yin and yang. Different sides of the same halfpenny.

Although the media in America trivializes the anger circling the globe concerning our aggressive push to capture patent rights on life forms, which will further impoverish the poorer developing nations, especially when it comes to food and food distribution, this debate is far from over.

When the 1999 Seattle WTO meeting was going on, with much ado, and tens of thousands of activists effectively disrupting the expected success that was not achieved, 'developing countries led by the African Group {proposed} that "all living organisms and their parts cannot be patented; and those natural processes that produce living organisms should not be patentable.'"[245] In addition, it was also proposed to "ensure the protection of innovations of indigenous and local farming communities; the continuation of traditional farming practices including the right to use, exchange and save seeds, and promote food security."[246]

Why should anyone have the right to 'own' a plant-line or an animal-line or even <u>you</u>! via your genome, or your plotted out gene-chromosomal make-up? John Moore, a cancer patient, had his cell lines patented by his own doctor![247] 'The human genomes of the populations of Estonia, Tonga, and Iceland have been bought and patented by private corporations.'[248] 'The cell lines of the Hagahai of Papua, New Guinea and the Guami of Panama are patented by the U.S. Commerce Secretary.'[249]

Monsanto and the Gene Giants have patented so-called 'terminator' seeds, which will produce plants whose seeds will not germinate. Then purchasing farmers will have to buy new seeds from them every year. Although now, Monsanto et al. have arranged contractual obligations in which the seed-purchasing farmer is not <u>allowed</u> to replant non-terminator seeds spawned from Monsanto et al. owned/sold crops.

As in the nightmare at the beginning of this chapter, Monsanto is indeed 'suing hundreds of U.S. farmers for "patent infringement" for the "crime" of having genetically engineered plants growing on their property without paying royalty payments' to their company.[250] The problem with genetic drifting of unwanted bio-engineered pollen cross pollinating non-genetically-engineered farmers' fields has been reported by Canadian CBC radio with canola plants 'all across the Canadian prairie.' "The genetically modified canola has, in fact, spread much more rapidly than we thought it would," said Martin Entz, a plant scientist at the University of Manitoba . "It's absolutely impossible to control."[251]

Farmers are fighting back by counter-suing Monsanto in North Dakota and Illinois for 'deliberately causing genetic pollution, and then turning around and suing innocent farmers who are victims of this genetic trespass.'[252]

Renowned visionary author Jeremy Rifkin comments that the biotech industry is "hoping there's enough contamination so that it's a fait accompli." A done deal. Something that has happened and cannot be undone. "But the liability will kill them. We're going to see lawsuits across the Farm Belt as conventional farmers and organic farmers find that their product is contaminated."[253]

Meanwhile, though, Tennessee cotton farmer Kem Ralph was sentenced to eight months in jail for saving seeds supposedly created by Monsanto. He also 'is being forced to pay $1.7 million dollars in damages to Monsanto.'[254] And the relatively famous case of Monsanto vs. Percy Schmeiser went to the Supreme Court of Canada, where the verdict went 5-4 in favor of Monsanto. Monsanto sued Schmeiser because 'their' canola seeds/plants were found on his land, which Schmeiser claimed was due to that 'impossible to control' genetic drift, mentioned above. However, neither party won any fine or amount of money from the other. A strange decision, as related by Brewster Kneen, co-editor of *The Ram's Horn*, discussed in the June 2004 issue of that British Columbia, Canada, publication.[255]

The latest and most terrible actions concerning all this are the so-called 'Monsanto Laws' already passed in fifteen states as of this writing. Because over 100 communities in the northeast USA, and counties like Mendocino and Trinity in California, have passed laws outlawing the planting of genetically altered crops within their borders, Monsanto et al. have quietly overriden them via 'legal' means. But not being hoaky or impractical, these Monsanto Laws have been passed with <u>statewide</u> application for Texas, Oklahoma, the Dakotas, Iowa, Florida and other states.

Voters and farmers, activists and consumers and backyard gardeners, who wanted to protect their crops, organic and not-necessarily-organic, from genetic drift of genetically altered pollen/ seed, were the citizens who worked to accomplish 'justice' at the ballot box. But money talks loudly, as we all know, and the deed was sneakily being done in state halls of congress without the media doing any significant reporting of this issue. Have you read about it in your local paper, or the *USA Today?* heard about it on your radio or TV?

I was going to tell you to watch for the furtherance of this tactic in states like Illinois, North Carolina, and the biggie state for agriculture, California. However, new <u>state</u> "Monsanto Law" passage seems to have ground to a halt by just minimal information being provided about the heinousness of these laws. So what did the Gene Giant corporations try to do with their wicked means to an evil end? They worked new language into the massive 2007 Federal Farm Bill for <u>all</u> of the USA that would have pre-empted "any state prohibitions against any foods or agricultural goods that have been deregulated by the U.S. Dept of Agriculture [USDA]."[256] In other words, if this version of the Farm Bill would have been passed, we would then have had one totalitarianly bad "Monsanto Law" for the entire nation that would have taken enormous energy, money, and time to repeal.

At last report, this language was removed from the 2007 bill, so aware individuals who fought against the "Monsanto Law" federalizaton can breathe a sigh of relief (?temporarily?).[257] Here are some interesting state laws that would have been compromised via pre-emption as cited on the organic consumers website concerning this scamming:

"Legislation adopted this year [2007] in the state of Washington, which prohibits planting of GE canola in areas near the State's large non-GE seed production. Brassica (cabbage, broccoli, and other such crops) seed producers pushed for this legislation, since GE canola can cross-pollinate with and contaminate natural cabbage seed. The Skagit Valley area in Washington produces $20 million in vegetable seed annually and is home to half of the <u>world's</u> cabbage seed production."

"Legislation in California and Arkansas that gives these states the power to prohibit the introduction of GE rice. The major rice growing states are particularly concerned after last fall's [2006] revelations that several unapproved varieties of GE rice had contaminated natural rice, resulting in massive losses for US farmers when export customers in Asia and Europe closed their markets to US rice."

"In addition, the vague language [introduced into the 2007 Farm Bill] raises concerns that states would be barred from taking action when food safety threats arise. For example, states could be barred from prohibiting the sale of e. coli-tainted ground beef if the meat has passed USDA inspection, as was the case in last week's [May 14, 2007] massive 15-state beef recall."[258]

Thus, you can see, we avoided a very irritating mess, engineered by forces alien to our national health and common sense. Our federal Congress, instead of smartly funding organic options for our food, still is "subsidizing genetically engineered crops, factory farms, and chemically-intensive agriculture to the tune of... $220 billion for the current five year Farm Bill...while"[259] only giving less than $5 MILLION per year to "organic research, promotion, and marketing."[260][261] Yet, thanks to such continuing anachronistic policies, U.S. farm exports have fallen 15% since biotech crops have come onto the market in 1996. That includes a $400 million a year drop in corn exports. And a 14.3% decrease in soybean exports to the European Union (EU); while Brazil, which is producing GE-free soybeans, has increased its niche to the EU by 10.7% to take up the slack.[262]

Here, about the genetically engineered world, so-called 'free trade' is not free. The WTO, primed by our own country, is still trying to enforce those Intellectual Property Rights to include life forms, onto the rest of the planet's member nations, whether they like it or not.

There are claims that genetically engineered foods will better feed the world. But, right now, it seems more like genetically engineered foods will feed the funding of the Gene Giants' corporate accounts.

For the Round-Up Ready soybeans that Monsanto produces, as one example, were engineered to fit their herbicide, 'Round-Up;' not the other way round. Apparently it is easier to genetically engineer a type of crop than it is to create a new herbicide or pesticide. According to the Pesticide

Action Network "about 73% of all GE crops planted last year (2000) were engineered to be used with weed killers – not to increase yields or to be drought tolerant, but to increase sales of special brands of weed killer."[263]

And did you know: as time has run along, with more than 90% of our soybeans being genetically altered, and the majority of these being of the 'Round-Up Ready' Monsanto variety, pounds per acre use of Round-Up or glyphosphate on average is now 2-5 times <u>greater</u> than herbicide usage on non-genetically altered soybean fields[264].

In Argentina, one of the big five countries producing genetically altered crops, so much Round-Up has been sprayed on the soil, 'the bacteria needed for breaking down inert vegetable matter' are 'being wiped out...dead weeds did not rot.'[265]

Denmark has officially banned Round-Up/glyphosphate effective September 2003. Reports show that 'the toxic chemical is not breaking down in the soil and, as a result, is polluting their water at a level that is five times what is considered safe for the environment and human health.'[266]

In addition, scientific information is adding up, implicating glyphosphate's adverse affects on animal/human tissue. The June 2005 scientific journal

"Environmental Health Perspectives" reported that this herbicide damages human placental cells at exposure levels ten times less than what Monsanto claims is safe. A study in the August 2005 journal "Ecological Applications" found that even when applied at concentrations that are one-third of the maximum concentrations typically found in waterways, glyphosphate/Round-up still killed up to 71 percent of tadpoles exposed.

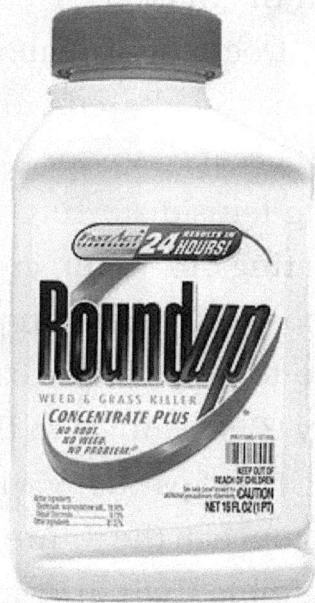

'Round-Up' herbicide, or glyphosphate, has been banned in Denmark, linked to non-Hodgkin's lymphoma, ADD, miscarriage, placental damage, yet it is still used on genetically engineered soybeans, corn, canola, cotton

Similar glyphosate studies around the world have been equally alarming. The American Academy of Family Physicians epidemiological research has now linked exposure to the herbicide with increased risk of non-Hodgkin's lymphoma, a life-threatening cancer, while a Canadian study has linked glyphosate exposure with increased risk for miscarriage. A 2002 study linked glyphosate exposure with increased incidence of attention deficit disorder in children. Despite these studies, Monsanto continues to advertise Round-Up, sprayed heavily on 140 million acres of genetically engineered crops across the world, as one of the "safest" pesticides on the market.[267]

Does this frighten you? Realizing that as genetically altered crops spread across our planet's landscape, so does the use of toxic cancer-causing fetus-distorting/killing chemicals? Contaminating our water, crossing into the wombs of our mothers? And companies like Monsanto continue to issue the Great Lie until Truth (temporarily??) succumbs to their uncaring sales pitch?

Perhaps there will be better ideas and better science in the future, when Monsanto and DuPont humble their arrogance. Meanwhile, what other reason can these companies muster to continue to push a concept like having Bacillus thuringiesis (Bt) produce its toxin in 'all commercialized genetically engineered insecticidal plants"?[268]

Organic farmers have used Bt in very limited amounts for decades to ward off pests like moths and nematodes.[269] Seizing upon the power of Bt, *why not use/insert Bt into every plant we make?* conclude the genetic engineers. Who cares about the organic farmers, and what they normally do?

Entomologists warned that Bt crops exuded '10-20 times the amount of toxins contained in conventional (non-GE) Bt sprays,' and that they 'are harming beneficial insects and soil microorganisms,' plus 'may likely be harming insect-eating bird populations.'[270]

With Bt plants producing the toxin 'in most, if not all, parts of the plant'[271] instead of refining the engineering to just the ideal part(s) of the plant, as they most likely will do in the future,[272] the predicted problems are arising.

Besides adversely affecting Monarch butterflies and beneficial lacewings and ladybugs; moths and roundworms have developed resistant genes to Bt. "As more crops with Bt genes are planted, it is only a matter of time before populations of Bt-resistant insects grow numerous enough to become economically troublesome to farmers hoping to control them." University of California at San Diego Biology professor Raffi Aroian's team of scientists 'found that a single mutation in a gene that disables an enzyme in roundworms' guts is responsible for imparting resistance to a range of Bt toxins.'[273]

Our organic farmers are losing an important

tool in their relatively harmonic contest with nature.

In addition, the pollen from the Bt crops, especially corn and cotton, 'can flow to wild and weedy relatives, with potential long-term ecological consequences.'[274] Weeds resistant to Bt, multiplying all around the prairies and hillsides. The nightmare as reality could happen all too soon.

Farmers are suing on the Bt front too.

But don't worry. The genetic engineers have another wonderful idea. They want to release genetically altered insects to mate and kill off female pests. The pink bollworm, a cotton pest, will have a green flourescent protein from a jellyfish inserted in it so it will glow in the dark and therefore, be 'traceable.' Eventually, radiant males then could be used and lead to 'development of female-killing genetic enzymes capable of eradicating the bollworm pest.' Aside from the possibilities of viral transfer of a 'modified "transposon" to a variety of insects,' who but alarmists would care about the enormity of complications that could occur in the exponential multiplications common in the insect population?

Never mind what happened with the mongoose and the snake population in the West Indies. (The mongoose was introduced, killed off the offending frightening snakes, and now the mongoose is the problem, killing off chickens and whatever else these predators need to eat to survive, without having any natural predators on those Caribbean islands.) Or the heinous cane frogs in Australia, where these amphibians too have no natural predators. After being introduced from

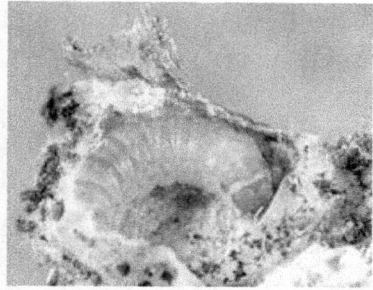

Cotton pest, the bollworm, is becoming increasingly resistant to pesticides, so why not mutate and introduce a glowing male variety altered with a jellyfish gene, that can also kill off females? Remember the mongoose and the snake in the West Indies, the mongoose killing off the snakes, becoming 'the problem' there after being promoted as the solution.

Hawaii to kill off pesky bugs, they have multiplied to such a degree, devouring various plants and animals, and are so detested by locals, that Aussies often enjoy riding out with their vehicles to run over the hordes of these tough skinned ugly varmints and feel them *squoosh* under their wheels.

Progress.

So often, that is the cry of the scientist and the businessperson, when questioned on the problems of the above described adventures with your food and your earthly existence. The big corporations do not matter-of-factly admit that it is not good to wipe out diverse types of plants and seeds and animals in their quest to unify their grip on their market share. Too bad if a 'monoculture' develops where one strain of wheat or corn is sold to farmers in a vast area of the USA. Along with the companion pesticide(s) to 'protect' said strain. The more the profiting companies sell of both crop and pesticide, the more money they make. Even if the wheat scab or fusarium head blight devastates the sitting duck-susceptible monoculture wheat crop. When 'pests of cropping systems are evolving faster than agricultural science's ability to develop the technology necessary for protection. Currently, cropping systems in...the world contain less biological diversity than at any other time in human history.'[275]

Less diversity, easier target for pests and pesticism.

But truly, there is hope. Even if corporations try to protect their 'inventions,' their biotech 'foods' that they patent, and limit the sharing of knowledge and germplasm because they want to protect their patents and investment, when the increase of knowledge could be common ground for the benefit of all forms of life. Instead of just those who prefer to own and profit from it. Eliminating species, narrowing the availability of precious seeds saved and selected over the centuries, creating a less and less varied culture, both biologically and socially. One dangerously susceptible to a catastrophic blight.

In a time when fast food is the norm, but Slow Food as a movement has developed as 'A firm defense of quiet material pleasure...{and} is the only way to oppose the universal folly of Fast Life.'[276] There are other answers to heedlessly accelerated biotechnology and food irradiation and callously degrading factory farming, with bio-imperialism not sharing the wealth equitably between producer, nurturer, marketer, and consumer across an unfairly WTO-globalized world in which we all must live.

New Zealand plans to go totally organic by 2020. Germany aims to be 20% organic by 2010. Again, in the USA, more than 50% of us prefer to ingest organic food. With organic food sales in the USA increasing by 15-20% annually every year since 1990, the trend is obvious.

Although the agribusiness industry has tried to

knock its way into the organic trade, corrupting idealized standards to do so, its arrogant tactics initially failed miserably. The proposals to include food irradiation, genetically engineered foods, sewage sludge treatment of crop fields, etc., common to corporate farming practice, caused over 300,000 people to officially protest to our Agriculture Department against their inclusion in new organic standards. As a result, we developed, for a few years, what our former Agriculture Secretary Dan Glickman called "the strongest and most comprehensive organic standard in the world."[277]

Even much of the organic community seemed satisfied, or relieved. Though there were some reservations. While the standards were set, so was a 'ceiling' erected precluding private food certifiers from using their seals to express higher levels of performance than USDA standards. This could have included the prohibiting of informing the consumer about how a food is handled or grown.[278]

Then, in 2004, the USDA announced 'it would no longer monitor organic labels on non-food products...[adding] that pesticides, animal drugs, growth hormones, antibiotics, and tainted fish-meal would be allowed on organic farms.'[279]

Ronnie Cummins, National Director of the Organic Consumers Association, stated that this "is the third time in six years the USDA has tried to degrade organic standards, and they were beaten back each time – especially this year [2004], as this is an election year."[280] [On May 26, 2004, the USDA re-affirmed that it would revert to doing its duty, after a flurry of citizen and organic community response re-turned the tide. However, the USDA still was refusing to monitor 'organic label claims on what they have narrowly...defined as non-food and non-agricultural products,' like 'vitamins, body care products, nutritional supplements, fertilizers, and even seafood.'[281] Yet, this too was overcome, after much hard work by the Organic Consumers Association (OCA) and its hundreds of thousands of supporters. A joint lawsuit had to be filed along with Dr. Bronner's Soaps to get qualified body care products, pet foods and nutritional supplements certified organic. Seafood did not make it into the certification.]

Then, while the media sleeps, individuals and executives not dedicated or caring about the sanctity of our food, agriculture, animals, or plants were slimed onto the National Organic Standards Board (NOSB), and into the compromised US Department of Agriculture. Not to be outdone, "Republican leaders of Congress attached a rider to the 2006 Agricultural Appropriations Bill to weaken the nation's organic food standards in response to pressure from large-scale food manufacturers."[282] This allowed more than 500 synthetic additives and processing substances to be used in organic foods without public review, from sulphur dioxide, a preservative, to ethylene, a chemical essential to the petrochemical industry. Young dairy cows could continue to be treated with antibiotics and fed genetically altered feed prior to being converted to organic production. Loopholes were provided for "non-organic ingredients to be subsituted for organic ingredients without any notification of the public based on "emergency decrees."" Have you read about this on the front page of your newspaper, or heard Katy Couric reveal the depths of its implications? Highly doubtful, right?

Now, to undo this travesty, an "Organic Restoration Act" has to be passed in Congress. OCA is working on this, and supporting "a thorough, carefully managed NOSB process...to review and approve all synthetic substances proposed for organic processing."[283]

Ronnie Cummins is concerned that, somehow, the USDA could be handed authority over organic standards, rather than the NOSB, which was designed just for this purpose. For a much smaller number of synthetics had been approved by the NOSB that were "supposed to be "sunsetted" after five years and then re-reviewed. This never happened.

Instead of the organic community and the NOSB proposing rule changes to the USDA that would be published in the National Register and then subjected to a full comment period of 90-180 days, Cummins worries that the entire Organic Food Production Act could be open to Congressional revisions.[284]

Slow Food

So, take another deep breath....and be thankful the world has been blessed at this time with the development of the Slow Food movement. With their insignia of a snail, Slow Food is intent on preserving and promoting food quality and diversity, rather than quantity and the quick lire. Founded in Italy, which leads Europe with 27% of the continent's organic food produced by that country, Slow Food is defending 'the purple asparagus of Albegna, the black celery of Trevi, the Vesuvian apricot, the long-tailed sheep of Laticauda, a succulent Sienese pig renowned in the courts of medieval Tuscany, and a host of endangered handmade cheeses and salamis known now only to a handful of old farmers,' amongst other food products and wines.

Slow Food also produces the *Gambero Rosso* guides to wine and restaurants, which have been compared to the Michelin guides in France. 'A top ranking in *Gambero Rosso*'s wine guide virtually guarantees that a particular vintage will sell out almost instantly.' Then there is Slow Food's biennial Salone del Gusto (The Trade Fair), which is Italy's 'largest food show, featuring some 550 food and wine producers. The Salone has become an almost obligatory event for thousands of the world's most important restaurateurs and wine and food importers.'[285]

With 'convivia,' or small units of ideally 50 members per unit, Slow Food now has established itself in more than 100 countries, and on every continent but Antarctica. Here in America, we have 170 convivia, representing nearly every state, with the numbers still growing rapidly[286]. After Slow Food's first U.S.A. conference in San Francisco during July 2001, local convivia have fanned out across the nation. Their mission is to save local foods, while being 'dedicated to stewardship of the land and ecologically sound food production.'[287]

Slow Food's 1989 International Manifesto declares 'Let us rediscover the flavors and savors of regional cooking and banish the degrading effects of Fast Food.'[288] Rather than be xenophobic (afeared of strangers, and their different cultures), Slow Food offers the alternative of 'an international exchange of experiences, knowledge, projects...{toward} developing taste rather than demeaning it.'[289]

Carlo Petrini, Slow Food's founder, states that

he is not averse to globalization per se, but to unsustainable 'homogenization and {the} high-speed frenzy of chain-store, fast-food life.' He uses the phrase "virtuous globalization" to describe the international network Slow Food is dedicatedly building.

The integrity of food and its producers is of prime concern to Mr. Petrini. He points out that many of what may be considered 'delicacies' to-day that Slow Food is protecting, actually 'were peasant foods that were brilliant strategies to stave off hunger and contain worlds of knowl-edge about intelligent use of the environment.'

A terrific example is the Piedmontese cow and its beef, (not to mention its greatly prized milk and cheese). 'According to US Department of Ag-riculture tests, 100 grams of Piedmontese beef contains 1.7 grams of fat, compared with 11.3 {grams} in standard kinds of cattle, and 95 calo-ries, compared with 251 calories in most beef.'

With the Mad Cow Disease epidemic blow-ing European (and now concerned American) meat eaters off kilter, Piedmontese beef is a safe healthy bargain at about $4 dollars per kilo (2.2 pounds) versus $3 dollars per kilo for more com-mon, but questionably safe, beef. Slow Food

stepped in to help 'orga-
nize a consortium of six-
teen livestock farmers.
Rather than urge them
to expand their herds
and cut expenses to be-
come more cost-effec-
tive, Slow Food encour-
aged them to agree to a
series of strict protocols
for natural and organic

The Piedmontese cow, as-
sisted in its survival and
traditional natural breeding
by SlowFood. Lower in fat
content, slower to the market, without the acceleration of growth hormones or
unsavory additives.

methods of feeding and raising the animals in order to produce the highest-quality beef.'

This worked when desperation set in and beef
consumption dropped 30% in Italy due to Mad
Cow paranoia.

Economically it works like this: "The average
Italian eats about 20 kilos of beef {44 pounds per
year}...if you pay...about 50 cents a pound more
for Piedmontese beef, that comes to about $18 a
year – an entirely manageable cost for excellent-
quality, safe meat."[290] All this despite traditional
methods taking 18 months to bring Piedmontese
cattle to slaughter, compared to 14 months when
the cattle are raised using food additives and
growth hormones.[291]

Can these and other sustainable healthy alternatives feed the still starving 800 million of us seemingly out of the loop of the world's food distribution network? (Total population of planet Earth: 6.8 billion, or 6,800 million of us human *Homo Sapiens*.)

The answer to this question is not prohibitive, or necessarily knee-jerk reflex high tech, ah, yes. Supply and demand, profit and loss, are the real driving forces inhibiting or promoting equitable worldwide food distribution. The United Nations (U.N.) states that the Earth has 1.5 times the amount of food people need currently. In fact, 'we have more food per person {now} than {at} any {other} time in history – 4.3 pounds per day'[292] per person. The U.N. also states that 800 million people need $500 to buy their food or to grow it annually, which totals $400 billion dollars per year.[293] It has also been noted that the industrialized nations are 'overproducing' meat and animal products, which consume more resources and funds than do non-animal products and foods.

Some figures you should know here: people can live with grains like brown rice and corn supplying their sustenance, – vegetarians know this – but <u>for</u>

every pound of beef produced, it takes 16 pounds of grain to be fed to the cattle![294] And for every fast food hamburger made from rainforest beef, John Robbins tells us 55 square feet of rainforest is destroyed. That's the size of a small kitchen.[295]

Brazil is now 'the world's biggest beef exporter,'[296] with four fifths of that beef coming from deforested land in the Amazon basin. In 2003, '25,000 square kilometers of Amazon rainforest was destroyed' (that's a bigger area than the whole state of New Hampshire), with 'clearances for cattle pasture doing ten times more damage than logging'[297] there.

According to a WorldWatch report, the real cost for producing a hamburger if it were produced by clearing forest in India, without the subsidies there: $200.[298]

As far as our precious commons commodity, fresh water goes, of which only '0.0001 percent is readily accessible,'[299] get these numbers: one pound of potatoes requires 24 gallons of water, 49 gallons are needed to produce a pound of apples, 815 gallons for a pound of chicken, 1,630 gallons for a pound of pork, and somewhere between 2500 to 5214 gallons of fresh water are used up to produce just one pound of beef![300]

Also, for every 'quarter pounder' you buy at the fast food burger stop, our USA loses 'five times the burger's weight'[301] in that precious resource, topsoil. Do you know how long it takes just to form <u>one inch</u> of new topsoil? Eric Davidson, Ph. D., senior scientist at Woods Hole, tells us 'anywhere from fifty to several hundred years.'[302] Dr. Davidson also tells us that once a groundwater source is depleted, 'it takes <u>thousands</u> of years to replenish'[303] it.

All right, all right, so you're one of those urban people in the majority in America today who think of the land as 'the space between cities on which crops grow,'[304] as forester-ecologist Aldo Leopold chided.

Estranged as you may be from nature, and how your food gets onto your plate, be aware that we Americans consume 23% of the world's beef, while only four percent of the world's population lives in our country[305]. So there is 'wiggle room' available to our meat hungry planet, and country. From the above numbers, it should be evident that we can produce more nourishment at a lower cost to both the pocketbook and the environment. Unfortunately, right now our minds are brainwashed to eat meat and plenty of protein, when we could nicely survive with a more moderate diet that would be much better in tune with serving our Earth's survival.[306]

For just as 85% of the Earth's water is consumed by 15% of the Earth's people, commercial food also tends to be distributed to the highest bidders. Perhaps, that is why 'less than 0.3% of total corn exports from the United States…went to the 25 countries listed by the Food and Agriculture Organization as the world's most severely undernourished.'[307]

Genetically engineered food advocates claim they can help feed the world, but right now many of their products are in the experimental stages. They may indeed be of great assistance, but their dangers may presently outweigh their benefits. Testing and labelling is mandatory, for, unlike a high-bouncing superball or a Pinto automobile, many novel organisms/plants/insects artificially introduced into the world will survive and self replicate. Once released, they are life forms, not inert plastic, and cannot be recalled into the test tube. They will intermingle and mate with other amenable life forms, sometimes causing extinction, or who knows what kind of environmental havoc.

Appropriate regulation and strict protocols are the unavoidable practical answers to the wave of trade liberalization and de-regulation motivating the world's food policies, especially relating to novel foods and food processing techniques.

More locally appropriate, sustainable agriculture seems to be a most promising way to feed those people on the outskirts of the globalized economy. Rather than be forced to buy seeds from Monsanto and Dupont and Syngenta that they once used for free by simply saving them, poor farmers and individuals in 50 developing countries were able to improve food production by 50-150% with 'low-cost locally available technologies and inputs.'[308]

Addressing a conference at St. James' Palace, Prince Charles of England said:

> "One of the most commonly raised arguments raised by those in favour of {genetically modified foods} is that they are necessary to 'feed the world.' But where people are starving, lack of food is rarely the underlying cause. There is a need to create sustainable livelihoods. I would argue for a more balanced approach. Sustainable agriculture provides a pointer to what can be achieved."

Research by Essex University showed that people working together in groups in 'poorest areas' 'made better use of local natural resources' utilizing sustainable agriculture techniques.[309] Saving water, regenerating soils by using manures, foregoing deep ploughing to prevent erosion, reclaiming unproductive land, minimizing the application of pesticides and fertilizers, are some of these techniques.[310] It's not quite organic, but it's on its way there.

So we do not have to sit there while the world seems to accelerate beyond our control when it comes to our food. The World Trade Organization policies are compounding the problem. But most of us want simple, safe, healthy, adequately labelled nourishment for ourselves and our families.

New technologies may help us, now and in the future, but we should not allow them to get out of hand, and conquer our centuries of wisdom garnered by our foremothers and forefathers. There are unforeseeable dangers possible that can be prevented if we move a bit more rationally and slowly.

Environmental and genetic pollution can happen too fast, and may not be reversible in all too many cases. We do not need more Mad Cow Epidemics or plagues of stupidly released genetically altered flying insects zipping around our planet.

Food is basic to all of us. We need to maintain control of its production and distribution with pragmatic insightful policies. Not be rushed like lemmings over the cliffside by profit-seeking research-warped missionaries claiming our seeds, obscuring our innocence.

What Can I And My Family Eat?

Don't be overwhelmed by all you've learned lately about genetically altered foods and irradiated foods and fish farms and factory farms, because the ideal answer is obvious: <u>Organic foods</u> are what you want you want to eat, and buy for your family!

Though organic foods will not always be perfect - - perhaps there may be the odd disreputable farmer or corporation - - for the most part, this is the easiest way to go. Yes, they tend to be more expensive, because the forces-that-be have pushed them into a 'niche' market. In the olden days, maybe just 70 years ago, 'conventional' farming and foods were essentially what we find to be 'organic' today. Then along came the pesticide and fertilizer corporations, allowing us to forget the lessons of the dust bowl disaster from the 1930's: don't plant just one crop on the same field over and over again, vary your crops, rest the field with grasses during some growing seasons, etc.

The treadmill of planting monoculture acres of the same species of corn or potatoes on the same field sprayed by increasing amounts of pesticides, became what too many pundits today call 'conventional' farming.

And, now, just as we are really ready to go full blown toward an organic future, the political and agricultural landscape is being tainted by a bio-tech tipsied crop blight.

But you live in a new millenium where ingredients have to appear plainly on most of the foods that you buy. Use your eyeballs to protect yourself from buying foods you don't want. Remember that the biggest genetically altered crop, soybeans, in 80 % of processed foods, may somehow sneak into your "natural" food or gourmet food store. Those knishes and that tofu chicken and un-egg and un-turkey salad today could be made with soybean oil or tofu, from soybeans, that are not organic. This also goes for foods containing 'lecithin' or 'soy lecithin.'

What to do? Get your hands on foods that are made from organic soybeans or organic soybean oil or organic soy lecithin. More and more companies are providing these.

As you now know what you know, you can tell

your local food makers about U.S. soybeans being more than 90% genetically altered; that such an ingredient could have bad effects on one's intestinal tract, for example; that why not get smart and use organic soybean oil and soybean products? You can call other out-of-town food producers too, and inform them, because most of them don't know as much as you do now. They want to sell their product; they really don't want to harm anyone with what they intend to be wholesome food created by their own hands or machines.

Of course, you may not want to be bothered. But then you don't want to get cancer, or be a party to causing your beloved child to be ill. You do have a responsibility to your family and/or your self.

And let this thought hum in your brain:

"It is not the voice that commands the story:
it is the ear."

from the great writer, Italo Calvino.[311] You demand what is to come into your body, though the loud voice and advertising of the big food merchandizers may try to sell you something you don't want to swallow down into your gut ecosystem.

So, remember that USA corn is 73% genetically altered and is in many foods like cereals, popular sodas, and processed foods/additives such as corn starch and corn oil and high fructose corn syrup. Plus, many restaurants and fast food chains use corn oil, and soybean oil, when they cook up what you order. Plus, many of our animals that we eat ingest feed that is full of genetically altered corn and soy, and then there is that cottonseed oil.

I realize that most people do not think of cotton as food. But the story is that though 40% of what we know as cotton is the fiber from the plant, that goes into our t-shirts and clothes, 60% is seed by weight. Much of this ends up as cottonseed oil in our potato chips, cookies, salad dressings, baked goods, and other processed foods. What's so bad about it?

83% of USA cotton is genetically altered, plus it's the most pesticide-laden crop on the planet. Though it is planted on but 2.4% of Earth's arable land, it accounts for 24% of the world's insecticide market. And five of the nine pesticides used on cotton are cancer causing.[312]

Milk and beef cattle eat 6-8 POUNDS of cottonseed per day![313] 'Pesticide residues from cottonseeds concentrate in the fatty tissue of these animals, and end up in meat and dairy products.' And lest we forget: 'foreign proteins, bacterial and viral promoters, and antibiotic resistant genes, which {us} humans have never really eaten before,'[314] may be contained in too much of whatever form of cotton we hungrily may chow down.

Three terrible facts about pesticides that we should know to wake up our brains about what we are dousing our food and environment with:

'A 1987 National Cancer Institute study found a nearly seven-fold higher risk of leukemia for children whose parents used pesticides in their homes and gardens[315]!'

14 MILLION Americans 'are routinely drinking water contaminated with carcinogenic herbicides and 90 percent of municipal water treatment facilities lack equipment to remove these chemicals.'[316]

I know you all don't hate birds. Next: 'It is estimated that pesticides unintentionally kill 67 MILLION birds each year.'[317]

So, stay away from cotton on your food. Don't encourage its use. A third of a pound of agricultural chemicals go into making your typical t-shirt inorganically.[318] Yet, yes, there is organic cotton out there.

Why are we living this way? allowing things like this to happen? Because most of us just are not aware. We're shopping and raising our kids and working, and our education does not usually include citizenship and a unified consciousness about our land, environment, food, and the animals who share the Earth with us.

Factory farmed food from ever larger 'Confined Animal Feeding Operations' or 'CAFO''s provide much of our meat and dairy, but the 'product' is not exactly the wholesome one we wish it could be. 5,000 or 50,000 or 400,000 chickens or pigs or cattle in packed-together feedlots is not the way we always used to raise our animals-for-meat. And many of us are getting sick from all the wastes released from these enterprises into our air and water. Plus, much of the feed for these animals is splotched with sewage sludge, as if this is all right.

But yes there is organic meat and poultry and pork in existence out there, to more safely consume. And while America ignores healthier food opportunities, the European Union has 'agreed to end all caged egg production in Europe by 2012, replacing it completely with free-range farming.'[319] America can follow the European lead toward more healthy, less cruel animal treatment and rearing. But people like you, who are reading this book, must be some of the ones pointing out the problems that could be hurting our bodies by permitting more chemicals, synthetic additives, and crap into our food pipeline. You have to inform your butcher and supermarket about what's going on and what you want to buy.

As for fish, you now know of the dangers from farmed fish and other seafood. At least, no transgenic fish have been given permits for production YET. But salmon appear to stand first in line, and then catfish awaits the nod. No, there is no organic standard available yet for fish, though you may see false signs and claims about this wedged into the ice cubes at your local fishmonger's.

Main tip here: re salmon: the Atlantic salmon variety is the one used in most salmon farms. But Chinook or king salmon, plus coho or silver salmon, also may be farmed. Pacific and Alaskan varieties tend to be wild. Wild should be more healthy, but, of course, the question of pollution-tainting is what drove us to buy farmed fish in the first place.

I think if you inform your salesperson or filleter about dyes being used to make fish muscle appear pink, and antibiotics and insecticides not being appreciated for your dollar, you can lead to the changes being made that you want, one store and businessperson's mind at a time.

Also think and hope that the "Country of Origin" labelling will actually go into effect for fish in 2008, as is currently scheduled, to help you with your seafood choices.

Obviously, buying local is the best way to know what you are getting. Plus, you also can deal with your neighbors face to face to dialogue about what's in the stuff they're selling and how they can make it safer and better for their customers who might be listening to what you are saying. Buying from your local grower or meat producer/seller also might be cheaper, and perhaps more practical, than always having to buy organic. Which we cannot do all the time.

As far as irradiated food goes, don't eat it unless you have to. Who knows if it ever will be commercializable? Ensure that your kids don't have to be served any of it in their school's lunches. Stand up for your kids, like all those other school districts have. Even Congress is aware of the ire of America's parents.

Some other foods that could be irradiated or

261

gene-altered that should be avoided would be: Hawaiian papayas, mangoes, mangosteen, zucchini, yellow-squash, canola oil (60% of US crop, and almost all of Canada's, where most of it comes from).[320]

Foods that could contain gene-altered ingredients: mayonnaise, margarine, veggie burgers, crackers, beer, colas, sodas, candies, fructose sugar, bread, egg-based products, dairy products from milk-cows injected with bovine growth hormone.[321] [I have recently noticed various dairies advertising that their milk is NOT treated with any artificial growth hormones. That is good. Now to verify that this is true…via governmental inspection…]

Ingredients to avoid that could be gene altered, not yet mentioned: maltodextrin, citric acid, lactic acid, dextrose, aspartame (Nutrasweet), rennet (used as enzyme in cheese).[322] For more information on foods to avoid, go to the website: www.seedsofdeception.com

And beware of those fad diets that push protein, and cause you to eat much of your food as factory farmed, pesticide and genetically altered-feed tainted unhealthy animal fare. The WHO says you really only need about 4 ½ percent of your calories from protein, not 50 or 70 percent. That comes to 32 grams of protein per day for a 150 pound person, which comes to 128 calories from protein per day.[323] Various other reputable boards and agencies may nearly double this amount of protein in their recommendations.[324]

Just think of those Irish folks starving and surviving on potatoes, when their meat sources went defunct. If they just ate potatoes all day as their only food, and ate enough to fill their bellies, potatoes are 11% protein by weight.[325] That must have satisfied their bodies' needs, or you'd just find Englishmen proudly striding across the Burren to Dublin and on up to Belfast these days. 'Twasn't an ideal way to suffice, there must have been vitamin and mineral deficiencies, but it does show you that you don't have to blow yourself out with animal protein, and go grumpily lo-carb to thin your under-exercised figure.

If you eat too much protein that your body can't use, your kidneys just excrete the excess anyway, as they do with excess vitamins B and C.

Yes, Hippocrates, the father of modern medicine, told us: 'Exercise is man's best medicine.'

So, get out there and run or walk or bike or swim your miles and eat your best foods.

And, don't forget the alternative of raising your own animals and plants in your own sacred garden or backyard acreage! Also, don't forget that there are more than 3000 food co-operatives across the nation that you can join to purchase what is right and healthful for your family.

Let me leave you with this information concerning organic designations: '100 percent organic' food must contain <u>only</u> organic ingredients; products labelled 'organic' must be at least 95 percent organic by weight; processed products with at least 70 percent organic ingredients may be labelled 'made with organic ingredients,' listing as many as three of those ingredients on the package front – however the non-organic ingredients that are <u>not supposed to be</u> <u>genetically altered</u> do not require being

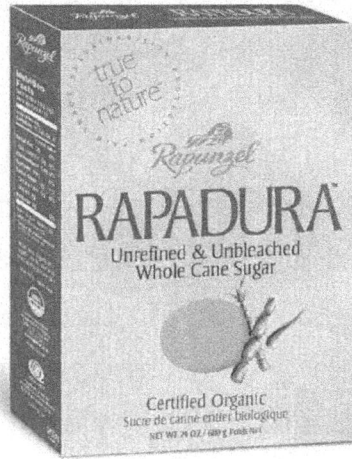

With its claim to be 'Certified Organic,' this box of cane sugar must contain at least 95% organic ingredients by weight to meet current USDA organic standards. Especially important with threat of 50% of non-organic granulated sugar becoming GMO with U.S. approval of GMO sugar beets.

tested for these[326]; those products with less than 70 percent organic ingredients may list the organic ingredients on their information panel, but not carry the phrase 'organic' anywhere on the front of the package.

Also, only the '100 percent organic' and 'organic' products are allowed to use the 'U.S.D.A.' seal on their products and in their advertisements.[327]

Again, the organic designation excludes any foods that are irradiated, factory farmed, animals fed animal wastes and body parts, foods that contain genetically altered substances, or are produced utilizing sewage sludge.

Magic numbers to call to reach your representatives on any of these food issues: your two Senators? 202-224-3121. Your Representative in Washington, D.C.? 202-225-3121. This is your great country. Speak up and inform those who need to know what you do. You're learning a lot to help your family and friends be healthy and wise. Now spread the word. That is why I wrote this book for all of you!

Chapter Three:
Star Wars and
Space Dominance

'The aerospace industry has stated that plans for space control, Star Wars, will be the largest industrial project in the history of the planet. The profit potential for the space weapons industry is astronomical.' – Global Space Newsletter 17, Winter 2006.[†]

"US Space Command: Dominating the space dimension of military operations to protect U.S. interests and investment....integrating Space Forces into warfighting capabilities across the full spectrum of conflict."[1] That's what our Air Force blatantly preaches in their 'Vision For 2020,' viewable on the internet (and page v) as the new millennium begins. Informing us that "The medium of space is recognized as the fourth medium of warfare." -- the others are land, sea, and air— within Earth's atmosphere -- "Joint operations

---

† Available at http://www.space4peace.org   in Section 4) Bush's Larger Plan For Space of Newsletter 17, Winter 2006.

Vice President Dick Cheney, and ex-Secretary of Defense Donald Rumsfeld in their 20th century days. As prime proponents of the Project for a New American Century (P-Nac), they blueprinted space weaponisation via 'missile defense' and the control of space, to be American goals for the 21st century. Meanwhile, the rest of the world declared there should be <u>NO</u> weaponisation of space via the <u>Outer Space Treaty of 1967</u>, and the <u>'Prevention of an Arms Race In Space'</u> resolution affirmed by 163 nations on Nov. 20th 2000.

require the **Control of Space** to achieve overall objectives."[2] {my italics}

Could this be . . . 'missile <u>defense</u>??'

Former Secretary of Defense Donald Rumsfeld and our onerous Vice President, Dick Cheney, spoke 'openly about the possibility of a 'new Pearl Harbor' that would 'catalyse the US people'"[3] to support their belligerent *Project for a New American Century* (P-Nac) back in the autumn of 2000, <u>before</u> they came into office. 'Space weaponisa-

US Space
Dominating the space dimension of military operations
Integrating Space Forces into warfighting

Control of Space

Control of Space is the ability to assure access to space, freedom of operations within the space medium, and an ability to deny others the use of space, if required.

The medium of space is recognized as the fourth medium of warfare. Joint operations require the Control of Space to achieve overall campaign objectives. The Control of Space will encompass protecting US military, civil, and commercial investments in space.

As commercial space systems provide global information and nations tap into this source for military purposes, protecting (as well as negating) these non-military space systems will become more difficult. Due to the importance of commerce and its effects on national security, the United States may evolve into the guardian of space commerce--similar to the historical example of navies protecting sea commerce.

Control of Space is a complex mission that casts USCINCSPACE in a classic warfighter role and mandates an established AOR.

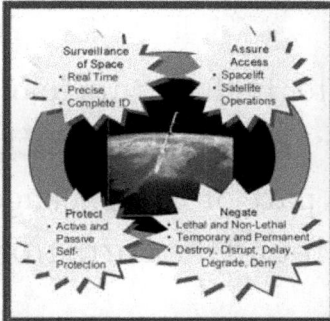

Surveillance of Space
- Real Time
- Precise
- Complete ID

Assure Access
- Spacelift
- Satellite Operations

Protect
- Active and Passive
- Self-Protection

Negate
- Lethal and Non-Lethal
- Temporary and Permanent
- Destroy, Disrupt, Delay, Degrade, Deny

The ability to dominate space

Control of Space Capabilities
- Real-time space surveillance
- Timely and responsive spacelift
- Enhanced protection (military and commercial systems)
- Robust negation systems

Page 10 of the U.S. Air Force's 'Vision For 2020,' telling us how important the Control of Space is to our 'classic warfighter role.'

tion via 'missile defence' was an essential part of the N-Pac programme.'[4]

'Missile Defence.'

Space weaponisation. Space as the "fourth medium of warfare." A "new Pearl Harbor." Hmm-mmm. . . . .

Today this is to include nuclear power, deployed into space[5], launched on rockets that too often explode, as part of the package to protect our <u>investments</u> around the Earth from the "have-nots."[6]

Does this sound a bit scary, and too unbelievable?

We must remember that America was the prime mover to <u>prevent</u> an arms race in space, developing the Outer Space Treaty in 1967 "to

Future Trends

Although unlikely to be challenged by a global peer competitor, the United States will continue to be challenged regionally. The globalization of the world economy will also continue, with a widening between "haves" and "have-nots." Accelerating rates of technological development will be increasingly driven by the commercial sector -- not the military. Increased weapons lethality and precision will lead to new operational doctrine. Information-intensive military force structures will lead to a highly dynamic operations tempo.

Accelerating rates of change will create challenges

From Page 6 of 'Vision For 2020.' Note sentence two on the left concerning the 'haves' and 'have nots.'

keep war out of space."[7] This occurred after the Russians launched their Sputnik, the first man-made satellite, back in 1957. Americans and humans across Planet Earth were frightened by such an astonishing accomplishment, and what it could mean. What if the Russians got control of space?! and could control what went on below with their superior knowledge of space technology? Would all the countries of the world then fall subject to the Russians and their totalitarian communism?

Well, now it is 2008, and the USA has taken over as the dominant power in space. In addition to remaining as the only 'Super Power' amongst all nations on terra firma. Since we happen to control the space dimension, we might as well damn keep it. We don't want anyone else up there, either, to seriously interfere in our realm, while we fantasize about strengthening our grip on what we have.

(The Vision For 2020 actually states: "...to assure access to space, freedom of operations within the space medium, and an ability to deny others the use of space, if required."[8]) (Underlining mine.) (See actual document and quote two pages back on page 269)

Though China, Russia and Canada 'annually repeat...the proposal, calling on the US to agree {upon} an international pact banning all weapons in space,'[9] we are preparing ourselves to reprehensibly become the first Earthly nation to dare shoot a weapon up into the space commons. Originally, we readied an 'N-FIRE' test satellite to be launched by NASA and the Pentagon via a Minotaur missile with an offensive 'kill vehicle' aboard. However, Congress killed the 'kill vehicle' option in 2004.

'N-FIRE' is an abbreviation of the 'Near Field Infrared Experiment,' with its supposed primary mission being 'to gather data on the exhaust of rockets in space... {This information then} will be used to help future space weapons differentiate more clearly between a target and its trailing plume'[10] of exhaust. On March 21, 2007

The N-Fire test satellite was to be the first weapon ever launched into space, the USA to violate the Outer Space Treaty of 1967 we originated to keep war out of space. Latest: its 'kill vehicle' temporarily killed by Congress.

the first N-FIRE was indeed sent up into space. The New Scientist reported that it will "gather information that could be used for a future missile defence system in space."[11]

Combine the implications of this paragraph with the latest version of U.S. National Space Policy, as advanced by the Bush administration. "Calling for the deployment of offensive weapons systems in space to "deter" and "deny" others the "use of space," Bruce Gagnon, co-ordinator of the Global Network Against Weapons and Nuclear Power in Space, infers this "will give the Pentagon the green light to put anti-satellite weapons in space that would be able to destroy other countries' satellites."[12] Which is close to being funded, as 'offensive counterspace systems,' that the Pentagon describes as being designed 'to disrupt, deny, degrade or destroy an adversary's space systems, or the information they provide,'[13] amidst the $649 billion approved for the Pentagon's budget in 2008[14].

Recall how aghast we have been that China could dare to be developing anti-satellite technology. Yet here we are, hypocritically following "Do-as-I-say, Not-as-I-do" behavior. And the Bush administration is not apologetic in proclaiming that we will "oppose the development of new legal regimes or other restrictions that seek to prohibit or limit U.S. access to or use of space." Mr. Gagnon translates this excerpt from our Space Policy document to mean "that the U.S. is now on record as being totally opposed to the development of an international treaty at the United Nations that would ban all weapons in space."[15]

Though you could take China's action as "a shot across the bow" to get our country to the negotiating table to ban weapons in space, as Theresa Hitchens, director of the Center for Defense Information, suggests. Such a "hard-power capability" development/manoeuvre being "a classic cold war technique."[16]

Although the N-FIRE's 'kill vehicle' has been removed from the space program for now, watch out for offensive-minded Congresspersons attempting to re-insert it. . . because they reckon it "can destroy passing missiles or, as ABC News reported in June {2004}, the satellites of the US's military and commercial rivals.'[17]

Don't forget the commercial component of the 'Star Wars' story, as so dubbed by the regrettable Ronald Reagan back in the 1980's. For the number one industrial export product today of our dear USA is not cars or steel or computers. No. It is WEAPONS! And know now that our "weapons corporations have been saying for decades that Star Wars will be the largest industrial project in the history of the planet Earth." Mr. Gagnon further informs us that "both Democrats and Republicans get the message and understand that their corporate sponsors want them to leave the door open to a new costly and destabilizing arms race in space."[18]

Is that what we the people want? spearheaded by an unpopular arrogant administration having an approximately 20% approval rating during most of 2007 - - though you might never know this by the way the major networks report the news for us.

Could the adverse influence and lobbying of the powerful weapons industry be why we have been behaving so uglily, especially since the new millennium began? Money and donations from weapons corporations; their wealthy CEO's hobnobbing with our politicians, lobbying our government, aided by the 55 percent increase in military spending since George Bush became President? "Thirty four weapons corporations' CEO's received record salaries in recent years," including the "top profiteers" since 9/11 earning: $200 million for the CEO of United Technologies, $65 million for the CEO of aerospace giant General Dynamics, $50 million for fellow-aerospace-giant Lockheed-Martin's CEO, and let us never forget Halliburton's CEO, getting paid $49 million. While the average army private, putting his or her life on the line, earns $25,000 per year.

According to Mary Beth Sullivan, "The U.S. arms industry is the second most heavily subsidized unit of the economy after agriculture. Arms exporters know they can rely on American taxpayers for billions of dollars annually to market and finance sales of their products. After all, taxpayers foot the bill for the weapons research and development work in the first place."

The obscene money/profits that we give to these corporations and their investors and CEOs and surviving workers comes back down the pipeline into our Congress and White House, leading "us" to benefit them and their industry by political/military/economic actions like acting unilaterally in invading Iraq - - so that today Iraq's "sovereign government" can directly purchase its $1 Billion of weapons directly from U.S. sources annually; scrapping the 1972 Anti-Ballistic Missile Treaty [ABM] in June 2002 so we will not be constrained from putting weapons into space, including nuclear powered weaponry; selling "arms to both sides in simmering conflicts"[19] around the globe; "providing nearly half the weapons sold to militaries in the developing world."[20]

As my colleague Robert Slutsky M.D. has often told me: "Wars will come and go, but the weapons remain, and they keep getting better and better!"

Thinking about that, the influence of our weapons industry, and how this might affect our government's decisions and policies, perhaps we should ponder the following quote from Barbara Tuchman:

> "A phenomenon noticeable throughout history regardless of place or period is the pursuit by governments of policies contrary to their own interests. Mankind, it seems, makes a poorer performance of government than of almost any other human activity. In this sphere, wisdom, which may be defined as the exercise of judgment acting on experience, common sense and available information, is less operative and more frustrated than it should be. Why do holders of high office so often act contrary to the way reason points and enlightened self-interest suggests? Why does intelligent mental process seem so often not to function?"[21]

Of course, in George Bush's case, we have to worry how much "intelligent mental process" exists. Especially pertaining to technical, scientific, and biological fields of knowledge. Knowing how heavily subsidized our arms industry is, should we be worried when Mr. Bush waxes enthusiastic to spend a TRILLION dollars or so[22] for a Mars mission when the TOTAL annual federal budget runs at about $2.9 trillion? Yes, most of the research and development for all these space weapons and travel propulsion will be paid for by the federal government. But, the mining of the minerals, and profits from weapons deployment will then be raked in by the Bechtels and Lockheed-Martins.

Getting back to the N-FIRE's, the Pentagon's February 2004 presentation to Congress revealed that the first N-FIRE, and the N-FIRE's to come, are but <u>initial</u> steps on a trail to deploy and ready 'whizz-bang'[23] space weaponry on into a US Earth-domination future. There will be tungsten (or depleted uranium) 'Rods from God' we can shoot from orbiting platforms at 12,000 feet per second that the Pentagon claims will be accurate to 'a range within just 25 feet, able to destroy even the most hardened targets.'[24]

Then there will be 'orbiting lasers, 'hunter-killer' satellites, and space bombers with the ability to swoop down out of the ether so as to target any uppity nation that displeases the world-Caesar in Washington,'[25] so Chris Floyd colorfully states.

This goes along with ex-NASA chief and former Navy Secretary[26] Sean O'Keefe telling 'the nation that from now on every {NASA} mission would be dual-use. By that he mean[t] that every mission would carry military and civilian payloads at the same time. This

Karl Grossman

is further evidence that the space programme has been taken over by the Pentagon.'[27] Karl Grossman, our Upton Sinclair of the new millennium, informs us in his mid-2004 article in *The Ecologist*.

Add into the mix the 2004 NASA mission internet quote:

'Today, only nuclear power can enable these scientifically vital, but incredibly challenging missions.'[28]

and you can see the horrific direction in which our leaders are taking us.

What's worse, peaceful as space could and should be, Rich Haver, [ex-]'vice president of major NASA and Pentagon contractor Northrup Grumman'[29] believes "space is the place we will fight in the next 20 years."[30]

Rich Haver, former VP of Northrop Grumman, ex-intelligence assistant to Donald Rumsfeld[31]

So, while space becomes the 'centre of gravity for the Department of Defense and the nation,' as projected in our Space Command's 1998 Long Range Plan (LRP); and our Space Command's Lance Lord dictates that "we must become a full-spectrum space-combat command," Bruce Gagnon observes that "space is viewed today as open territory to be seized for eventual corporate profit."[32]

First and foremost, this includes our one and only moon. Reality to the contrary, President Bush's rah rah vision of putting Earthlings on the moon and Mars, that most of you have probably heard about, is not some innocent exploration without any ulterior motives. By setting up military (and civilian) colonies on these orbs, and on stations in space, at exorbitant expense to our government and people, for the benefit of the few, the privileged, the indulgent defense contractors we should, but don't quite know enough about, we, the USA, will then be able 'to control the "shipping lanes of the future,"'[33] as Bruce Gagnon describes it for us. Plus mine precious elements, like 'the rare helium-3, which would be brought back to Earth to fuel supposedly cleaner {nuclear} fusion-power reactors.'[34] (But beware the dangers of tritium pollution with nuclear fusion, your author interjects.)

As a specific example here, "former astronaut and engineer Harrison Schmitt has created a corporation to mine the moon" for helium-3. Tom Taylor, vice president of Lunar Transportation Systems Inc., headquartered in Las Cruces, New Mexico, tells us "while it's a little early to speculate,

Former astronaut Harrison Schmitt has created a corporation to mine helium-3 on the moon. "It's worth about 5,000 Saudi Arabias." But space propulsion scientist John Brandenburg is not talking about oil here, he's referring to potential profits for a nuclear fusion fuel. Major problem: how to transport the tons of helium-3 back to Earth, while controlling (militarily esp.) one of the main space "shipping lanes of the future," from the moon to our Earth. Besides the unspoken about contamination problems & practicality of nuclear fusion itself.

helium-3 is worth about $12 billion per 2,000 pounds — if we could mine it on the moon, it would change our entire nuclear industry...If other countries get there first, I fear that our nation will drop into some lesser status."

'From a pure resource perspective, mining helium-3 could turn the U.S. into the top power producer in the world,' John Brandenburg, a senior propulsion scientist at Orbital Technologies Inc. in Wisconsin and a former scientist at Sandia National Laboratories, enthusiastically proclaims. "Once you get helium-3 on the moon, the moon becomes the new Persian Gulf," Brandenburg informs us. "It's worth about 5,000 Saudi Arabias."[35]

Researchers at the Princeton University Plasma Physics Laboratory have estimated that some one million tons of helium-3 could be obtained from the top layer of the moon."[36]

How to get the stuff back to Earth is a major obstacle to overcome, as is the morality of commercially mining the moon and other planets. For the 'Moon Agreement' of 1979 states '"neither the surface nor the subsurface of the moon" or "other celestial bodies within the solar system" shall "become property of any state...organisation...or...person."'[37]

Oh, fortunately or unfortunately, depending on your viewpoint, the USA never signed onto this United Nations document expressing the international desire to reserve space and the planets and their natural satellites/moons for 'the common heritage of mankind.'[38] Perhaps the advice of ex-Nazi "Major General Walter Dornberger, who was in charge of the entire V-1 and V-2 missile operation for Hitler's Germany,"[39] still rings in the heads of our 'deciders?'

Testifying before Congress in 1958, Dornberger, one of ~1500 "top Nazi scientists smuggled into the U.S. under Operation Paperclip after World War II [said] that America's top space priority ought to be to "conquer, occupy, keep and

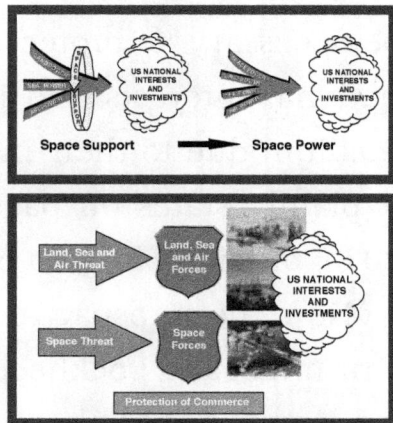

The emergence of space power follows both of these models. Over the past several decades, space power has primarily supported land, sea, and air operations--strategically and operationally. During the early portion of the 21st century, space power will also evolve into a separate and equal medium of warfare. Likewise, space forces will emerge to protect military and commercial national interests and investment in the space medium due to their increasing importance.

Page 4 of Vision For 2020, document available in its entirety at http://www.crestofthewave.com/showcase/docs/vision_2020.pdf

Major General Walter Dornberger, German Nazi smuggled into USA after WWII, provided Hitlerian spirit and direction leading to our current NASA space program's offensive priorities.

utilize space between Earth and the Moon.""[40] Sound a bit Hitlerian? [More on our Nazi-Space Dominance history later in the chapter.] Adolph Lives! in our ambitious space program to control who might and might not go anywhere from planet Earth. Thanks to those Nazis we incorporated into it, who helped formulate it, give it its spirit, starting back in the 1940's.

We, then, the USA and our military, will gloriously be out there in the fourth medium of warfare, acting as space police. Because it will be necessary "to protect military and commercial national interests and investment in the space medium due to their increasing importance," as it plainly states on page four of the Vision For 2020, provided for you on the opposite page.

Now, the aerospace corporations like Northrop Grumman and Lockheed Martin and General Dynamics may reap the profits of our space adventures, but won't it be kinda expensive? Mostly for us exhausted taxpayers?

Yes. "Soitainly," as Curly of the Three Stooges would say.

"NASA said it would cost $104 billion just to return to the moon for a first visit, but has declined to give estimates for the total cost of a permanent base."[41]

$104 <u>BILLION</u> Dollahs!!! <u>BILLION</u>!??!! And that is without late deliveries, technology malfunctions, and cost overruns! Like with the International Space Station, that was originally supposed to have a $10 billion price tag, but has actually cost upwards of $100 billion. And the durn thing is still not completed![42]

The latest proposal festering in our space corporations' radioactive blueprints to get us to the moon is one that seems ridiculously stupid, and failed miserably, flung into the nuclear dumpster of bad ideas back in 1972. But with the profit trough spreading wider and deeper under President Bush's divine guidance, why not fund a massively expensive nuclear rocket program? Nah, who cares that one of these might explode on take-off, contaminating the state of Florida essentially forever, and maybe also half of the USA's east coast? Besides sending radioactive debris like small particles of plutonium as 'fallout' into the wind currents circling the Earth so all our children might have an equal chance at contracting lung cancer, and other radionuclide-induced cancers and leukemias?? Not to mention innocent non-American children and adult citizens of other countries.

But, don't you see that chemical-powered boosters are not strong enough to get big payloads to that moonbase we want to electrify with nuclear reactors near the moon's sunnier south pole? Why, just heat up that hydrogen aboard our rocket with the fission of uranium, and "blast it out of the thrust nozzle at extremely high speed!" Oh, um...yes, well, the "hydrogen reaction mass became radioactive as it passed through the reactor." Was that supposed to happen?

Should we be told that the "reactor tended to come apart and fire itself out of the exhaust?" Thanks to those NASA budget cuts of 1972, this rocket research programme was unceremoniously terminated way back then.[43]

But money talks, and plenty of it can possibly go into newer versions of old failed impractical eternally-polluting programs. Michael Griffin, NASA's new director, has given the nuclear-happy Project Prometheus the assignment "to develop surface-based nuclear power systems for a proposed moon base colony."[44]

Have the pundits at Fox News, or Katie Couric told you about Project Prometheus? The nuclear bong-banger that Mr. Bush expects will put us over the top with his administration's daft drive 'to give the US monopolistic control over the heavens?'[45]

Though it was 'quietly unveiled' in February 2003, you should know that this NASA/Pentagon program already was 'underway to build nuclear-powered spacecraft and other atomic systems for space use.' ('*Quietly* unveiled,' because two days earlier the Columbia space shuttle 'had fallen to Earth, killing all aboard.')[46]

Prometheus' Crew Exploration Vehicle is scheduled to begin development and testing in 2008, with a manned mission planned to occur before 2014. For 2020, a "flight with a four-astronaut crew" to land on the moon for a short visit, may very well occur. This would be a prelude that "NASA envisions [to] people living on the moon for six-month intervals beginning in 2024."[47]

Hear NASA's announcement circa November 2003:

Froglike 'Crew Exploration Vehicle' in the foreground may be your next US spacecraft. Its first manned mission could occur before 2014.

> "Project Prometheus recently reached an important milestone with the first successful test of an engine that could lead to revolutionary propulsion capabilities for space exploration throughout the solar system and beyond."[48]

Sounds great, huh!!? But 'in space NASA's new 'high-power electric propulsion ion engine' would be powered by a 'small nuclear reactor.'[49]

Main point here is that these nuclear reactors and dangerous space toys have to be launched somehow into space. There have been too many rocket failures, shuttle destructions, plutonium powered-Snap-9A and Cosmos 954 reconnaissance satellite uranium spillages/contaminations that have occurred already so that we should know better by now.

Those Titan rockets you might have heard of, because they are one of our main vehicles to lift our satellites into orbit, have a record of 'one catastrophic accident for every ten launches!'[51] Not very good if you have plutonium or uranium aboard, is it?

'The Soviet Union's worst space nuclear accident ever' unfortunately spread a 600 kilometer swath of nuclear debris across Canada in 1978. 110 pounds of uranium aboard the Cosmos 954 reconnaissance satellite 'had probably been vapourised and dispersed globally.'[50]

Plutonium, and its close cousin from whence it usually transmutes in nuclear power plants (or other forms of 'generators' of nuclear energy), uranium, are so toxic, just one microgram of plutonium is enough to cause lung cancer, once the offending radioactive particle silently settles into your body's breathing apparatus[52].

You should know that Europe's Organisation for Economic Cooperation and Development,

289

along with the Swedish National Institute of Radiation Protection (NIRP) blamed the 1964 Snap-9A accident as 'the main source of Plutonium 238 in the environment'[53] in a report they jointly issued. 'A worldwide sampling programme carried out in 1970 showed Snap-9A debris to be present at all continents and all latitudes.' is quoted from that report.[54]

Oh, that was in 1970, like 35 years ago, you might say. I don't have to worry any more about that old news, now do I now?.....

But you should have the knowledge that plutonium 238 has a half life of 87.7 years according to our Environmental Protection Agency.[55] That means that wherever teeny micrograms (one microgram = one <u>millionth</u> of a gram; 454 grams make up one pound) of the stuff drift and mix and unite with other elements and objects and bodies and protoplasm anywhere on Earth, the toxicity and lung cancer-capability/danger continues to be a problem for 10-20 'half-lives,' or 877 to 1754 years. Is that acceptable? Is that comprehendable for you? When we know that just <u>one</u> <u>microgram</u> can cause lung cancer over a fifteen to thirty year period, on the average[56]? Plutonium is a radionuclide you cannot smell, taste, feel or see at such miniscule yet dangerous toxic doses. And some of its nanoparticles are still floating as 'fallout' above our atmosphere today, from that 1964 accident. In addition to the fallout from our nuclear testing, and other nuclear mis-adventures. . .

Lockheed-Martin and Boeing received the first Project Prometheus contracts. Massachusetts Representative Edward Markey tried to redirect the Project Prometheus money toward cleaning up hazardous waste sites. 124 Representatives in the House agreed that that would be a much better idea than funding a nuclear future for our space program. A Time/CNN poll had

---

Numbers to Help You (Once Again)

One trillion dollars >> $1,000,000,000,000
equals
one million million dollars
(or one million millionaires' money)

For plutonium ----> one microgram
equals a millionth of one gram
454 grams = one pound
one pound = 454,000,000 micrograms =
454 <u>million</u> micrograms
enough to give lung cancer to
454 <u>MILLION</u> people
@ 1 microgram per cancer

'more than three fifths of Americans oppos{ing}'...President Bush's moon-Mars colonization proposal/ boondoggle.[57]

Rep. Edward Markey

However, in our republican democracy, majority rules, and since 309 Representatives officially disagreed with Mr. Markey's idea, we better keep an open ear for Chicken Little's granddaughter warning us some nuclear poison may be falling out of the sky for us to unawarely inspire into our lungs when one of our cock-eyed space weapons disintegrates or explodes somewhere beyond our senses, or our media microscopes (besides fallout coming down to earth with rain or snow).

History is such an interesting perspective by which to correct our short term memory. Many of us may think at this point, that "Oh, this has been going forever." That we've had weapons in space for many years. But, no, this is not true. "Plans for potential space weapons were vetoed by the Clinton White House." When President Bush gave the formal go-ahead to develop space-based _weapons_ back in mid-2005, that was a sea change in our space policies.

Theresa Hitchens, the Vice President of the

Center for Defense Information, warned us then that this 'would trigger a new arms race in space.'[58] Evidence China's NERVE to develop anti-satellite weaponry, that has unsettled us to no end.

Never mind that we have the nerve to develop such technology ourselves. Laura Grego, 'a space weapons expert at the Union of Concerned Scientists tells us "We're legitimizing the idea of attacking other people's satellites, and we have the most to lose. This technology is diffusing rapidly. To be the masters of space you'd have to not allow anyone else to launch into space. But you can't blow up everyone's launch pads."'[59]

As far as numbers go, we have about 450 operating satellites, Russia has about 80, China about 45, and the rest of the world about 275. These include commercial (most common), government owned, and military, with less than 25 civil owned satellites in the overall total. These satellites are integral to our functioning today, helping with weather data, ship location, instantaneous credit card authorization, telephone and television transmissions, education and medical training, rescue operations, besides military and other uses.[60]

Please keep the following words from the Outer Space Treaty in mind should you ponder the

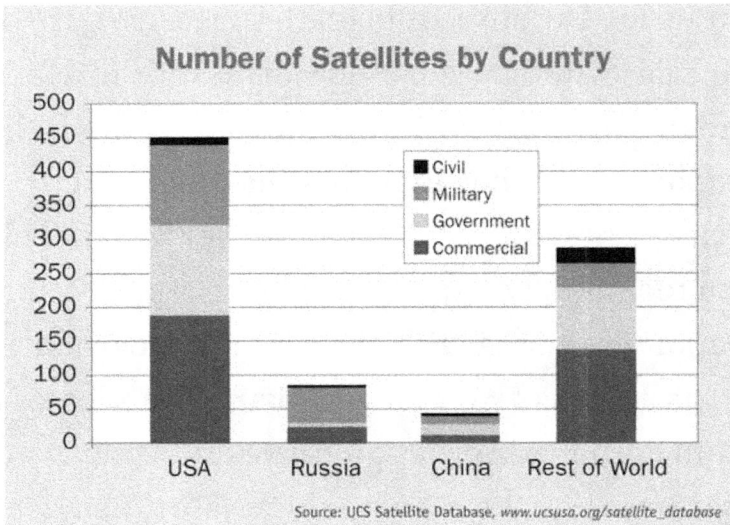

Number of Satellites by Country

Source: UCS Satellite Database, www.ucsusa.org/satellite_database

Which country has the most to lose if USA legitimizes attacking other nations' satellites? [Chart used by permission from Union of Concerned Scientists.]

picture of what is happening in space today, and what could and should be happening, if we do not want to allow our militarism to destroy our civilization:

> The exploration and use of outer space, including the moon and other celestial bodies, shall be carried out for the benefit and in the interests of all countries, irrespective of their degree of economic or scientific development, and shall be the province of all mankind.

Grounding yourself with this knowledge of our space politics and history, perhaps you should now be informed that President Bush II made 'missile defense' his 'top agenda item in countless meetings with NATO allies, Russia, and China.'[61] In fact, Condoleezza Rice had prepared 'a major speech {naming} missile defense as the Bush regime's national-security priority.'[62] BUT, the day she never got to give that speech was: September 11th 2001!

However, September 11th 2001 proved to be fate's gift to Rumsfeld and Cheney as that 'Pearl Harbor' they hoped would 'catalyse the US people'[63] to support their P-Nac programme, that I mentioned at the beginning of this chapter. Thus, now we have the 'global war on terror'[64] ever obscuring our loss of democratic freedoms, while space weaponization and military incursions like those into Iraq and Afghanistan (utilizing 'more than 50 military satellites to direct U.S. missiles and bombers to their intended targets'[65]) proceed to the glee of our current leaders and their zealous supporters.

For the current economic reality, with globalization and the World Trade Organization legalizing the claims to any country's natural resources by mostly first world corporations, "will only continue, with a widening between the "haves" and "have nots,"" as it says on page 6 of that 'Vision For 2020.' Furthermore, from that same page:

"Increased weapons lethality and precision will lead to new operational doctrine. Information-intensive military force structures will lead to a highly dynamic operations tempo."[66]

'Star Wars,' as Ronald Reagan tabbed it, is gluttonously fluffing out its feathers, filling in its multi-billion dollar expense vacuum. Soon, we could be firing nuclear-powered laser beams from space stations at vulnerable 'enemy' targets down on this Earth below. If our political hawks and special interest agents continue to fund their way past our stumbling instincts, we will soon be preparing to attack and fight the entire world beneath our orbiting weaponry at whatever provocation we deem requires a response.

The challenge extends to space

Space Trends

Space systems, commercial and military, are proliferating throughout the world. Space commerce is becoming increasingly important to the global economy. Likewise, the importance of space capabilities to military operations is being widely embraced by many nations.

Indeed, so important are space systems to military operations that it is unrealistic to imagine that they will never become targets. Just as land dominance, sea control, and air superiority have become critical elements of current military strategy, space superiority is emerging as an essential element of battlefield success and future warfare.

Bottom half of page 6 'Vision For 2020,' giving reason for the USA taxpayer to fund the challenge to accomplish 'space superiority' 'as an essential element of battlefield success and future warfare.'

You should know that on November 20th, 2000, 163 nations reaffirmed the overwhelming international sentiment *against* such aggressive policies by signing the "Prevention of an Arms Race In Space" resolution at the United Nations. Only the USA and Israel, plus the USA-dependent Micronesia islands, abstained from the voting.[67]

Nevertheless, George Bush's administration wants to persist in funding a poorly tested charade that would pretend to protect America from incoming missiles. $9 billion per year has been the magic budgetary number for the Missile Defense Agency for the last few years (though many say, add another fifty percent to that amount per year).[68] (NASA's 2008 official budget is $17.3 billion.[69]) Experts project that $120-$150 billion, <u>plus</u> other operational and support costs, would be the bill for USA taxpayers through 2015, when we are concerned about using 'limited resources to defend the U.S. against weapons of mass destruction.'[70] Like perhaps ten pounds of anthrax, carried in a cooler that is left open, when the perpetrator leaves the ship that transported him or her to one of our ports; or perhaps a 'dirty' nuclear-imbedded bomb strategically detonated by someone who picked up the uranium from Rus-

sia's disarrayed "'Home Depot' WMD infrastructure,"[71] an ideal black market shopping venue for
terrorists. (WMD stands for 'Weapons of Mass
Destruction,' as if you could ever forget such a
term in this day and age.)

The entire federal budget projected for 2008
will be $2.9 trillion, or 2,900 billion dollars.

If we lock ourselves into "funding an industry
that makes missiles, and anti-missiles, and creates policies to promote the use of missiles, and
more spending on missiles,"[72] as presidential candidate, Representative Dennis Kucinich of Ohio
warns us, how will we get ourselves off the treadmill? Something very important to consider....

For once we implement the beginning of the
'missile defense' plan to deploy ground-based interceptors to shoot those incoming Scuds and intercontinental ballistic weapons out of the sky, won't
the next stage be the space-based platforms and lasers? Note how our Air Force looks at the issue circa their 1996 *New World Vistas: Air and Space Power
for the 21st Century* : "A natural technology to enable
high power is nuclear power in space.... Setting the
emotional issues of nuclear power aside, this technology offers a viable alternative for large amounts
of power in space."[73] (My underlining.)

"Setting the emotional issues of nuclear power aside"?? George Bush, Dick Cheney (ex board member of TRW[74]) and their weapons corporation cronies are ready to do that, as I have annotated throughout this chapter so far. The present Bush administration is manically pro-nuclear, pushing for the building of new nuclear plants in America, highlighted by their 'Nuclear Power 2010' program 'streamlining' approval of the ultimately polluting technology that had not seen a new nuclear plant ordered in our country since the 1970's due to adamant citizen opposition and nuclear contamination danger. Many in the Bush administration thought it was OK to use 'mini-nuke' bunker busters, at least one fifth as powerful as the atomic bombs that destroyed Hiroshima and Nakasaki, Japan—of course, to be deployed by USA forces exclusively. No other country or terrorist would dare to touch such a

Both Dick Cheney and his wife Lynn have been board members of corporations that stand to profit from waging war in space: TRW and Lockheed Martin, respectively.[47b]

weapon if we decided it would OK for us to do so. Right? (Thankfully, Congress recently decided not to fund any more research or production of this radioactively-wrong weapon.)

Now you also know of plans to use nuclear power to electrify colonies on the moon.

And down here on Earth, we're deregulating nuclear waste releases (see Chapter 5), while President Bush enthusiastically signed a bill allowing the shipping of high level radioactive wastes through 43 states by rail and truck, to Yucca Mountain in Nevada, with its myriad of earthquake faults, imperiling the water supply for one of our fastest growing cities, Las Vegas. Not to mention the danger of a transportation accident, or *did* anybody say 'terrorist?' Nah, it's impossible they could even THINK of highjacking one of these shipments. Impossible.

Who would want to entertain the likelihood that a nuclear powered generating device could explode on the Kennedy spacepad as the Challenger did? Although such an explosion was supposed to have a 1 in 100,000 chance of happening back in 1986 <u>before</u> the Challenger disaster, NASA shortly thereafter changed malfunctioning expectations for launched shuttles to occur in <u>1 out of every 76</u> launches![75]

Imagine 24.2 pounds of plutonium being blown to bits of microscopic particles to eventually settle in our lungs and cause lung cancer, if it had been the NEXT Challenger mission that had exploded in May of that same year of 1986.[76] That was the nuclear payload of the Ulysses space probe, to be on board that then-postponed launch. (The Ulysses probe was eventually launched in 1990, without catastrophe.) Plutonium, as I continue to remind you, is the most toxic element known to man, 24.2 pounds of it theoretically being enough to kill every human being on Earth if each of us inhaled a 2-4 micron sized dose into our lung.

And don't forget the seldom told tale that Al Qaeda's initial plan for September 11th 2001 was to crash those planes <u>into nuclear power plants</u>. But they decided against this 'for fear it would "get out of hand."'[77] Chernobyl in America?...Can you imagine?.....

The Theoretical Basics Of Missile 'Defense'

To get basic now on so-called 'missile defense,' the Trojan Horse we are promoting to gain domination of space, the fourth medium of warfare - - I mentioned 'interceptors' (back on page 298). The idea with them will be to shoot a bullet with a bullet. A missile starts coming at us, we shoot up an interceptor, it hits the missile bullseye, and POOF! our wonderful superior technology conquers the enemy once again!

Actually, a non-nuclear 'kill-vehicle' will be employed atop our proposed three-stage booster missile that will be based in an underground silo. That was what our 'interceptor' originally was proposed to be. It is designed to operate at a range <u>above</u> 130 kilometers, or about 80 miles above the surface of Mother Earth. Our atmosphere surrounds the Earth to a height of approximately <u>75</u> miles.

Once released from its booster, the 'kill-vehicle' will use infra-red light seekers and small side-thrusters to home in and 'direct[ly] impact' its target.[78] (The idea of using hydrogen bombs in space, or other related nuclear options, were not utilized for a variety of reasons, not least of which were adverse effects on communications and guidance satellites, and adverse public opinion.[79])

Sound intriguing, and technologically mundane for us in the Star Trek era? Alas, the real story is that the testing of these interceptors has been es-

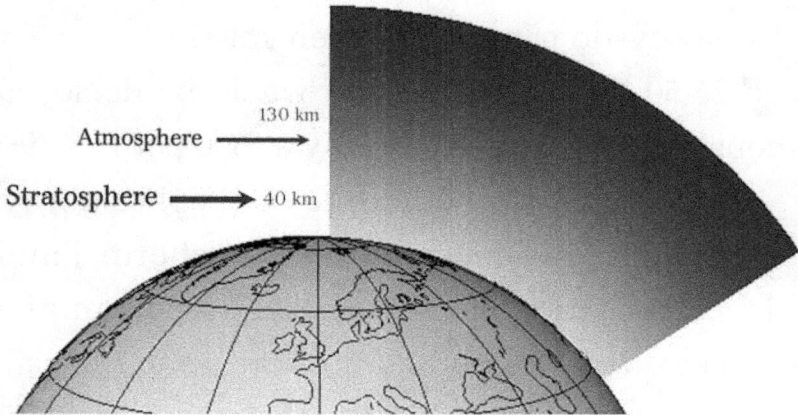

130 km

Atmosphere →

Stratosphere → 40 km

Interception of incoming enemy missile would target the 'mid-course' of its flight, when it would fly in the vacuum just beyond Earth's atmosphere. Usually running ~30 minutes.

sentially bogus -- we don't know if they will work well in actual conditions. Plus, what IF we had twenty or forty of them installed at Fort Greely, Alaska or Hokus Pokus, North Dakota? [Or perhaps in Poland or the Czech Republic, or in any of the 26 member nations of NATO that "will spend 75 million euros over six years to purchase command and control systems for missile defense?"[80] - - (Yes! another bright venue earning more MONEY for our space/weapons contractors!)]

Decoys are the matter that is not being properly addressed. As physicist Lisbeth Gronlund of the Union of Concerned Scientists tells me, any country that acquires or develops the capability to launch an intercontinental ballistic missile (ICBM), also will utilize accompanying relatively inexpensive decoys to camouflage their missile. On average 25-50 decoys per missile would be deployed, though there could be as many as 500 per missile.

These decoys, whether they be replica decoys or metallized balloons or shrouds of thermal multilayer insulation or chaff or electronic jamming devices[81], would look like the warheaded missile we need to hit, to our detecting devices. So each decoy would require to be hit, in addition to the actual missile, because we couldn't tell which was which. And just one interceptor may not do the job per target missile/decoy, so say two or three interceptors may be required per decoy! In other words, for three detected missiles we would need at least 150 interceptors to be fired. We'll not have that many, even by 2010. For Congress is bridling at the multi-billion dollar cost of buying weapons that do not meet the criteria for 'fly-before-you-buy' procurement laws.[82] That is, show that they work first, then spend the taxpayer's money.

There is talk of better 'discrimination' techniques in the future for discriminating between missiles and decoys, via both radar, and the interceptor itself. Possibly, the N-FIRE test satellites may help here – if they work. Radar technology is getting increasingly sophisticated, but timely integration with the interceptors in tracking missiles is still not where it should be to differentiate between decoys and the travelling missile that could be threatening us as it travels through space.

As for the interceptors doing the detecting, $4.6 billion may be requisitioned for a ten year contract 'to develop progressively more complex countermeasures <u>to test</u> how well interceptors can discriminate between warheads and decoys....The March 2004 General Accounting Office (GAO) report states that "a notable limitation of system (Ground-based Missile Defense, or GMD) effectiveness is the inability of system radars to perform rigorous target discrimination," and that "the kill vehicle (interceptor) itself must perform final target selection."

According to a Carnegie Endowment report: "knowledgeable scientists doubt that the problem of mid-course discrimination can be solved effectively in the near term, if ever."[83]

OK, 'mid-course' is the portion of the route of the missile where it leaves our atmosphere, and until it re-enters it. Time length: about 30 minutes. This is where we are concentrating our efforts on striking potential in-coming missiles. It is also the usual region where the decoys would be released to smarm up our detection system.

Now there is talk about hitting a missile in its 'boost phase,' just after the missile lifts off from its launch pad. Then, probably no decoys to flutter or reflect or get in the way! One interceptor, maybe two, or three, and blow that missile to smithereens!

But why is there always a problem?! The time is much shorter (up to 5 minutes for liquid-fueled rockets[84]); the range in distance of the interceptors also will be limited. Long-range missiles stationed inland in China or Russia, for example, could not be effectively reached or intercepted.[85]

Then there is the nightmare of an interceptor going awry. What if it strikes a crowded building or an unanticipateable densely populated civilian area? Or what if it hits the target missile, but does not destroy the warhead? Which then ends up landing somewhere, totally off its intended track, in some innocent bystander country? With a nuclear explosion or anthrax plague released as a result?

No problem! We'll just have some defense contractors build Aegis destroyers armed with smaller-ranged interceptors aboard to surround our next exaggerated missile threat, China, to the tune of $1 BILLION per ship. Yes, they'd also be good to check that axis of evil country, North Korea, if we can get the ships close enough to that ultramilitary land. And we can dock them, and send them out, from Japan's Yokosuka base. Maybe Japan would even buy a few? (Why, they have! Four at last report.[86]) To help boost the profits of our poor underappreciated weapons corporations! And those of the politicians and potential warrior laborers of Bath, Maine, and Pascagoula, Mississippi, where the Aegis's will be made.[87] We'll practice and practice, and pay and pay and pay, until they work. Don't we taxpayers, and hopers for peace, love it!?

Oh, these interceptors will only have a range of 1000 kilometers instead of the 10,000 kilometers required for integration into the National Missile Defense (NMD) system?[88]

Oh, that is only 625 miles? Well, remember that satellite that was falling back to Earth that we destroyed in early 2008? Yes, that was hit with a weaker-ranged relative of one of these interceptors! At about 150 miles above the Earth.[89] A grand test of our National Missile Defense prowess! OK, there were no decoys to confuse us. But, you have to admit that it did work! All

307

the more reason to continue spending those billions of dollars on Star Wars, rather than education or health care or the economy!

By the way, we do have an official interceptor test run scorecard. Out of twelve flight intercept tests, six have been 'successful'[90] in well scripted scenarios. Fifty percent. Fifty-fifty. Alas, these tests:

> 'employed the same unrealistic target missile trajectory, known in advance, and flown at low speed and altitude. The simple target missiles have been rigged with transmitters that exaggerate their signatures to a surrogate transponder/FPG-14 radar combination for mid-course tracking....nine of the twenty remaining programmed tests have been canceled, and others postponed...'[91]

states the Carnegie Endowment Special Report of March 25, 2004. Decoys were not used in a realistic scenario. Nor were several aborted runs counted in the tally.[92]

Bottom line: the entire Star Wars fantasy is still basically that. A fantasy. In functional practical terms. Many more tests have to be run, decoys and other countermeasures have to be allowed into the picture, a dependable booster rocket for the interceptors has not yet been developed, etc.

That Missile Defense 'shield' that you probably heard about? It would work like some protective bubble or force-field circle/sphere to surround the borders of our country from ocean to ocean... and maybe include Alaska and Hawaii...?...to save us from missiles that would bounce off its impenetrable plastic surface, or be vaporized by the lasers or missiles/interceptors we would produce? Like in the sci-fi movies!...?

Well, no, not exactly. The fantasy you may have swallowed. as being the ultimate defense against some serious massive attack from say Russia, or China, versus the relatively basic start-up plan President Bush is pushing billions of our tax dollars into for this decade and the next, are worlds of magnitude apart.

In 2007, there are about 30,000 nuclear weapons on our seemingly shrinking planet, most of which are still Russian and American owned.[93] Even though the Cold War is supposed to be over, Russia has most of its 5,800 active nuclear warheads[94] aimed at the U.S. Some targets just in New York City: each of its three major airports, its major bridges, Wall Street, four major oil refineries, major rail centers, and of course, power stations.[95]

The U.S. maintains about 5,700 active offensive nuclear warheads in our arsenal.[96] Though most of them used to be pointed at Russia's command centers and missile silos, very recently we switched the targeting points to the middle of the ocean. That way, in case of a computer glitch that erroneously says Russian missiles have been launched and are coming our way, U.S. missiles would not be initially fired at Moscow or St. Petersburg.

However, according to former Assistant Secretary of Defense Philip Coyle III, the U.S. has not deleted the co-ordinates of the Russian targets from our computer systems. Within seconds we could just switch our bulls-eyes back to where they have been, if we had to.[97]

With 2,500 of our nuclear warheads set on hair trigger alert, apocalypse always hovers near our over-militarized civilization.[98] For example, back in 1983, on the Soviet side, their early warning system indicated five U.S. missiles had been launched at the Soviet Union. "Disobeying orders, Stanislov Petrov, a lieutenant colonel in the Soviet Strategic Rocket Forces, decid[ed] against informing his superiors, correctly thinking a system malfunction had occurred.

"I couldn't believe that all of a sudden some-one would hurl five missiles at us," Petrov told Mosnews.com in 2004. "Five missiles wouldn't wipe us out. The U.S. had not five, but a thou-sand missiles in battle readiness.""[99]

Mr. Petrov is one of the heroes we should acknowledge for saving our planet. He might not have killed 'the enemy' or died as a result of 'friendly fire,' but humanity can continue on its self-destructive course a bit longer because of people like him.

According to the unlikely combination of former Vietnam-war-era Secretary of Defense Robert Mc-Namara, and the grandmother of the anti-nuclear movement, Dr. Helen Caldicott of Australia: if we detect what we think is a real missile attack coming at us, our Strategic Air Command has three min-utes to decide that this is not some static or tomfool-ery caused by hackers or wildfires or solar reflec-tions off clouds or even the ocean! Meanwhile, the President has to be reached within ten minutes, to receive a thirty *second* briefing on possible counter-attack options. Then there are three minutes to de-cide what to do, what targets to launch our missiles at. Within 30 minutes thereafter, the missiles could strike Russian soil and silo, etc.[100]

What could maybe fifty interceptors do against 5,800 incoming enemy missiles? you might be asking yourself. Each with about 25-50 decoys, thus requiring 26 X 5,800 = <u>150,800</u> total interceptors if we would conceivably be doing it right, using the lower number of 25 decoys per missile, plus the one missile itself to equal 26.....Too impossible, huh? Too insurmountable a problem?

Instead of starting up a new arms race in space, one might wonder, why not work on diplomatically disarming ourselves of these apocalyptic weapons. Each of them has about twenty times the potency of the bomb that we dropped on Hiroshima. We could incinerate ourselves and burn our Earth to a tragic crisp if we don't commence using our wisdom to get off the nuclear inertia express. The Berlin wall came down back in <u>1989</u>!

Americans should be aware that our current administration continues with the antiquated thinking that our nuclear forces must be able "to defeat any aggressor," as it states in their 2001 Nuclear Posture Review.[101] In other words, win a nuclear war!

Military reasoning postulates an answer for any problem posed. Plan to defeat a possible aggressor, if there is an attack upon us. However, in the real world, Nuh Uh, we cannot <u>DEFEAT</u> any nuclear aggressor. For the most likely nuclear scenario would see the U.S. and Russia mutually destroying each other, landing their thousands of missiles on the other's territory within 30 minutes, plus the subsequent rounds that *might* be launched, devastating and irreparably contaminating the surrounding environments of our atmosphere, continents, and seven seas.

We have to <u>avoid</u> such a possibility, as we have managed to do so far. We have to work towards peace, not war. Our technology is too powerful today, to ignore what we could unintentionally do to our children and our human family. Iran, we reckon, wants to join the nuclear club. North Korea recently became a member, successfully testing its first nuclear bomb. India and Pakistan are dangerous cousins in the Gang of Nine countries who are nuclear-bomb-capable.

Though, actually, with all our mega-projects shoring up our biggest budget deficit ever, at $445 billion for 2004[102] (previous highest budget deficits were $374 billion in 2003[103], and $290 billion in 1992;[104] $244 billion is projected for 2007[105]), consider what 'Robert Walpole, National Intelligence Officer for Strategic and Nuclear Programs, stated in Senate testimony on Feb. 9 2000:

We project that in the coming years, US territory is probably more likely to be attacked with weapons of mass destruction from non-missile delivery means (most likely from non-state entities) than by missiles, primarily because non-missile delivery means are less costly and more reliable and accurate. They can also be used without attribution.'[106]

So, as predicted, post September 11th 2001, the blight of cheap means of terrorism threatens America's civilization. We worry that Al Qaeda or some other cell of lunatics might spread a toxic gas -- sarin or VX? -- like they did in the Tokyo subway. Or maybe a bit of botulism to poison us as we eat, instead of using it to youthen our facial flesh? Or would they use some other low tech, inexpensive, more easily accomplishable assault on us rather than firing a massive intercontinental ballistic missile, risking relentless inevitable retaliatory bombardment to the point of decimation of their 'rogue nation?' What about bazookas? or those smaller shoulder-fired missiles?

Our former Secretary of Defense, Donald Rumsfeld, whom the Washington Post called the "leading proponent not only of national missile defenses, but also of U.S. efforts to take control of outer space,"[107] before taking office in 2001, chaired what some called the 'Space Commission,' and others called the 'Rumsfeld Commission.' The report issued by this commission, besides stating 'We cannot fully exploit space until we control it,'[108] focused on the emerging missile threat to the USA from Iraq, North Korea and Iran.

Although utilizing the report to show why we needed some sort of missile defense program begun, no matter what the practicality of its function, based on foreseeing that each of these three countries could develop -- or purchase -- modest long range missile systems, accommodating small payloads, over the next decade; the commission also revealed that more likely, and more quickly, we might see them having shorter-range ballistic missiles that a country like Iraq could launch covertly from a commercial ship off of our shores.[109]

What the commission did <u>not</u> consider as a group was "the vulnerability of the US to biological weapon attack from ships off shore, from cars or trucks disseminating biological weapons, from unmanned helicopter crop dusters, or from smuggled nuclear weapons or nuclear weapons detonated in a US harbor while still in a shipping container on a cargo ship; but these capabilities are more easily acquired and more reliable than ICBM's,"[110] a commission member divulged.

'The Soviet Union was reported in the late 1980's to be developing new types of chemical and biological submunitions for a variety of delivery systems, including short-range ballistic missiles.'[111] 'Submunitions' or small bomblets, perhaps a <u>hundred</u> per warhead seems to be the magic number, that could contain anthrax or nerve gas instead of one nuclear bomb. Easier to disperse and do plenty of damage in their own right.[112] Excellent for countries or terrorists that have not as yet developed sophisticated nuclear capabilities.

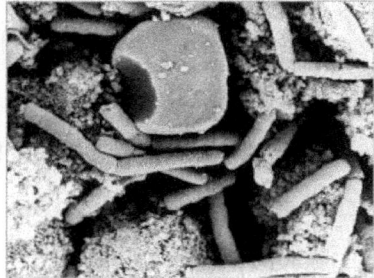

Anthrax germs: The ones that look like worms under the electron microscope

The submunitions could have all sorts of dispersal contraptions to release them from their warhead individually at different angles, at different speeds, so each bomblet would have to be dealt with by <u>at least</u> one of our interceptors or kill-vehicles, if released from an ICBM in its midcourse. That would mean 2-3 interceptors again, per submunition, with 100 submunitions didging and dodging at perhaps 7 kilometers per second in the weightless vacuum of outer space.[113] Doing your basic multiplication, that translates to 200-300 interceptors being required per missile releasing submunitions. By 2010 maybe we'll have 100 or more interceptors? inadequately tested. Does this sound re-assuring? spending, or wasting, your hard earned money like this?

With a short range missile, the time would be too brief to intercept, as something like a Scud would release its submunition payload one minute or so after lift-off.[114]

For countries whose missile technology is not as sophisticated as ours, the spreading out of where the bomblets land would more than make up for the poor accuracy of their developing missile system technology. It would be better than putting everything in one basket, one warhead, also, because dispersal would be better from one hundred bomblets with more total surface area, than one more vulnerable large 'lump' of the same amount of anthrax seeding its spores from just one dispersal point.

But don't concern yourselves, our missile defense proponents claim our in-flight tests have shown success against submunitions. Well, perhaps, my fellow

Americans, though highly doubtful. Like so much of our testing for so-called 'missile defense', our November 30th 1993 scenario apparently seemed typically staged to help the program look good, rather than scrupulously test our defenses against a realistically created attack. A target missile carrying 38 canisters filled with water to simulate some chemical weapon submunitions was intercepted.[115] However, the submunitions were not dispersed early in the flight as they certainly would have been in a real situation. 'Instead the canisters were all clustered together in a single package, which makes no sense from the point of view of an attacker facing a missile defense.'[116]

Of course, if a terrorist wanted to, or could, would he bother with any sort of missile, if he could gain access to something like anthrax? With its spores that are so potent, 300 million lethal doses could exist in just six grams. (454 grams make up one pound.) If a concoction of that minimal amount could be dispersed in an ideally sized aerosol, the entire population of the USA could be wiped out. That's where something like cropdusters or just that one cooler holding those ten pounds of the germ left open in New York harbor, could kill an awful lot of people.

We know about this because the USA was a pioneer in testing military uses of anthrax. For our M143 bomblet that carried just six grams of the germ's spores in a slurry form. 6 trillion anthrax spores that little 8.6 centimeter bomblet carried. (That's about 3 inches in diameter.) The lethal dose of anthrax has been reported to be 10,000-20,000 spores inhaled.[117] ~ 100,000 doses can fit on the head of a pin!

However, Dr. Matthew Meselson, Nobel Prize-winning molecular biologist from Harvard, who studied the 1979 Sverdlovsk, Russia, accidental release of less than one gram of anthrax which killed 68 people, suggests that there may indeed be no threshold number of spores required to initiate the deadly disease process. In other words, but one potent spore could possibly cause the disease. The type, form and potency of the anthrax more likely would dictate how many spores it would take to infect an individual.[118]

For example, some of the anthrax mailed in those fatal envelopes to individuals in Florida and elsewhere earlier in the decade, were reportedly of the genetically engineered variety. Perhaps this type was more potent than other varieties more commonly found in the world on contaminated animal hides.

Although we never fired the M143 at any adversary, we still have more than sixty laboratories in the USA licensed to handle anthrax. Why, you might ask, if we are so civilized and would never use biological weapons ourselves on any combatant or enemy nation, do we have what must be hundreds of people playing with this ominously dangerous germ? That <u>some</u> terrorist may have gotten his or her hands on right here within the borders of our country to produce those killer mailings? And may do so again...

Perhaps, our government would say we have to figure out how to defend ourselves against these microbes. That is why we must continue to study them. For defensive reasons.

Meanwhile, the Bush administration demagogues warn us we also have to continue worrying about Iran and North Korea's missile capabilities. Even if North Korea may only have 10 missiles in its entire armamentarium. And it's main 'Nodong' missile with its small two and a half cubic meter volume nosetip capacity[119], sounds like it needs Viagra. The <u>threat</u> from these 'rogue' nations is reason enough why we have to have missile defense. At the cost of billions. Making our defense contractors very happy to help. As we channel some of our greatest minds into the industry for which we want to enhance our space dominance. While the Osama bin Ladens of the world figure out some simpler more economical means to terrorize Americans. And our military affirms that they are needed in space because, though "we would like space to be a peaceful medium, the fact of the matter is history says that it may not always be that way."[120]

So we have what few Americans are aware of: the National Reconnaissance Office (NRO) that basically runs and monitors our satellites of all types, including those that spy on other nations. Annual budget: approximately $7 billion per year. Which is more than the annual budget for the CIA! But according to Robert D. Steele of oss. net, of all the information taken in "we process roughly 10% of our collected imagery, 6% of our Russian signals, 3% of our other major targets, and less than 1% of our rest-of-world signals."[121] The light's always on, but how often is anybody home? doing the work, figuring out what we have collected?

Priorities, priorities!

But did you know that things are getting so crowded in space, there are already more than

110,000 manmade objects orbiting the planet, ~9000 of these larger than a softball. These include bolts and metal debris, to defunct satellites and rockets.[122] And some of our entrepreneurs want to put gigantic advertising billboards out there, visible for potential marketees down here on Earth! Others, like the publicly traded SpaceDev corporation, want to stake private property and mining claims to asteroids like Nereus, dispatching a 'prospector' device to find minerals on the orbiting body.[123]

Since "typical impact velocities" are greater than 20,000 miles per hour, 'space junk' now poses a navigational hazard.[124] So much so that the USA and Norway are have erected a giant radar station, Globus II, in Norway's Arctic, to ostensibly monitor orbiting debris.[125] (Though I have been informed this more likely is being used to monitor Chinese and Russian satellites in space.)[126]

"Even millimeter-sized objects can cause considerable damage"[127] and "probably more than 95% of the objects...are not catalogued because they are too small to be reliably detected by Space Surveilance Network detectors,"[128] reports the Department of Defense. NASA "replaces pitted orbiter windows after most flights" of space shuttles, but the National Research Council warns that very serious accidents with space debris could result "in the loss of life or the vehicle"[129] involved at anytime in our Star Wars future. Newsday reports that space junk has <u>doubled</u> since 1990.[130]

Just imagine a weaponized space involving a war scenario. Space junk would litter the heavens, and some of it would rain down on us Earthlings, as Murphy's Law[131] dictates. And possibly be NUCLEAR! if our current rabid-dog leaders have their way.

Though in January 1978 Chicken Little, Senior, might have rightly alerted Canadians about the Soviet Union's Cosmos 954 nuclear powered satellite real-world-crashing about our northern neighbor's Northwest Territories. Radioactive debris had survived atmospheric re-entry and spread '110 pounds of highly enriched ura-

Globus II radar station, erected in Norway's Arctic to ostensibly monitor 'space junk' travelling at 20,000 miles per hour --but actually more likely monitoring Chinese and Russian satellites

nium'[132] over a 600 kilometer (~375 mile) path.[133]

As recently as November 16, 1996, the break-up of the Russian space probe left half a pound of plutonium deposited along the border of Chile and Bolivia. Remember half a pound of the ultimately toxic plutonium could result in a quarter million cases of lung cancer, if dispersed in particles small enough to fit into our alveoli in our lungs.[134]

Then there are the hare-brained space missions and fly-by's that the USA has launched, including the Cassini, with 72.3 pounds of plutonium aboard, and the Climate Orbiter that crashed into Mars in September 1999 due to lack of mathematical unification of metric and English numbers by erudite scientists!

The Cassini was sent up in a Lockheed-Martin manufactured Titan-4 rocket, then was to buzz the Earth on a return, accelerative 'fly-by' of our planet to better slingshoot it toward its final destination, Saturn.[135] Travelling at 42,300 miles per hour, Cassini was to pass 700 miles above the Earth, though <u>NASA</u> stated in its Final Environmental Impact Study that if the probe inadvertently re-entered our 75-mile high atmosphere, without any heat shield, it would break up, releasing plutonium, and "approximately 5 billion of the....world population of the time....could receive 99 percent or more of the radiation exposure."[136] 72.3 pounds of plutonium can cause a lot of deaths and contamination, as you now know.

Luckily, the Cassini did not veer too close to Earth, getting close enough to Jupiter to photograph that remarkable planet and some of its moons in mid-2004, but NASA's Climate Orbiter, just five weeks after the Cassini fly-by, did <u>not</u> safely pass over Mars. The two separate teams of scientists involved at Lockheed-Martin and NASA's Jet Propulsion Laboratory were excruciatingly embarrassed because they had used different scales of measurement at their respective facilities, never converting the English units to metric units when calculating their altitudes.[137] How could that happen? you might ask. What is the likelihood of such an error?

Well, might we say it was just one of those infamous human errors? which simply cannot be quantified.

Though we can quantify the number of Titan-4 errors. As stated earlier, one in ten launches of Titan-4 rockets have exploded on, or soon after, launching. In fact, the three Titan-4's launched right after the Cassini went up, all catastrophically destroyed themselves. Floridians and Earthlings of all nationalities should be grateful that none of these had plutonium on board, as did the Cassini.[138]

Remember the previously sited NASA projection of one in 76 shuttles/rockets malfunctioning?[139]

Yet, eight future USA space missions projected to 2015, all needing to be launched by rockets which could blow up at the above rates of likelihood, are being designed to have nuclear powered generators.

Meanwhile, the European Space Agency's Rosetta probe, launched on March 2, 2004, bound beyond Jupiter, to rendezvous with the Churyumov-Gerasimenko Comet on a 7 BILLION kilometer journey[140], is powered by "high efficiency solar cells."[141]

Why are we sticking stubbornly to nuclear power, while Europe goes solar with her space probing? Could be because we want to keep the nuclear option open in our scheme to dominate space, and we want to keep the technology 'dual use,' optimally integrated. Not like the Mars Climate Orbiter and its lack of integrating metric and English units.

Though the USA, following a 1964 space accident that dispersed plutonium all about the planet, also became the pioneer in developing photovoltaic solar energy, which is <u>now the power system on all U.S. satellites.</u>[142]

Yet, just like those athletes thinking they have to keep up with the McGuires and Alzados via using steroids, we hear that maybe the European Space Agency [ESA] thinks it should keep up with the USA by using Radio-isotope Thermoelectric Generators [RTG's] to power their future deep space missions. Uh-huh, RTG's are nuclear powered, with somewhere between 20 and 75 POUNDS of plutonium-238 producing the electricity, exploding on the launchpad, or someplace else crucial, or not. And, why, blimey, won't they be purchased from the USA? Makes perfect sense, doesn't it?

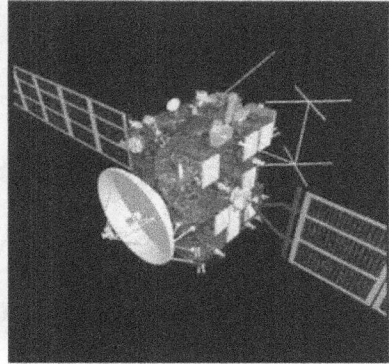

Rosetta space probe, launched by European Space Agency, March 2 2004, utilizing <u>solar</u> power for its 7 BILLION kilometer journey. Yet the newest incarnation of USA's NASA insists on using nuclear power, to be consistent with its 'dual use' policy for both military and civilian options for our space projects.

Plus, mark that you heard it here about those new <u>plutonium powered batteries</u> NASA and our

Department of Energy [DoE] want to produce at the Idaho National Laboratory. These 'could eventually wind up in everything from space-based killer satellites to battlefield computers. "The primary driver for us to start production is for national security requirements."' So says 'Tim Frazier, director of the DoE's nuclear power systems programs.'[143]

Stratcom in Omaha, Nebraska, Where The Next War Will Be Planned, Launched and Co-ordinated

Now for you horror fans, we have our Strategic Command [StratCom] center in Omaha, Nebraska doing things you probably never woulda thunk. Deep in the heartland StratCom's "satellite surveillance system keeps a close worldwide eye on anything suspicious and shady" in conjunction with "its international network of 'listening stations.'" Being the command center for our nuclear arsenal, where our 'warrantless wiretaps' were brainchilded, and our National Security Agency [NSA] has been made a 'Component Command' thereof, how much doubt would there be that the

Bush/Cheney empowered StratCom wouldn't simply adore radioactive gadgets like plutonium powered batteries aboard the satellites they use to monitor American citizens and anyone else/anything else they choose to monitor? At a recent "'Strategic Space and Defense' trade show and arms bazaar in Omaha" an "industry sponsor" called StratCom "a laboratory for the future of warfare."

'Having foregone all semblance of its purportedly 'defensive' role, StratCom today serves as the command center for offensively waging the administration's international 'War on Terror' with conventional as well as nuclear weapons,' in addition to being given the responsibility 'to launch a first-strike against any perceived threat to America's national security anywhere on the planet within two hours.' Tim Rinne of Nebraskans for Peace tells us.

He further reveals that "StratCom's long-range plans call for securing space exclusively for the U.S. and its approved allies. This strategic goal is already fixed, and will go on regardless of which party controls Congress or whether a Democrat or a Republican Congress occupies the Oval Office. StratCom is fast becoming a law unto itself."

Stratcom in Omaha, Nebraska, "fast becoming a law unto itself," where "the next war...will be planned, launched and co-ordinated."

This is most important to know, since the Bush administration has padded StratCom "in quick succession" with control of U.S. Space Command, so-called 'Missile Defense,' "combating Weapons of Mass Destruction" plus 'Full-Spectrum Global Strike.' And do NOT forget that included here is Space Command's "C-4ISR" missions, which are Command, Control, Computers, Communications, Intelligence, Surveillance and Reconnaissance.[144] All back there in cornhusker land, Omaha, Nebraska, where Mr. Rinne warns us "the next war... will be planned, launched and co-ordinated."[145]

And to further chill you, effective since 2007:

"STRATCOM's satellites will be allowed to keep an eye not only on foreign foes but on you and me as well. This spring [2007], the government for the first time granted the Department of Homeland Security and other domestic law-enforcement agencies access to real-time, high-resolution images and data from military intelligence satellites as they pass over America's cities and countryside.

Indeed, after her conference talk, Brig. Gen. Jennifer Napper, deputy commander for USSTRATCOM's Global Network Operations told reporters, "The FBI and CIA are in our operations center 24/7.""[146]

Couple that information with what Senator Frank Church wrote back in 1975 when he was chairman of the Select Committee on Intelligence investigating 'the government's massive and highly secretive National Security Agency' (NSA)'s capability:

"That capability at any time could be turned around on the American people and no American would have any privacy left, such is the capability to monitor everything: telephone conversations, telegrams, it doesn't matter. There would be no place to hide. If this government ever became a tyranny, if a dictator ever took charge in this country, the technological capacity that the intelligence community has given the government could enable it to impose total tyranny, and there would be no way to fight back, because the most careful effort to combine together in resistance to the government, no matter how privately it was done, is within the reach of the government to know. Such is the capability of this technology... I don't want to see this country ever go across the bridge. I know the capability that is there to make tyranny total in America, and we must see to it that this agency and all agencies that possess this technology operate within the law and under proper supervision, so that we never cross over that abyss. That is the abyss from which there is no return. ..."

Senator Frank Church, concerned in 1975 about our awesome spying technology being to a threat to our democracy if tyrannically turned against the American people.

When Sen. Church made this statement the NSA was not authorized to spy on American citizens'[†] as it now is allowed to do.

The USA's Nazi-Riddled Space History

Very perplexing, isn't it? that this could have eventuated in our land of the free and home of the brave? Alas, now it is time for you to learn more specifics about our Nazi-riddled history that may still be infecting the spirit of what we are doing and plan to do with the heavens above, and the Earth below.

World War II ends. 1945. German scientists who produced the Scud-like V1 and V2 rockets are turned over to Richard Porter for interrogation. Porter "represented the General Electric Corporation, which held the Army contract for the first long-range ballistic missile under development in the United States."[147] Instead of punishing these Nazis, Professor Jack Manno described the goings on as being "like a professional sports draft....Nearly one thousand German military scientists {were adopted by the USA}, many of whom later rose to positions of power in the U.S. military, NASA, and the aerospace industry."[148]

Wernher Von Braun became the most prominent Nazi star to instill his genius into our budding space program. In fact, he and his team were the heart of the George C. Marshall Space Flight Center at Redstone Arsenal in Huntsville, Alabama. Von Braun was appointed Marshall's

---

† 'The Corporate State and the Subversion of Democracy,' by Chris Hedges, May 31, 2008 in TruthDig.com also available at http://www.common-dreams.org/archive/2008/05/31/9331/

director, and after ten years in that position became Deputy Associate Administrator of NASA! in 1970.[149]

Then there was former German Major General Walter Dornberger, who we mentioned before, championing us to conquer and occupy space between Earth and the Moon.[150]   This USA-Nazi space program draftee produced a planning paper in 1947 projecting "a system of hundreds of nuclear-armed satellites all orbiting at different altitudes and angles, each capable of re-entering the atmosphere on command from Earth to proceed to its target." The Air Force began working on this concept under the name 'Nuclear Armed Bombardment Satellites' or 'NABS.'[151]

Dornberger also produced a variation of this NABS idea in which he "proposed an anti-ballistic missile system in space in the form of hundreds of satellites, each armed with many small missiles.   The missiles would be equipped with infrared homing devices and could

Wernher Von Braun, German Nazi military scientist drafted by USA to eventually become Director of our Marshall Space Flight Center; and in 1970 appointed Deputy Associate Administrator of NASA.

be launched automatically from orbit."[152] The Air Force studied this variation, labelling it 'BAMBI' for 'Ballistic Missile Boost Intercept,' which became the seminal "idea that would reappear in the space-war dreams of the Reagan administration in 1983."[153]

Professor Manno laments the tragedy of "extending and accelerating the arms race"[154] into space, the world losing its opportunity to cap the genie before peace could secure the realm surrounding our planet.

Manno also wrote in 1984 rather prophetically: "Even if militarists succeed in arming the heavens and gaining superiority over potential enemies, by the 21st century the technology of terrorism - chemical, bacteriological, genetic and psychological weapons and portable nuclear bombs - will prolong the anxiety of constant insecurity."[155]

The Nazi scientists, Professor Manno sees as an important "historical and technical link, and also an ideological link[156]...{to our} space program of today...{with} its roots deep in the strategy of world domination through global terror pursued by the Nazis in World War II."[157]

The aim of missile defense or Star Wars or

whatever title one wants to give our space program, according to Professor Manno "is to put all the pieces together and have the capacity to carry out global warfare including weapons systems that reside in space."[158] A new factor is the growth of the global economy, recently sped up by the creation of the World Trade Organization in 1995 (see Chapter Four), with those promoting it wanting to attain "control over the process of globalization."[159]

Don't think that Ronald Reagan or either of the Presidents Bush or former Secretary of Defense Rumsfeld or the Republican party are alone in their aggressive push for U.S. control of the space above planet Earth. The Democratically controlled U.S. Congress authorized "Military Space Forces: The Next 50 Years" in the mid-1980's. This book advocated military bases on the moon, nuclear power in space, and preventing other nations from becoming military space operatives by 'controlling attitudes.'

"The basic objective would be to deprive opponents of freedom of action, while preserving it for one self." the book proclaimed on page 48. (Which phrase, 'freedom of action' in space, President Bush has been using these days to rationalize why we should be able to continue acting as we are, rather than co-operating with other nations to prohibit anti-satellite tests and development of space weapons.[160])

Using the 'gravity well' advantage of operating from the moon - - "Put simply, it takes less energy to drop objects down a well than to cast them out...{U.S. forces could} enjoy more manoeuvering room and greater reaction time."[161] Not having to be restricted by the Earth's much more powerful gravitational pull, as compared to that of the moon. Nor the Moon Agreement of 1979, prohibiting such appropriative/military behavior.

Reprehensibly, this same U.S. Congress-authorized book - - about which Les Aspin, Democrat from Wisconsin, and Secretary of Defense under President Bill Clinton, approvingly stated: "No other military space study puts all pieces of the puzzle together."[162] And to which people like Senator/Astronaut John Glenn, and Senator/former Senate Armed Services Committee Chairman Sam Nunn of Georgia, Representatives John Kasich (Republican) of Ohio, Ike Skelton (Democrat) of Missouri, and many others affixed their honorable signatures - - proclaims that U.S. "armed forces might lie in wait at that {ideal moon site} location to hijack rival shipments"[163] of materials mined by other nations!

Too many of our Congressional leaders from both sides of the political aisle willingly, enthusiastically, jumped on the bandwagon approving space piracy!

And what about this, with our terrible anxiety about being victims of anthrax poisonings, from the same U.S. Congress authorized book:

"Self-contained biospheres in space accord a superlative environment for chemical and biological warfare....Clandestine operatives could dispense lethal or incapacitating chemical/biological weapons agents rapidly and uniformly through enemy facilities."[164]

Shame on us!

I understand that the military has to brainstorm ideas to be creative and overcome enemy manoeuverings. Many of these ideas are ridiculous and crazy, reaching various stages of fruition. But our history does not look very good when it comes to the documents that our government has commissioned and approved in the name of the United States of America, and its people: US! you and me! as discussed above.

There used to be the ABM (Anti-Ballistic Missile) treaty with Russia that aided maintaining the peace between our nations. BUT ALSO 'PRO-HIBITED THE TESTING OF SPACE-BASED AN-TI-BALLISTIC MISSILES!'[165] Its abandonment in the light of an escalated arms race in space, bestows upon both nations the curse of <u>increased in</u>-security as we march ahead with what President Bush labels as 'missile defense.'

Evidence here on Earth, Russia's recent [July 2007] dropping out from the Conventional Forces in Europe Treaty, largely because of the Bush administration unrelentingly pushing to deploy radar and interceptors in Europe under the veil of 'missile defense,' ostensibly for fear of Iran's nuclear ambitions. "Russian officials regard the project as unnecessary because they believe Iran is many years from developing long-range missiles. And, more critically, military officials here believe the system can - - and probably will - - be used by the United States to peer deep into Russian territory."[166]

After all the work our diplomats have done to reduce the threat of nuclear destruction, ending the 'Cold War,' and the 'Arms Race,' here we are going again. On Earth as it is starting to be in Heaven.

Oh, you might also be surprised to know that "on Jan. 4, 2007, George Schultz [former Secretary of State], Henry Kissinger [the infamous egotistical Republican policy-master], Bill Perry [former Secretary of Defense], and Sam Nunn [former Senator] wrote an op-ed in *The Wall Street Journal* calling for the elimination of nuclear weapons. Three weeks later, Mikhail Gorbachev responded with his own statement supporting the urgent call to action."[167] We really cannot practically use these weapons, can we? Unless we want to start a nuclear war. Such an alliance of hawkish Americans and Mr. Gorbachev realizing how foolish the continued coddling of the nuclear option is, should stun the observant citizen in us.

Honestly Evaluating Our Space Program
Honoring The Outer Space Treaty of 1967
Keeping Offensive Weapons Out Of Space

For now, testing of our military thinkers' public relations first stage toward full-spectrum space dominance has to be adequate to show high effectiveness of new weapons under varying conditions, with enough acceptable tests to also give us a high level of confidence that these expensive gizmos will actually work, will actually protect us. That means that countermeasures like multiple decoys; radar-impermeable metalized balloons; nitrogen cooled shrouds over warheads; clouds of wire chaff; flares to generate confusing infra-red signals that could mask the relatively cooler warhead(s) being searched for by radar or space-based infra-red sensors on satellites like the N-FIRE; manoeuverable warheads; electronic radar jammers employable on both warheads and decoys; plus tens of submunitions that could carry anthrax or VX nerve gas; must be encountered and defeated in realistic scenarios.

In other words, we cannot depend on tests that have just one decoy and one warhead, with the shape of each, or the relative infra-red signal of each, known beforehand as in the test of October 2nd 1999.[168] What the Union of Concerned Scientists (UCS) suggests in their 'Countermeasures' report is that an independent 'red team' should act out the enemy role, employing 'the most effective countermeasures that the emerging missile states could field. It is clearly important that the countermeasures that are developed and tested are not "dumbed down" to make the job of the defense easier.'[169]

In addition, there must not be a "rush to failure."[170]   'Testing must be outcome-driven, not schedule-driven. There must be an opportunity to assimilate the results of one test before rushing headlong into another. Program managers must carefully distinguish testing done to learn and testing done to verify.'

'Thus, it is important that the NMD (National Missile Defense) program be insulated from congressional and administration pressures for unrealistic testing and deployment schedules.'[171]

Three or four or five tests simply will not be enough to determine how or if we should go ahead with the whole schmorgasbord. 100 or more intercept-type tests seem to be a more pragmatic number to the Union of Concerned Scientists' panel (UCS), at a projected cost of at least $5 billion dollars.[172]

Notwithstanding all that, UCS concludes that 'any country capable of deploying a long-range missile would also be able to deploy countermeasures that would defeat the planned National Missile Defense system.'[173] This statement comes from the UCS Executive Summary report, which does preclude President Bush's changing/evolving approach to missile 'defense.' Which will encompass the previous Clinton administration's basic plan of layers of defense, adding the danger-ously destabilizing offensive weaponry described in this chapter, intended for space-deployment, into the equation.

As the second version of the Rumsfeld Com-mission, chaired by President Bush's former Sec-retary of Defense, reported in January 2001:

> 'In the coming period, the U.S. will conduct op-erations to, from, in and through space in support of its national interests both on the earth and in space....{The Commission urges that the U.S. Pres-ident} have the option to deploy weapons in space to deter threats to and, if necessary, defend against attacks on U.S. interests.'[174]

If you wonder why most of what you hear and see via your favorite media outlets may be sur-prisingly uncritical of 'missile defense,' not get-ting down to such specifics as those revealed in this chapter, a big reason could be that people like Diane Sawyer, Dan Rather, Tom Brokaw, Jim Leh-rer, Barbara Walters, and the Washington Post's and New York Times' top editors are members of the ∼3000 member-strong Council on Foreign Relations, which has existed since 1921. Many consider this elite organization as representative of 'The Establishment' or the 'power structure' of our country, with bankers and politicians from Dick Cheney to Colin Powell to George Bush and Bill Clinton also counted amongst its membership.[175]

This Council's published 1998 report 'Space, Commerce, and National Security,' written by Council Military Fellow, Air Force Colonel Frank Klotz, espoused: "the most immediate task of the United States in the years ahead is to sustain and extend its leadership in the increasingly intertwined fields of military and commercial space. This requires a robust and continuous presence in space."[176]

Our major networks, and their leading talking heads, fed on the above approach to space and so-called 'missile defense,' should not be expected to surgically dissect major problems with deploying weapons, including those of the nuclear variety, in space to save our civilization.

While 'missile defense' and our goal of embellishing our space dominance may be quietly funded at nine billion dollars per year (plus fifty percent) under the entertainment media radar, we will continue to hear the red and orange hued warnings from our officials in our War on Terrorism. Will it be North Korea or Iran that might attack our cities with a missile in the future, or will there be the unthinkable accidental launch of Russian or our own missiles desolating the planet before we use our smarts to disarm and take these weapons off hyperactive hair trigger? while we build up our 'Space Force,' developed on equal footing with our Army, Navy, Marines, and Air Force, as some governmental hawks and space industry vultures wish us to do?[177]

Yet more likely, a less expensive, lower tech, more functional way to terrorize our country and the world will be used by enemies of our way of life. A bloated, poorly thought out, inadequately tested, runaway space program will only make those who wish to destroy us, create more focused means to defeat an overloaded behemoth of unco-ordinateable technology. A few grams of anthrax or botulism, relatively simply developed counter-measures to radar and infra-red sensors - - tech-nologies devised by the USA in our checkered past - - seem probable to haunt our immediate future, employed by some antagonistic sub-superpower entity framed within our own hazy paranoia.

To really save civilization as know it, and want it to be for our children and future generations, we should seriously consider the words of Kofi Annan, former Secretary-General of the United Nations:

"Above all, we must guard against the misuse of outer space...The international community has acted jointly, through the United Nations, to ensure that outer space will be developed peacefully...We must not allow this {20th} century, so plagued with war and suffering, to pass on its legacy, when the technology at our disposal will be even more awesome. We cannot view the expanse of space as another battleground for our earthly conflicts."[178]

The big thing at this point is that no weapon has been deployed in space, YET. If all the facts come out, as portrayed, for example, in this chapter, perhaps no arms race will occur in space. Since you now know the true story about so-called 'missile defense' and its actual offensive nature, maybe you could do your part to inform your

Kofi Annan, former Secretary-General of the United Nations

friends, family, government representatives, media employees, and anyone that you can, about this potential escalation toward apocalypse. Before it is too late. There is room for reason and diplomacy in our human community. Even with George Bush and Dick Cheney pontificating to the pre-emptive contrary.

Chapter Four:
Surrendering U.S. Sovereignty to the World Trade Organization

"I see in the near future a crisis approaching that unnerves me and causes me to tremble for the safety of my country. Corporations have been enthroned and an era of corruption in high places will follow, and the money power of the country will endeavor to prolong its reign by working upon the prejudices of the people until all wealth is aggregated in a few hands and the Republic is destroyed."

—U.S. President Abraham Lincoln, 1864[1]

Running for President in 2004, Dennis Kucinich said the <u>FIRST</u> thing he would do, if elected, was withdraw the U.S. from the World Trade Organization (WTO). Why would he think that would be more important than fixing up the turmoil in Iraq, or creating more jobs, or putting a halter on the mad scientists in their Monsanto and Syngenta gene-alteration laboratories compromising our food supply?

Could it be that we and the other 151 nations that have signed on to the GATT agreement (<u>G</u>eneral <u>A</u>greement on <u>T</u>ariffs and <u>T</u>rade) Uruguay Round establishing the WTO, should be ultimately worried about the sanctity of our own laws now?

That in secret, usually three men can decide that any of our federal, state or local laws are WTO-illegal and have to be changed, or we must pay the consequences? That our Supreme Court or our Congress no longer reign superior to the WTO and its undemocratic dispute panelists' decisions? That almost every challenge to any country's laws in WTO-court has deemed those laws 'illegal barriers to free trade?'

This can apply to almost anything from any local or federal law prohibiting sale of goods made by child labor, to transportation of goods, to environmental standards on what we have scientifically discovered to be toxic levels on a pesticide or an element like arsenic, to our use of eco-labels, to <u>eliminating any limits</u> on blasting food with irradiation, to whatever could be considered as 'services.'

Wallach and Woodall describe 'services' suscep-

Under WTO rules, laws placing restrictions on products made with child labor would be WTO-illegal, and be grounds for massive punitive sanctions.

tible to the almighty WTO as 'everything that you cannot drop on your foot.' Here's their list of what this means to you: 'retail stores, banking, insurance, energy, telecommunications, maintenance and repair, construction, mining, toxic waste processing, tourism, museums, libraries, food preparation and restaurants and hotels, laundry and cleaning, and transport.'[2]

I know you can't believe that so many things could now be controlled by some 'trade' entity whose center is located far away across some ocean (in Geneva, Switzerland, Europe). But I hate to tell you that the list isn't finished! The WTO can also decide upon their concept of the legality of our laws relating to "essential public services," such as education, hospitals, social security, mail delivery, police and prisons, and water and sewage systems.'[3]

If some foreign hotel chains want to build new hotels on land that some Florida or California town has decided has had enough development, perhaps for environmental reasons, or to stop rampant urban sprawl, well, too bad. Our USA laws here are now subject to challenge and being ruled WTO-illegal.

How about this one: once a 'service' sector is opened to what is called GATS, or the General Agreement on Trade in Services, as one of 18 WTO major agreements[4], 'new government services or private not-for-profit monopolies can never be established again!'⁵ For once a 'service' sector allows its first privatization at any level of government from federal down to the village level, under GATS and WTO rules, you've entered yourself and your entire country on a 'one way' street that you cannot back out of anymore. That 'service,' whether it be social security or health care or electricity can never go back to the way it was, unless 'all WTO member countries whose service providers are affected'⁶ are compensated. Which is just about absolutely prohibitive in a real world WTO-world.

Think about Enron and Global Crossing and our deregulation of electricity as one shining USA example of this, and GATS and the WTO expanding this deregulation policy for the whole planet. Or Canada trying to go back to their unique coast to coast public health care system after 'the government of Alberta introduced legislation allowing private, for-profit facilities to offer overnight care,'⁷ jimmying open the window for GATS saying 'That's it, sucker. You can never go back to exclusively public health anymore under your single-payer Medicare program.'

Our Clean Air Act has been affected by the WTO, our offshore-earnings income-tax-break law has been ruled WTO-illegal, our national parks could be overrun by concessions that we already have decided would ruin them, non-profit elder-care or childcare mandating would run afoul of

GATS and the WTO. And on and on and on and on. . . . .

How have we let this happen? That by subjecting our 'service sector to WTO disciplines, almost no human activity from birth (health care) to death (funeral services) remain outside WTO's purview?' Is this the New World Order? Would the WTO be the all-powerful overlooking Big Brother that George Orwell feared when he was writing his classic novel *1984*?

I know most of you probably never thought about the reality of all this, but here you have perhaps the most important international issue driving the politics and economics of the entire world's future. Corporations want access to the natural resources of the world without any drain on their possible profits. Health, labor, or environmental laws are seen as needless impediments, 'externalities,' that routinely should be scraped into the refuse bin as 'illegal barriers to free trade,' from heartless WTO dispute panel tabletops. Yes, tariffs as trade charges on goods coming into our nation would be great to eliminate theoretically, as that should then lower the price of foreign-made televisions and cars and toys that we Americans could buy.

Just imagine if tariffs could be eliminated around the globe, lowering the financial burden on smart-shopping consumers everywhere. 'Free trade!' is the cry of the venture capitalist and the salesmerchants of mutual funds as dreams of limitless wealth spread from the yellow brick road winding out into our new millennium...

Well, some local businesses in various countries, including ours, might be driven out of business by those giant corporations with vast amounts of capital at their disposal. OK, yes, you could argue, these would be too weak anyway, survival of the fittest, pip! pip! social Darwinism and all that. But the U.S. and most first world countries have historically protected their fledgling businesses, even helping them to grow, so some of them have become the powerful behemoths they are today.

That's the way it *was*. But now, the poorer third world countries are not allowed to do that, because under the current WTO regalia, 'harmonizing' laws and regulations so one size fits all about the entire Earth, local laws can't protect their important new (or old) industries like before. Foreign corporations can come right on in and have to be given 'national' treatment, be treated just like Mom and Pop Jones' Clothing store, even if it's some unscrupulous polluting hard-to-prosecute megacorporation stationed in China, or Dusseldorf, Germany.

Remember that 70% of the U.S. economy is made up of 'services,' while over 60% of the world's economy is.[8] Think about this as one haunting example of the WTO problem: big corporations like Nestle and Coca Cola and Pepsi want unlimited access to fresh water for commercial sale. Corporations don't really care if they suck your acquifer dry. They have to make as much money as possible, as quickly as possible, to satisfy their investors. If they go too far, they have well-financed legal teams to clean up the monetary and public relations mess. Corporations are not inherently created blessed with morals.

When there still was Enron, as an icon of unscrupulousness if there ever was one, Rebecca Mark, the CEO for their water division, Azurix, announced that her goal was "to fully privatize the global water market."[9] Estimated value of such *agua*: $800 billion, as per the World Bank.[10]

Did you hear about what happened in Bolivia when Bechtel Corporation jumped in under a privatization-or-else order to privatize the water of Bolivia's third largest city, Cochabamba? Water became so unaffordable for the people of that city that they marched in the streets, leading some people to actually be killed over this issue! Eventually, public outrage was so great, Bechtel was booted out of Cochabamba, but there still remains a 25 million dollar suit from that huge American-based corporation to recoup its possible lost profits.

And get this: Bechtel is one of the world's richest corporations, garnering $14.3 billion in receipts for the year 2000. At the same time, the country of Bolivia's entire national budget was but $2.7 billion!![11]

And don't think what happened in Cochabamba, Bolivia can't happen in the USA. The city of Atlanta had 'the biggest water contract in the nation,'[12] for $430 million with United Water, which actually is a 'subsidiary of [a foreign corporation,] the

Bechtel's takeover of Cochabamba's water system, outrageously raising prices for water for the extremely poor citizens of Bolivia's third largest city, led to riots, deaths, and eventual ouster of the national government.

French utility Suez.'[13] But the city government was troubled by the company not fulfilling 'its contractual obligations.'[14] Atlanta decided to terminate the contract. Tune in to see what happens if Suez induces France to take this brash move by our-Georgia's capital city to WTO court.......

Now you might be saying, if this all true, why haven't I heard more about this threat to my country's sovereignty, besides the news that NAFTA is wonderful (but what about all those jobs lost to Mexico?), and capital should freely flow across all borders (but not pirated movies and other patented 'intellectual property' like Monsanto's and Dupont's gene-altered seeds)? It could be that the increasing concentration of media ownership, down to five corporations controlling 75% of what Americans see and hear, is depriving us

General Electric, owner of NBC, would not be expected to broadcast nuclear news that could hamper potential sales of its nuclear reactors

of essential information. Because of interweaving conflicts of interests of board members, and other corporate investments Disney [owner of ABC, ESPN] and General Electric [NBC] and Viacom [CBS, MTV] and Newscorp [Fox] and Time-Warner [CNN] have made and want to make in the globalized marketplace adjudged under the auspices of the WTO.....

But let's go back to the 1980's and the GATT agreement. Formerly, this was the big international trade agreement facilitating the lowering of trade barriers, and those bothersome tariffs. Initially, the idea was to lower duties on manufactured goods—'tariffs'--so these goods would cost less for the consumer to buy, and the producer to make and export. GATT was more benign than its current incarnation back then, because though the old GATT may have had about 100 countries in its fold, any decisions made, or disputes settled, required UNANIMOUS approval from each member nation for any decision to be 'executed.'

Then came along Ronald Reagan and George Bush I, and their ultra pro-corporate profligees. Over several years they composed a new radical version of GATT that, amongst other things, would establish the 'self-execution' of any dispute decisions, to grease the wheels for corporate interests. That meant that if a decision was reached by a three person dispute panel in secret, it would <u>automatically</u> go into effect, or be 'self-executing.' The onus was to be reversed: the only way to <u>prevent</u> the decision from going into effect would be that <u>ALL</u> member nations would have to vote <u>against</u> the decision, <u>including the nation that brought the challenge to the dispute panel!</u>

The new devised rules and regulations were so outrageous, there seemed no way in Hell they could pass through a bipartisan Congress and be approved. But then there arrived the middle-of-the-road, semi-Republican, forever-polling-for-approval Bill Clinton, and his Vice President Al Gore, to do the dirtywork of pro-corporate radicals like Carla Hills (US Trade Representative from 1989-93) and the Republicans. What would happen was, NAFTA (North Atlantic Free Trade Agreement), nicknamed 'baby-GATT,' could be offered as a trial balloon, and if that made it under the radar of America's consciousness, then the GATT's next version, called the 'Uruguay Round' -- because it was ultimately negotiated down in that South American nation -- could follow, be approved as a similar agreement, establishing its World Trade Organization (WTO) entity as the supreme governing body championing 'free trade' for the entire Earth!

Somehow, Ross Perot emerged as the loudest voice speaking against the ills of NAFTA (which involves three countries: Mexico, the U.S.A. and Canada). On the fateful night of November 9, 1993, corporatists held their breath while the Larry King show began, watched by millions here in America and around the world, with Al Gore and Perot about to debate the pros and cons of NAFTA. Unfortunately for the planet, Perot appeared discombobulated, getting lost in that point about the patent for Tennessee whiskey, and Gore won the day, the battle, and ultimately the marketplace for corporations wanting more secure access to all GATT-signatory nations' natural resources, services, and whatever else they could profit from, once the next step would be taken.

For once NAFTA was adopted thereafter by all three North American nations, the debate

After Al Gore and Ross Perot's one-sided TV debate on November 9, 1993, NAFTA was adopted for the USA, Mexico and Canada, and the WTO was established in 1995. Now we have to be concerned about a North American Union (NAU) being quietly laid upon the foundation of NAFTA, undemocratically threatening the sovereignty of each of NAFTA's three signed-on nations. See page 402, To see the actual debate on youtube http://www.youtube.com/watch?v = GhwhMXOxHTg

in the media about adopting NAFTA's big brother, GATT's Uruguay Round, seemed moot. Our media refrained from portraying the scope and importance of the GATT-WTO story, spewing out its sports, gossip, and superficial extravaganzas to capture the advertising dollar for its sponsors. Demographics. Shiny new cars captivating the beautiful models drooling over driving them. Business as usual. So that in 1995, with Americans basically in ignorance of the implications, our Congress approved the crowning of the WTO as the kingly arbiter of all trade disputes about the globe below.

Nowadays, new laws possibly affecting trade have to be 'least trade restrictive,' some laws will have to be changed, and some laws will never be made that should be made. Otherwise, sanctions in the millions and billions of dollars, as imposed by the WTO, via their undemocratic dispute decisions behind closed doors, threaten our own sovereignty today and on into the years beyond. Of course, this affects all 151 nations that have signed on to GATT's Uruguay Round.

This goes for our farm subsidies, our dolphin-safe tuna laws, our internet gambling laws, our product safeguard regulations, our determinations of unsafe pesticide concentrations in our foods, our patenting of life forms and medicines, and our own threat to force genetically altered 'Frankenfoods' onto the tables of angered Europeans.

We speak of democracy, and how nations must be 'free' to enact openly debated policies that will make the world a better safer healthier place for our children and grandchildren. But the WTO hovers ominously over the parliaments and piazzas of the planet, when it comes to finally putting down what's best for us all on paper or parchment. . . .

Protecting The USA Against Asian Beetles and Borers

Try this one on for size: as our trade escalated with China and Hong Kong, we discovered that this insect pest, the Asian longhorned beetle, was arriving with the televisions and the textiles. It turns out that these beetles have no native predators here on this side of the Pacific Ocean. They like to bore into trees, infesting them. Once they're beneath the bark, it's too late to use a pesticide.

The entire tree has to be destroyed, as 'the beetles can spread and kill trees quickly once they enter an area.'[15]

Infestations of Asian longhorned beetles have been discovered from Los Angeles to Cincinnati to Sinking Springs, Pennsylvania, to Brooklyn, New York.[16] Thousands of trees had to be destroyed by our Animal and Plant Health Inspection Service (APHIS) to prevent further spread of the dreaded bug.[17]

To curtail the problem, APHIS decided that we had to require all solid wooden packing materials like pallets and crates to be "fumigated,

The Asian longhorned beetle is a species without predators in America. Once it reaches our shores, usually in unprocessed wooden packing crates from Asia, and attacks our trees, our trees must be destroyed before an uncontrollable infestation occurs. Estimates of cost control run to $138 BILLION for a nationwide program. WTO rules prevented us from issuing timely regulations to keep these bugs out of the USA.[18]

heat treated, or treated with preservatives"[19] if they came from China or Hong Kong (this was in 1998, before Hong Kong became re-ruled by China once again in 1999). We had to do this because we found these materials were the means of transporting the beetles into America.

We all should know that the second greatest cause of species extinction today is invasive species from elsewhere, visiting and establishing themselves in non-native environs, wiping out unsuspecting native species as they multiply.[20]

Like the Asian longhorned beetle settling into American treetrunks -- while it also happens to be considered as a pest in its native China, Hong Kong and Japan.[21]

Similarly, zebra mussels from Europe tenaciously colonized Lake Erie and other waters, threatening the survival of native *unionid* clams.[22] And what about gypsy moths, which originally came to Medford, Massachusetts with Professor L. Trouvelot in 1868? One of his bottles of these insects fell out a window, spilling the brood into his garden. No one expected them to be such a problem[23], but today, New Yorkers and Vermonters, to those of us in the Pacific northwest,[24] see these cursed critters making nests in our oak trees, defoliating them, and sometimes killing them off.

Shouldn't we be allowed to make our own laws to protect ourselves and our environment? Modern transportation being what it is, all kinds of insects and pests may stowaway to undetectedly invade our shores to wreak havoc in increasing numbers as the future rolls along. When we find an obvious problem, we naturally have to solve it before the situation gets out of hand. Common sense.

The voracious gypsy moth, transported to Massachusetts by a Professor L. Trouvelot in 1868 to 'breed a hardy silkworm.' [25] Escaped and spread across the USA thereafter. An 'outbreak' of gypsy moth caterpillars can eat all the leaves on an oak tree in one week.

Yet this proved to be very difficult, in light of what Hong Kong/China tried to do, relating to our regulation of their potentially infested crates. In 1998, Hong Kong, as a WTO member, registered a complaint with the WTO Sanitary and PhytoSanitary (SPS) Committee about our regulation. This was not quite a 'challenge,' but was a step on the way to raising one.

Since the WTO, working to lower trade barriers, already had passed their own "rules restricting quarantine and other Sanitary and PhytoSanitary measures,"[26] the onus was on us. We didn't want Hong Kong's complaint to progress into a formal challenge, where we would find ourselves on the wrong end of one of those dispute decisions, and have to pay out millions of punitive sanction dollars. So we had to postpone implementing our regulation.

We did utilize diplomatic channels to ensure a compromise with Hong Kong, and finally did implement a wood packaging rule in 2004. But we had waited half a decade to summon the legislative fortitude to do this. Meanwhile, the emerald ash borer, another Asian insect that also had been arriving in our country within those wooden packaging/pallets, helped kill off 25 MILLION of our ash trees by the beginning of the year 2008!

Baseball fans may be the most aware of this as a large group, because announcers have pointed out that new bats are being made of maple and other woods. But these are not as durable or flexible as ash. APHIS now is considering importing wasps from the emerald ash borer's natural Asian habitat to attack the bugs.[27] Let us hope *these* potential predators do not wreak havoc on our environment, if they are transported to America based on current testing versus the emerald ash borer.

Here we have a very damaging, annoying example of how our knowledgeable powerful

The Emerald Ash Borer/ash baseball bat. Another invasive species arrives in USA via wooden packaging crates. Responsible for killing 25 MILLION ash trees in USA! Blame the WTO for baseball hitters having to use those splintering dangerous maple bats.

country can be held hostage to behave like a paralyzed pawn in the New World Order of things. By slavishly kneeling to the demons of trade, we hesitated doing what should have come naturally and democratically: proceeded with the timely crafting of legislation to protect ourselves from invasive species with no predators within our borders.

The 'Dolphin-Safe' Tuna Story
Plus
The Gutting Of Our Sea Turtle Law

As one of the leaders of the world, the USA, and its citizens, realize that sometimes actions have to be taken to keep the Earth safe for our interdependent biosphere. Animals, plants, water, air, need to be protected and healthy if we, the people, want to continue living and pursuing happiness under the shining light of liberty. With our technologic powers ever increasing, our planet seems to be shrinking in size, with more and more of our natural resources carelessly used up or wasted.

Yes, political behavior too often to the contrary, most of us do realize how important our blessed diverse forms of life are, and that they deserve their place on Earth, while also being necessary to perpetuate our mutual long-term survival.

Specifically, many of us gathered that something had to be done to stop the killing of seven <u>million</u> dolphins over a thirty year period by a tuna catching technique called 'purse seine' fishing.[28] We also discovered that up to 55,000 sea turtles were being killed and dismembered annually by shrimp net fishing.[29] Since we are intelligent responsible beings, we Americans, as a country, decided to pass laws to protect these creatures. We realized that shrimping kills more sea turtles than all other human threats to turtles combined.[30]

We decided to utilize a 'dolphin safe' designation for tuna not caught in purse seines, and excluded tuna caught, either domestic or foreign, by the purse seine technique from the 'dolphin safe' designation.[31] We required a 'TED' device to be installed inexpensively ($50 to $400 each) in shrimpers' nets so that turtles could escape out the bottom of these nets, avoiding drowning, while maintaining the efficiencies of shrimp catches.[32]

Millions of children and adults were behind

A 'TED' device that allows turtles to swim free from a shrimper's net.

these laws, developed over time, with lengthy and widespread democratic debate.

However, both laws were challenged in WTO court, and effectively overturned as being illegal barriers to free trade. What else would you expect? 'With only two exceptions, every health, food safety or environmental law challenged at the WTO has been declared a barrier to trade.'[33] Remember also that as Public Citizen global trade experts Lori Wallach and Michelle Sforza inform us: "Far from being a neutral arbiter, the singular and explicit goal of the dispute settlement process is to expand trade in goods and services."[34]

Dispute panelists are chosen with only trade credentials, without any mechanism "for ensuring that individuals serving as panelists have any expertise in the subject of the dispute before them."[35] Neither are there any requirements for "panelists to consult with experts."[36]

Very worrisome when you have prototype trade panelists like Kym Anderson saying things like: "Environmental and labor concerns can provide a convenient additional excuse for raising trade barriers."[37] (my underlining)

Well, we can appeal unfair decisions, can't we? you might ask. Not outside of the WTO, you can't. The WTO decision is supreme, beyond and above all other Earthly democratic legal realms and institutions. And almost all appeals to WTO dispute panel decisions within the WTO's own Appellate

Body have failed, including our own appeal of their sea turtles verdict.[38]

As a result, the USA has had to change our law concerning turtles and shrimp fishing nets. Which we had enacted in compliance with the Convention on Trade in Endangered Species.[39]

The dolphin-safe tuna story is a little more involved, with twists and turns of intrigue. Legally, we have a Marine Mammals Protection Act, ratified and signed into law in 1972.

After an environmentalist signed on board as a cook and videotaped the slaughter of hundreds of dolphins; viewing of the incriminating, repulsive images led to millions of children writing to Congress imploring us adults to 'Save The Dolphins.'

In 1988 Congress passed the dolphin-safe tuna provisions of an amendment to our Marine Mammals Protection Act.

However, in 1991, with a pre-WTO version of GATT in place, Mexico challenged the USA in GATT court concerning said dolphin-safe tuna law, and won. Recall that back then, GATT's rulings

CHUNK LIGHT TUNA IN WATER AND SA
NET WT. 4 LBS. 2½ OZ.

were not yet 'self-executing.' Mexico had to round up unanimous 'consensus' from all the GATT countries to change our law. So the fruits of their victory would not be consumed for several years, their spectre lingering in the background...[40]

Meanwhile, NAFTA was coming up for a vote in the USA, the golden goose country for all trade deals and expenditures. Assurances about sovereignty being unimpaired were made high and wide. Though some of us should have been more wary, had it been more widely reported that the Director General of GATT, Peter Sutherland, had warned that "Governments should interfere in the conduct of trade as little as possible," in a March 3, 1994 speech in New York City promoting U.S. approval of the WTO.[41]

During 1994 the European Union took the USA to GATT court because they wanted to export processed tuna from Pacific fish stocks caught via the purse seine technique. They won. Both the Mexico and European Union challenge-rulings claimed that our law was not 'necessary' to protect dolphin health because the USA could have used other measures that would not have contradicted the almighty GATT.[42] ??Whatever those measures could have been??....??....???

Oh, well, Renato Ruggiero, Secretary General of the WTO, was soon to tell us that environmental standards "are doomed to fail and could only damage the global trading system."[43] So why be surprised that once NAFTA was safely approved, and the WTO set with its self-executing powers, that Mexico in 1995 threatened a WTO enforcement case against our country?

Renato Ruggiero, first Director-General of the WTO.

President Clinton scrambled to comply with the WTO to eradicate our long fought for, and implemented law. It would require an act of Congress to contravene the will, and legally enacted right, of millions of Americans. An amendment had to be arranged to the Marine Mammal Protection Act.

Mr. Clinton went to the fraternity of anti-environmental Congressmen to recruit soldiers for the scurvy cause. He found Representative Randy "Duke" Cunningham, a California Republican, and Senator John Breaux, a Louisiana Democrat, amongst others, to front the deed. They introduced a bill that would weaken the Marine Mammal Protection Act so the USA would not have to pay millions of dollars in trade sanctions to Mexico and the European Economic Union by being WTO-non-compliant. The bill was nicknamed the "Dolphin Death Act" and was initially repulsed in 1996.[44]

However, Al Gore stepped to the bat and produced significant RBI's in the form of a different version of the bill, which was passed in 1997. The Mighty Casey, Barry Bonds, Babe Ruth, <u>Al Gore</u>!!!! once more. He and his accomplices did such a magnificent job that the tuna caught in the purse seine nets can even be labelled 'DOLPHIN SAFE!' now. All it takes is <u>one</u> monitor standing on deck of what is often a football field length of industrial style fishing boat, observing the miles and miles of nets encircling the tuna (?dolphins?) to confirm that no dolphins were killed.[45]

How would you like to be that well-meaning monitor? Standing on the bridge, walking the higher decks of the vessel in question, maybe even climbing the crow's nest, relentlessly vigilant of the distant nets tightening around their catches......

Imagine the pressure from the crew, the fishing company. Would you be a well-meaning monitor, or one picked to see-no-evil and rubber stamp the process? How can the USA really monitor each catch for compliance faithfully? Before, the law had applied on a country-by-country basis, not a catch-by-catch basis. That meant that any country in question, like Mexico for example, was responsible for certifying that their 'entire fishery did not use encirclement techniques'[46] that could be killing innocent dolphins, for them to get their tuna imported into the USA.

Today, what we have to do is rely on the pro-ducers themselves saying their catch is OK, 'dolphin safe,' the fox in charge of the chicken coop, the polluter self-regulating his toxic discharges into the now-algae-red river water, heck with the dolphins, they're external to the profit sheet, so deems the WTO. Because, nope, that old law just wasn't 'least trade restrictive' enough. Just like Thailand's anti-smoking law wasn't, nor was our nervy regulation concerning those long-horned Asian beetles.

Further information on your trade dispute panelists: beside science or consumer concerns, or public health or consideration of human or labor rights not playing a role when deciding upon whom shall be the members of any dispute panel, panelists do not have to reveal any possible conflict of interest in the case they could be judging. They are only required to 'self-disclose' their backgrounds. Especially if such disclosure would be "administratively burdensome."[47] This pertains to 159 names on the Dispute Panel Roster, more than 90% of whom are male, at last report[48].

For our three man dolphin-safe tuna case dispute panel, we scored three officials from land-locked countries deciding in secret on our law affecting what goes on in our surrounding salt waters, worked for by millions of Americans through an extended democratic process, involving *years* of contemplation and give and take.

POOF!! our law was negated, in practical terms.

Yes, it is true that the WTO and its zealots give lip service to ensure increased 'transparency,' meaning openness concerning negotiations, disputes, and decisions. Making the WTO sound like a dignified body of justice, filled with integrity. Our former chief Trade Representative, Robert B. Zoellick, during his reign, campaigned about the globe for "transparency" to "become a hallmark of all future trade negotiations and trade agreements..." He avowed that "It will counter corruption and reveal the protection of special interests."

As Zoellick was pushing for public arena approval of the Free Trade Area of the Americas (FTAA), extending NAFTA-type rules to the rest of the western hemisphere, he bespoke that "The Western Hemispheric democracies agreed to make public the preliminary negotiating text of our agreement – to open the process and contribute to a more informed public debate on trade."[49] This speech was given on May 15, 2001, three weeks after the secretive FTAA meeting in Quebec City, Canada, conducted behind the Berlin-type walls that were hastily erected to protect the visiting ministers. No such text was made public before or during the Quebec meeting. In fact, not until six months after that meeting was the information released to the world public. So much for transparency in the deregulated world of 'free' trade. Or as some call it: 'corporate-managed trade.'

Although our chief Trade Representative reportedly chided the WTO to proceed with transparency, we must remember this: according to the WTO's Dispute Resolution Understanding (DSU) the only specific operating rule is 'that all panel activities and documents are <u>confidential</u>.'[50] Meaning all panel proceedings 'operate in secret, documents are restricted to the countries involved in the dispute, due process and citizen participation are absent, and no appeal outside the WTO is available.'[51]

Somehow, we have allowed the almighty WTO to stand supreme over all forms of law it decides to decide upon. There is no higher court. The world, with all its democratic aspirations and idealistic encouragement, has sacrificed its sovereignty to an autocratic institution with no actual responsibility to the people of the Earth. The corporations instead gain the benefit of WTO-orchestrated trade, and they are really just a paper entity without any flesh and blood national rootings. If they are being prosecuted in a country, they can simply pack up their belongings, their computerized records, closing their factories in Detroit and Philadelphia and Los Angeles, and trot off somewhere else beyond our confining national boundaries, where they will be gratefully accepted, paying lower wages, having less environmental requirements to satisfy, perhaps setting off a challenge in WTO court just before they leave to cover their derrieres.

A comic example of doublespeak and double-talk, plus a sprinkling of hypocrisy re 'transparency' from our own Mickey Kantor, USA chief Trade Representative in 1994, when all the GATT-WTO Uruguay Round negotiations were aflutter:

<u>Before</u> WTO approval, testifying before Congress:

"The Uruguay Round Agreements provide for increased transparency in the dispute settlement process."[52]

<u>After</u> WTO approval, Kantor "said he strongly supported the idea of making the WTO more accessible to most Americans and noted that he has been pushing WTO Members to agree to end rules keeping most WTO proceedings secret. He reported that important progress was made in the last few weeks on this point, and said he would be glad to brief the _committee behind closed doors_."[53]

Ha Ha! Not very funny. Not when it's the real world. I know. Sorry.

Yet there is one kind of transparency the WTO and corporations seriously slobber over: "transparency in government procurement."[54] In accordance

Mickey Kantor, US Trade Representative and Secretary of Commerce during the Clinton Presidency. Strong force in implementing NAFTA and the WTO.

with another WTO major agreement, the AGP, or Agreement on Government Procurement, this could mean all signed-on countries' governmental bodies down to the village level would have to faithfully list all the things they procure or buy, with the WTO. Big Brother WTO watching over the lesser institutions: the governments of the world. Because then, when all the nations of the world that are members of GATT, sign on to the Agreement on Government Procurement (AGP), the corporate gorge-feast can rage voraciously with minimal interference.

Trillions and trillions of dollars could be the precious prize. Where governments then would <u>not</u> be allowed free reign to make their own rules and regulations on how they spend their own money, and are muzzled by the WTO concerning what restrictions and parameters they could lay down for their purchases. Like requiring companies competing for a contract to maintain health care benefits for their workers, or using the least toxic practices in producing the desired goods.[55]

Some specific numbers for you to see the light?

<u>Health care</u>, valued globally at $3.5 trillion dollars per year (U.S. federal budget currently $2.9 trillion total); <u>education</u>, valued globally at more than $2 trillion per year; <u>water</u>, valued globally at $800 billion per year.

Sounds unbelievable? Big Brother-WTO making a list and checking it infinite times, recording it in His computers for future use - - typing in each and every government purchase that must be mandatorily reported....

Currently, it is mostly the <u>developed</u> countries that are amongst the 40 that have signed onto this Agreement on Government Procurement travesty of democracy. Yes, the USA is one of the signees of the AGP.

One of the AGP rules henceforward is that no 'performance requirements' for government purchases are allowed. If a company hires slave labor in Slovakia, or Myanmar/Burma, or encourages soldiers to dispossess people from their homes to make a gas pipeline, forcing them to work without enough food and water in the hot sun, where some die, and those who refuse are shot and/or raped, that provides no grounds for a government under the AGP to eliminate or penalize such a company concerning gaining a government contract in the state of Massachusetts or the town of Carboro, North Carolina.

For AGP countries like us, none of our governments are now permitted by the WTO to consider "environmental, human rights or labor practices as criteria"[56] for how we treat our foreign or domestic bidders for public contracts.

The Burma/Myanmar Boycott Illegalization Story

We all know about Nelson Mandela and the Union of South Africa, and the boycotts in America and worldwide that brought down the white supremacist apartheid government in that country. That was in the 1980's.

But do we now know about what happened when we tried to do the same thing for Burma after the military dictatorship in that Asian nation refused to step down after they lost 80 % of the seats in their parliament in a 1990 election?

Aung San Suu Kyi, the leader of the newly elected government, was placed under house arrest, and many elected legislators from her National League for Democracy party also were arrested. The brutal military of Burma, which prefers the nation now to be called Myanmar, accelerated its practices of forced labor, relocating civilians, killing and beating those who fought back, using subjected Burmans as 'porters' to walk in front of soldiers to detonate landmines and act as human shields during combat, so that in a nation of 42 mil-

A young Nelson Mandela

Aung San Suu Kyi
(pronounced 'Gee') of
Burma/Myanmar

lion people in this modern new millennium, more than one million have been displaced from their land. Half of this number now reside in concentration camps.[57]

Companies like Unocal, based in Los Angeles, used the military to help them build their gas pipelines, but denied any guilt for how the military behaved. Our current Vice President, Dick Cheney, while CEO of the Halliburton oil company 'himself inked a deal to build a pipeline between India and Burma in 1996.'[58] The bottom line was money, and perpetuation of power to make more money: a universal story in this imperfect world that is all we have.

Ms. Kyi "called for Americans to "use your freedom to protect ours."" She encouraged "people's boycotts" that deny the political legitimacy and foreign exchange needed by {Burma's} military government. "You cannot divorce economics from politics," she said…{expressing her 'rationale for} opposition to investment that strengthens the power of the military and "prolongs the agony of my country."'[59]

Americans responded with measures similar to those used against South Africa's apartheid government. Counties, cities, towns, 'terminated pur-chasing contracts with companies doing business in Burma.'[60] But it was the state of Massachusetts' selective purchasing law that was challenged by the European Union and Japan in WTO court.

As a result, a multinational corporation front group entitled USA*Engage (Halliburton acting as a 'vital member'[61] of their campaign) took up the gauntlet domestically, and, facilitated by WTO echoes in the power halls of our own vari-ous governments, claimed in several U.S. courts, including finally the U.S. Supreme Court, that the Massachusetts law violated the U.S. Constitution by 'interfer{ing} with the executive branch's ex-clusive authority to conduct foreign policy.'[62]

Unbelievably, Massachusetts, and thus all other government bodies that had acted to help Burma as they had helped South Africa previously, lost the battle. This despite our spirit of independence, the Boston Tea Party, and Massachusetts' Attorney General Thomas F. Reilly reminding us that: "For more than two hundred years, citizens of Mas-sachusetts and other states have used boycotts to support the "natural, essential and unalienable rights" of people around the world," quoting article I of Massachusetts' constitution of 1780.[63]

Japan and the European Union had tempo-rarily dropped their challenge in 1999, while USA*Engage did their dirty work. However, down in Maryland, in 1998, the latter's WTO challenge

Boston Tea Party – Enraged citizens of Boston hurling crates of tea overboard, during their rebellious historic Tea Party.

would cast a shroud over a similar story taking place on the floors of Maryland's Senate and House. Maryland was about to authorize a law barring the state from signing any contracts with Nigeria's government or companies doing business there.[64] Nigeria is the relatively wealthy oil-rich west African nation that has <u>executed</u> poets and human rights workers for protesting against the practices of <u>oil companies</u> like Shell Inc.

It looked like the Maryland law was certain to pass, but the Clinton administration saved the day once again. It dispatched a State Department bureaucrat to testify before a Maryland legislative committee, warning of violating WTO rules against such free-thinking sanctions leading to embarrassing global trade complications.

The threat of another WTO challenge like the one that Japan and the European Union were making against Massachusetts scared enough legislators to defeat the Maryland law by one vote.[65]

Oil companies and multinational corporations loved this, but the plight of man about the planet, and our own right to make our own 'unalienable' laws as the leader of a democratically progressing world, has suffered - - perhaps mortally. A decade after the Supreme Court struck down the Massachusetts law, Noble Peace Prize winner Kyi is still under house arrest, while her country continues to be terrorized by a vicious military dictatorship. As another attempt was made to prosecute oil companies for complicity in murder and forced labor under our old Alien Tort Claims Act, our then-Attorney General, John Ashcroft, instead of standing up for Burmese villagers facing death, displacement, decimation, and enslavement, argued that such 'suits interfere with U.S. foreign policy and undermine America's war on terrorism.'[66]

Where are our priorities? Are non-economic motives for our laws irrelevant in the 21st century? Is that why nobody says anything about the terrible 'Plan Puebla Panama,' as dubbed by Mexico's President Vicente Fox? This will be one horrific component of the much-heralded Free Trade Area of the Americas (FTAA) that President Bush is still unabashedly championing.

The game is: Asia and its low paying sweatshops

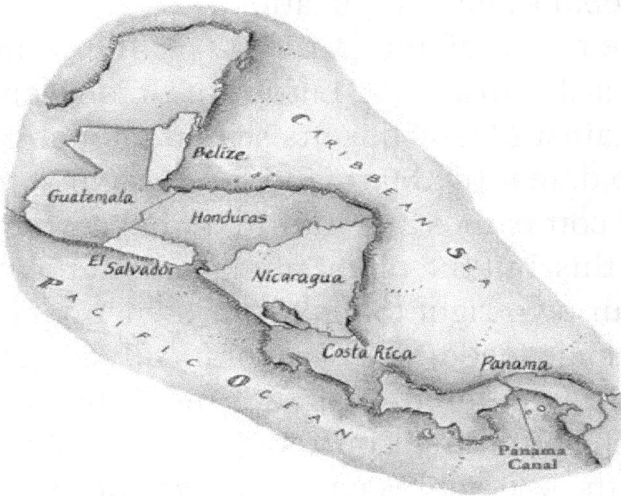

Central America, the object of commercialization and privatization via 'Plan Puebla Panama' and the CAFTA agreement.

want to sell their goods to Europe and the eastern United States. Central America can offer 'expendable' labor and rich resources to facilitate this, while being the ideal route through which said goods can travel and be redressed, if necessary.

Problem is, the aging antiquated Panama Canal cannot handle the upcoming inevitable globalized lode that would have to trundle through its locks. ACERCA (Action for Community and Ecology in the Regions of Central America) foresees:

> 'cross-isthmus dry canal mega-projects [that would] involve the construction of massive deep-water ports on both coasts [Pacific and Atlantic], capable of hosting the largest ocean freighters.

These ports will be connected by high speed rail
lines and highways. This massive transportation
corridor will open the region to further exploitation of the region's forests, minerals and oil and
lead to the development of extensive networks of
maquiladora sweatshops (where components manufactured in Asian factories can be assembled into
finished products.) The dry canal megaprojects will
also involve the construction of industrial shrimp
farms, oil refineries, smelters, and vast industrial
development, leading to wide swaths of ecological
and cultural devastation along the Isthmus.[67]

ACERCA expects that the diverse indigenous
people that have lived in Central America for
thousands of years, and have kept the land upon
which they live as communal property, will be displaced unceremoniously to 'urban slums located
adjacent to sweatshop factories.'[68] Financed by
the World Bank, etc., roads will be built through
rainforests, and clearcutting can be accelerated
apace. ACERCA asserts that 'land privatization
is key' to globalization/FTAA goals of controlling
and plucking 'spectacular fresh water reserves, 30
million low-wage workers...important petroleum
assets, 34 million hectares of virgin timber'[69] for
wondrous profit. And with 12,000 or more U.S.
troops in such an "unstable" region, there is the
worry about the use of these troops 'to crack down
on activism'[70] against the appropriation of indigenous lands.

Painfully, the excuse may again be our 'War on Terrorism.'[71]

Basically, the FTAA appears to be an improved NAFTA/GATT on steroids. It "intends to "enhance compliance" with existing WTO rules."[72] Like facilitating corporations in "demanding compensation"[73] for laws or government policies that "actually or potentially"[74] cut into their profits! Be these pollution controls or public health laws. Or WTO anti-dumping policies that put the burden on the country dumped on, rather than on the dumper of sub-normally priced goods intending to put local producers out of business, or overflooding the market with, say, unwanted genetically altered foods. Usually the poorer country is the one dumped on, by the richer better-resourced, over-subsidized one, and does not have the capabilities, money, trained lawyers, etc., to fight the dumping.[75]

If you might wonder sometimes why you see so many Hispanic people working, or looking for work, in your town, think NAFTA and at least 1.5 MILLION Mexican farmers and farm workers who lost their livelihood.[†] According to Public Citizen's Global Trade Watch expert, Lori Wallach, this is one of NAFTA's worst results, in addition to a 20% drop in real median wages for Mexican workers.[76]

[†] Stephen Lendman states that this number could rise to 10 million small farmers, and their families, being displaced from their land 'eventually' 'by one estimate.' Expected result: northward migration from decimation and poverty, into the USA, through no fault of their own. See: http://www. globalresearch.ca/index.php?context = va&aid = 6456

With the passage of CAFTA (the Central American Free Trade Agreement) further lowering the boom for seven more Latin nations, Mexicans are worried that jobs will flow downstream out of their country to even poorer economies than their own. And more displaced Central American Latinos may show up on your streetcorners, building your houses, harvesting your fruit.

Plus, with NAFTA/CAFTA/FTAA, there is the abominable provision making it all too easy for corporations to directly sue any member government for billions of dollars in sanctions, should democracy dare impede their business ambitions. Before the wheels were tefloned to this extreme, only a country could sue another country in GATT court.

A prime example here would be the Harken Oil $57.5 BILLION arbitration claim for "expected profits" from offshore oil drilling, versus the government of Costa Rica. Harken never addressed concerns of harm to the Talamanca region, 'one of the richest marine ecosystems on the planet,'[77] from oil spills and other problems possibly eventuating from their prospecting, so Costa Rica rejected Harken's application for environmental permits. If CAFTA had been in place, Costa Rica could have been subject to probably lose a Harken-initiated suit, as most suits are lost, in secret trade court decisions. Costa Rica's Gross Domestic Product only totals $51 BILLION.[78] Understand that this was one big factor why Costa Rica had not adopted CAFTA, until doing so by a very narrow margin during the 2007 presidential election year, and why CAFTA was one of the most important issues in this election.

Also note that 'CAFTA goes even further than NAFTA - - specifically allowing multinational investors to challenge government decisions about natural resource agreements, such as mining and offshore oil contracts.'[79]

Indigenous people, whose cultures have existed for centuries, sustained by their equilibrium with, and respect for their environment, are enraged that CAFTA and the FTAA and NAFTA want to consume them and their natural resources, only to spit them out like hapless woodchips after a clearcut gorge-out. Here is a viewpoint that should help you gain insight on this from Subcomandante Marcos of Chiapas, Mexico ('Pemex' is Petroleum of Mexico, a government owned company):

> "You have now entered by one of the three existing roads into Chiapas: the road into the northern part of the state, the road along the Pacific coast, and the road by which you entered are the three ways to get to this southeastern corner of the country by land. But the state's natural wealth doesn't just leave by way of these three roads. Chiapas loses blood through many veins: through oil and gas ducts, electric lines, railways; through bank accounts, trucks, vans, boats, and planes; through clandestine paths, gaps, and forest trails. This land continues to pay tribute to the imperialists: petroleum, elec-

tricity, cattle, money, coffee, banana, honey, corn, cacao, tobacco, sugar, soy, melon, sorghum, mamey, mango, tamarind, avocado, and Chiapaneco blood all flow as a result of the thousand teeth sunk into the throat of the Mexican Southeast.

These raw materials, thousands of millions of tons of them, flow to Mexican ports, railroads, air and truck transportation centers. From there they are sent to different parts of the world - - the United States, Canada, Holland, Germany, Italy, Japan - - but all to fulfill one same destiny: to feed imperialism. Since the beginning, the fee that capitalism imposes on the southeastern part of this country makes Chiapas ooze blood and mud.

. . . in Chiapas, Pemex has 86 teeth sunk into the townships of Estacion Juarez, Reforma, Ostuacan, Pichucalco, and Ocosingo. Every day they suck out 92,000 barrels of petroleum and 517 billion cubic feet of natural gas. They take away the petroleum and gas and, in exchange, leave behind the mark of capitalism: ecological destruction, agricultural plunder, hyperinflation, alcoholism, prostitution and poverty. The beast is still not satisfied and has extended its tentacles to the Lacandon Jungle: eight petroleum deposits are under exploration. The paths are made with machetes by the same campesinos who are left without land by the insatiable beast. The trees fall and dynamite explodes on land where campesinos are not allowed to cut down trees to cultivate. Every tree that is cut down costs them a fine that is ten times the minimum wage, and a jail sentence. The poor cannot cut down trees, but the petroleum beast can, a beast that every day fells more and more into foreign hands. The campesinos cut them down to survive, the beast cuts them down to plunder.

The tribute that capitalism demands from Chiapas has no historical parallel. Fifty-five percent of national hydroelectric energy comes from this state, along with 20 percent of Mexico's total electricity. However, only a third of the homes in Chiapas have electricity. Where do the 12,907 kilowatts produced annually by hydroelectric plants in Chiapas go?"[80]

Besides Mexico and the nations of Central America, be aware that countries like Bolivia, Venezuela, and Argentina also have many indigenous cultures patchworking their population, that do not want to accept extermination at the hands of the current in power/in vogue western capitalist model that is devouring the Earth. Specifically, the FTAA is being opposed by these Latin American nations, though the media portrays the story as if it is 'communists' and unrepresentative troublemakers who are causing 'free trade' its steamrollable problems.

"Corporate managed trade" is what Public Citizen's Sforza and Wallach call this grand new millennium version of 'free trade' in their book Whose Trade Organization?[81] Otherwise, would there be increasing restrictions on trade in so-called 'intellectual property,' for example? Plus all these other rules that, overall, favor corporation protection at the expense of the people of the Earth?

Yes, it does make absolute sense to protect mov-

Lori Wallach, leader of the fight against corporate managed trade disguised as 'free trade'

ies, books, artwork, etc., by copyright and other means, but patenting seeds and genetically altered life-forms so corporations have exclusive rights of control over these entities? And corporate-claimed water rights so they can sell their bottled patented privatized water? Is this what 'free trade' is all about?

What about traditional herbal medicines that countries like India have protected by laws passed so that they will always be available for common uncommercialized use? Will 'free trade' and the WTO prevail in overruling the sovereignty and humanity of such common sense? Force countries like India, with its 5000 year history, to succumb to modern corporate appetites to claim and patent and plunder what local people have had the right to use for free for centuries, in the name of 'free trade?'

Deregulation, 'liberalizing' trade laws (HEY! We're grandly championing the use of the 'L' word!????), aiding the free flow of capital: these are the phrases we have to swallow as the WTO expands its dominion over our planet. But our natural resources have to be managed wisely, not just be helpless targets for corporate profits. Shouldn't

389

each country maintain its sovereignty to do what is correct for its people and environment?

Why should the WTO, and the corporations this supranational organization shamelessly favors, be given increasing free reign over the future of our planet? Will most Americans remain in the dark about the goings on within this powerful body, based in Geneva, Switzerland?

How many of us are aware that our federal law prohibiting internet gambling has been successfully challenged by the Caribbean island of Antigua in WTO court?[82] Or our law allowing income earned abroad by corporations to be <u>exempted</u> from U.S. taxing, has also been successfully challenged in WTO court by the European Union, even after we amended it?[83] Or our huge cotton subsidies of $3-4 billion per year to mostly massively sized corporations, have also been successfully challenged, by Brazil, China, Australia, Canada, Chad, India, the European Union and several other countries, in WTO court in 2004[84]?

Yes, it is true that so much money leads to driving down the price of cotton, while the five biggest corporations get $84 <u>MILLION</u> in corporate welfare (over a five year period) from mostly unsuspecting hard-working USA taxpayers[85]. Yes, those 3 or 4 billion dollars are more than the total income of some African countries dependent on cotton exports to feed their people.[86]

While we mention cotton here, you might as

well know these other interesting things about it: the USA is the world's greatest cotton exporter, though China <u>produces</u> more cotton than we do. As discussed in Chapter Two, cotton accounts for 24% of the world's insecticide market, though it uses but 2.4% of the world's arable land.

So, who cares that 1/3rd of a pound of agricultural chemicals are typically used to make just one of your cotton t-shirts? you might snidely say. Remember that five of the nine pesticides used on cotton are cancer causing. You think cotton and clothing - - that's the association most of us assume. But that's not the one and only.

You now will get to know that YOU consume the cottonseed oil in so much of our food and food products, and the fact that our beef cattle and milk cows eat 6-8 <u>POUNDS</u> of it per day.

With the genetically altered influence infiltrating America's fields and water, 83% of our cotton is now genetically altered.

There are organic alternatives, but cotton is to American agriculture as sheep are to New Zealand and wine is to France[87]. The money and subsidies are entrenched in our country's budget culture since way back into the 1930's.

Who is going to dare be the politician to change this, while most of us unaware Americans are the ones who are getting hurt, along with people the world over who are ingesting what they would probably prefer not to be ingesting?

Put this all together, my dear countrywomen and men, and see how you are being deceived daily by how everything looks so good on your TV......

Meanwhile, the most serious WTO challenge of all, in my opinion, is the one our own environmentally unfriendly, and science-unfriendly embarrassment of a president has fired at the European Union concerning genetically altered foods. In spite of moratoriums and over 80% of Europeans not wanting any part of 'Frankenfoods,' the USA via our trade warriors launched a formal WTO complaint on the part of our biotech corporations, and the farmers they are supplying with seeds they have altered and patented. We want Europeans to open their borders to our genetically altered foods! NOW! Before science might find the time to prove these products definitively and finally unsafe.

Fears of health dangers to stomach linings, allergic reactions, unknown repercussions from inadequately studied novel foods and life forms, contamination of fertile safe soon-to-be organic fields by superweeds and pesticide-engineered organisms, is not what the doctor ordered to most Europeans.

But George Bush is arrogantly stomping upon the sensibilities of those who want to be cautious about releasing experimentally bred crops into their countrysides and food stores. It is a big political move that was temporarily muzzled while we built up the case for our war involving Saddam Hussein and Iraq. Most Europeans are very angry at this bullying affront to their sovereign desires on this issue. Meanwhile, we Americans hear practically nothing about this on our televisions and radios, though it is one of the most serious issues on the world stage today.[88]

We *speak* as the champions of democracy and freedom, yet, because of our intimidating corporate influences, we *act* as one of the loudest most powerful voices in the WTO *threatening* democracy and freedom. The biotech industry wants immediate access to the European market. The issue is complicated. Not just a trade issue. The European Union (EU) states they will decide "in a free and responsible manner...how to approve and regulate"[89] genetically altered organisms. That "the EU will always aim at responding to the legitimate interests of its citizens, not to narrow economic interests."[90] (See Appendices for more on this.)

But the easiest way to win the battle is to bash your way into WTO court, where you know you can conquer any sovereign claims to what a country's-worth or a continent's-worth of people have democratically decided is right. This will and has rebounded to affect us within our own borders back here in the USA. Though the USA has often utilized the WTO to attain certain victories, our country, and all the countries of the world suffer, because of how this unique trade organization has been propped above the world's other subordinate mechanisms of justice.

Our more dim-minded leaders may think the WTO is the great shining laser for globalization's inevitable conquering of all four corners of the Earth. But really, it respects no governments, the best wishes of their people in the long run, nor the commons of the planet that no corporation deserves to own or supervacuum clean for the benefit of that immortal idol: Profit.

Corporations are what gain from this great miscalculation by our fawning politicians, all too many with their brains bent and seared by the mirage of invincible empire globally warming up our atmosphere en route to an inconceivable apocalypse.

Will Americans ever be informed so they can see what is happening to our revered land, air, sea, and our microsopic protoplasm itself? Or, more likely, will most of us Americans passively accept the puzzlingly unsatisfying current status quo that is rapidly eroding the sustainable survival of a possibly healthy democratic world as we now know it?

Here is a valuable retrospective and prospective view from Public Citizen in Feb. 2007 on what has happened during 'The WTO decade,' and what might be expected in the future, to give you a unique angle on so-called 'free trade,' focusing on South Asia and China as an exemplary region:

'During the WTO decade, economic conditions for the majority [of the world's people] have deteriorated....The number of people living in poverty has also increased in South Asia, while growth rates and the rate of reduction in poverty have slowed in most parts of the world - - especially when one excludes China, where huge reductions in poverty have been accomplished, but not by following WTO-approved policies (China became a WTO member only in 2001). Indeed, the economic policies that China employed to obtain its dramatic growth and poverty reduction are a veritable smorgasbord of WTO violations:

high tariffs to keep out imports and significant subsidies and government intervention to promote exports; an absence of intellectual property protection; government-owned, operated and subsidized energy, transportation and manufacturing sectors; tightly regulated foreign investment with numerous performance requirements regarding domestic content and technology transfer; government-controlled finance and banking systems subsidizing billions in non-performing debt; and government-controlled, subsidized and protected agriculture.

Many of these same policies are those employed by the now-wealthy countries during their period of development.'

It's not as if the status quo is working for most people in the rich countries either. During the WTO era, the U.S. trade deficit has risen to historic levels - - from around $100 billion (in today's dollars) in 1994 (the year before the WTO went into effect) to nearly $800 billion in 2006. The U.S. trade deficit is approaching 6 percent of national income - - a figure widely agreed to be unsustainable, putting the United States and the global economy at risk. Soaring U.S. imports during the WTO decade have contributed to the loss of nearly one in six U.S. manufacturing jobs. U.S. real median wages have scarcely risen above their 1970 level, while productivity has soared 82 percent over the same period, resulting in declining or stagnant standards of living for the nearly 70 percent of the U.S. population that does not have a college degree. And for the first time in generations, the United States is headed for net food-importer status, having seen monthly agricultural imports outpace exports in August 2006. The United States lost 226,695 small and family farms between 1995 and 2003, while average net cash farm income for the very poorest farmers dropped to an astounding -$5,228.90 in 2003, a colossal 200 percent drop since the WTO went into effect.

Although trade and the status-quo model's failure were important in many 2006 U.S. congressional races, the bottom-up public pressure that has altered trade policies in many nations has not risen to a level in the United States that translates into significantly altered negotiating positions. Thus, while a majority of the U.S. public is losing under the Bush administration's trade agenda, the U.S. WTO position continues to be that of the narrow commercial interests that have bankrolled the administration's campaigns and those of the GOP majority in Congress.' [any underlining is mine][91]

Incidentally, Greg Palast reports that Chinese goods may be 'dumped into Mexico to be hauled northward as duty-free "Mexican" products,' along the new "NAFTA highway," if the North American Competitiveness Council (NACC) gets its way. The Council met the week of Earth Day, 2008 in New Orleans with the three NAFTA nation leaders, including President Bush, to discuss issues beneficial to the corporate members. Included are 'Wal-Mart, Chevron Oil, Lockheed-Martin and 27 other multinational masters of the corporate universe,'[92] as Palast puts it.

Although agreements may be made 'harmo-

Chinese goods - - duty-free - - may start entering the new NAFTA highway that is only supposed to include goods of Mexico, Canada and the USA, if corporate mega-business leaders quietly get their way. . . on a road that could possibly lead to an unvoted upon North American Union - - whose rules could supercede the democratic laws of the three above NAFTA nations.

nizing' our, and Canada's and Mexico's, standards down to the lowest common denominator - - often those of Mexico, yet too often those of the USA (e.g., forcing the allowance of GMO crop cultivation in Mexico; elevation of pesticide residues allowed on certain fruits and vegetables; claiming meat and milk from cloned animals is safe[93]) - - for anything from 'health, labor rights, oil drilling, polluting and so on - - in other words, any regulations that get between The Council [NACC] members and their profits,' Canadian Maud Barlow reportedly informed Palast that 'the US Ambassador to Canada told her the legal changes wrought in New Orleans will not be put before the three national Congresses for a vote. "We don't want to open another NAFTA." So, they'll skip the voting stuff. Democracy is so, like, 20th Century,'[94] are Palast's words.

Lurks The Danger Of An Unvoted Upon Establishment of a North American Union (NAU)

Connie Fogal, leader of the Canadian Action Party, doesn't speak so lightly of what has evolved out of NAFTA on our North American continent. Her article discussing the subject has been selected as a co-winner of the number two Project Censored "Most Censored" News Stories of 2007-2008 Awards.'

Speaking of the North American Union (NAU), breaking down borders, and 'harmonizing' laws of Mexico, the USA, and Canada she states:

> Since March, 2005, under the direction of three senior cabinet ministers of each country, about 100 working groups of unelected officials from government and industry have been meeting at taxpayer expense deciding on and directing the implementation of the restructuring of the apparatus of governance and the form of rule over the people. Their command goes out down the chain of bureaucracy expending vast amounts of taxpayer dollars implementing the changes in our border crossings, in our airports, on our airplanes, in our skies, on and to our roads and highways, in our personal identification systems, in our health, in our vaccines, over our food supplements, in our pesticide safety levels, in our schools and universities, in the exploitation of our natural resources - - our rivers, lakes, oil, gas, in our environment, in the arms industry, in the manufacture and use of depleted uranium, in the exploitation of and experimentation on our indigenous people and our military personnel, in immigration, over our right of Habeus Corpus, in our right of due process, our right to assemble and our freedom of speech, etc., etc.

Ms. Fogal claims that 'the fourth formal step in implementing the NAU' is the Security and Prosperity Partnership Agreement (S.P.P.) which 'is deliberately not a formal international treaty.' For she believes 'as a legal treaty...it would never have flown because it would have been exposed to scrutiny. It remains a work in progress agreement of incredible treacherous magnitude.'

What Ms. Fogal suggests is curbing 'the exercise of power in the executive branch of government... This need is of paramount importance because it is the key to why and how three nations (Canada, USA, Mexico) are being dismantled.'

America's 1950's president General Dwight D. Eisenhower, who warned us to beware of the 'military-industrial complex' in his final speech to the nation. Apparently now we have to beware of the military/industrial/financial complex threatening our national sovereignty and that of Mexico and Canada also, via a North American Union being quietly undemocratically installed upon our hemisphere, headquartered in Washington, D.C.

One last frightening sentence to share with you from Fogal's document: 'The S.P.P. is a treasonous metamorphosis of our federal and provincial government bureaucracies into formal instruments to implement the agenda of the shadow government – the military/industrial/financial complex exemplified by the Canadian Council of Chief Executives who in turn are dominated by the U.S. Council on Foreign Relations, and the U.S. military apparatus.'

Former U.S. President Dwight D. Eisenhower's warning about the dangers of the military-industrial complex from his final address to the nation is referred to by Ms. Fogal. Is her perception of what is occurring reality or paranoia, one has to wonder.[95] Most of you have probably heard nothing about the North American Union or the S.P.P. or the proposed unit of currency for the NAU, the 'amero.'[96]

So for you, and me, here is a bit more detailed view from Stephen Lendman, from his co-winning (#2, with Ms. Fogal) Project Censored article for 2007-2008:

> S.P.P. was formally launched at a March 23, 2005 meeting in Waco, Texas attended by George Bush, Mexico's President Vicente Fox and Canadian Prime Minister Paul Martin.[97] It's a tri-national agreement hatched below the radar in Washington containing the recommendations of the Independent Task Force of North America. That's a group organized by the powerful US Council on Foreign Relations (CFR), Canadian Council of Chief Executives (CCCE), and Mexican Council on Foreign Relations.[98] It advocates greater US, Canadian and Mexican economic, political, social, and security integration with secretive working groups formed to devise non-debatable agreements that, when completed, will be binding beyond the power of legislatures to change. It's also taking shape without public knowledge or consideration.
>
> From what's already known, S.P.P. unmasked isn't pretty. It's a corporate-led coup d'etat against the sovereignty of three nations enforced by a common hard line security strategy already in play separately in each country. It's a scheme to create

Mexican President Vicente Fox, American President George Bush, Canadian Prime Minister Paul Martin meeting in Waco, Texas March 23, 2005 launching the purposely unpublicized Security and Prosperity Partnership Agreement (S.P.P.) - - which is intended to be a crucial formal step to implement a North American Union (NAU), establishing priorities for the benefit of corporations and the military without public consideration.

a borderless North American Union under US control without barriers to trade and capital flows for corporate giants, mainly US ones. It's also to insure America gets free and unlimited access to Canadian and Mexican resources, mainly oil, and in the case of Canada water as well. It's to assure US energy security as a top priority while denying Canada and Mexico preferential access to their own resources henceforth earmarked for US markets.

It's also to create a fortress-North American security zone encompassing the whole continent under US control in the name of "national (and continental) security" with US borders effectively extended to the far reaches of the continent. The scheme, in short, is NAFTA on steroids combined with Pox Americana homeland security enforcement. [99]

Mr. Lendman goes on to reveal that the North American Competitiveness Council (NACC) Greg Palast mentioned above was only created as recently as March 2006 'at the second annual S.P.P. summit in Cancun, Mexico....In February, 2007' it laid out 'a set of S.P.P. priorities' that Mr. Lendman claims included the following, most to be accomplished by the end of 2008, plus:

some longer range ones targeting 2010.
-- implementing planned land clearance projects, meaning less for the people and more for corporate predators;
- putting in place more business-friendly border security practices, meaning militarizing the border;
-- further simplifying NAFTA rules-of-origin requirements, meaning no restrictions on regional trade even for unsafe products;
-- simplifying the NAFTA certification process and requirements aiming at their total elimination;
-- ending the consumer-protective US Animal and Plant Health Inspection Service (APHIS); [note the May 2008 discontinuing of pesticide tracking

national survey by USDA[100] , of which APHIS is a sub-agency.]

-- removing regulatory standards and practices that impede trade even if doing it harms consumers;

-- working toward a goal of uniform global regulatory standards and practices regardless of the consequences or concern about national sovereignty;

-- easing cross-border tax burdens forcing consumers to pick up the difference;

-- cooperating in identifying common financial regulatory concerns, then work to eliminate them;

-- completing a coordinated Intellectual Property Rights (IPR) Strategy aimed at protecting them and keeping their prices high;[101]

Besides an expected recommendation about cutting red tape, and saying that "regulations impede the efficiency and competitiveness of businesses in all three countries," Lendman reminding us that 'regulations, in fact, serve (or should serve) to protect consumers, not harm them,' Lendman goes on to note that NACC recommended:

energy integration specifically emphasizing Canada's vast oil sands that make its overall reserves second only to Saudi Arabia.

[Mr. Lendman also informs us that] Canada aims to triple its oil sands production by 2015 to three million barrels daily to feed America's insatiable energy appetite these resources are earmarked for. [102] Mexico's oil is also targeted, but the report hides NACC's aim for state oil company PEMEX to be opened to private investment saying only while the country is "blessed with abundant reserves, (it) faces major challenges in attracting capital" needed to realize their potential. NACC wants Mexico to "increase the competitiveness in (its) energy sector" without saying it wants it privatized so foreign investors can plunder them for profit. [103]

Then there is the North American Future 2025 Project, that Mr. Lendman states involves 'former American political heavyweights...including Sam Nunn, Zbigniew Brzezinski, Harold Brown, William Cohen, Henry Kissinger and others.

The agenda involves preparing a final report to the US, Canadian and Mexican governments by September 30 [2007] expected to recommend the benefits of integrating the three nations into a single political, economic and security bloc.

What's known has activist groups upset including the Council of Canadians and Coalition for Water Aid. They're protesting what they say amounts to a sub rosa effort for corporate interests to control Canada's huge fresh water supply, estimated at one-fifth of the world's total. They want Canadian energy and other resources, too.' [104]

The water from Chiapas, '40% of Mexico's total and the reason Coca Cola is dying to get hold of it,' is also mentioned by Mr. Lendman as another corporate goal for plunder. He states that the

Mexican government wants to remove the Mayan people from their native Lacandon jungle to accomplish this.

Walia and Oka address the issue of expropriating indigenous lands and cultures, displacing people who have lived sustainably in their local environs for centuries - - interweaving this with the SPP's approach to cheapening labor expenses and the buggaboo of 'illegal' immigration:

> The exceptional freedom and mobility of corporations and businesspeople [as provided by the SPP] is dramatically contrasted with proliferating restrictions imposed on marginalized communities. Ironically, border controls are deployed against those whose very recourse to "illegal" migration was destroyed by the license afforded to corporations by free trade agreements to ravage entire economies and displace entire communities in the South [e.g., Mexico, Central American nations - especially via Plan Puebla Panama]. Similarly, the focus on resource extraction and development in the SPP will work to further dispossess and displace indigenous communities.

> The SPP intensifies neoliberalism through an increased reliance on labor flexibility as a means of increasing profits. For example, the SPP undermines labor laws through employment of contract and part-time labor, as well as through the enforcement of exclusionary citizenship through Temporary Foreign Worker Programs. The expansion of guest worker programs allows for capital interests to increasingly access cheap labor that exists under

precarious conditions, the most severe of which is the condition of being deportable. Given their unstable legal status, governments and businesses are able to hyper-exploit migrant workers by denying them basic rights afforded to citizens. They also maintain the sanctity of the fortified national security apparatus and the racist regime of border imperialism by legalizing the "foreign-ness" of migrant workers. [105]

Utilizing the Canadian 'model,' whose detrimental effects you can probably easily imagine, or have experienced in your own life:

....[With] rapid expansion of temporary guest worker programs that are required to ensure cheap labor in light of the repressive migration and security controls...Canada's Seasonal Agricultural Worker Program is seen as the "model" to implement despite widespread documented abuse including being tied to the "importing" employer; facing deportation if workers assert their rights; and exploitative working conditions including low wages, long hours, substandard housing, and overt discrimination. [106]

Perhaps we should be more perceptive in our condemnation of those 'illegal aliens' then? Especially when we berate them and their border crossings into the USA, while one million Americans now live as expatriates in Mexico as retirees and residents of exclusive gated communities, like those in Cabo San Lucas at the southern tip of Mexico's

A quarter of all American expatriates, one million in number, live in Mexico today. Gated communities insulated from native Mexican neighbors exist in places like Cabo San Lucas at the tip of southern Baja peninsula, in contrast to maquiladora polluting industries located along Mexico's northern border, and hostility to Mexicans and Central Americans displaced to USA from their native soil due to NAFTA/corporate trade policies.

Baja peninsula. One quarter of ALL American expatriates have chosen Mexico as their new home, and that number will be increasing as we follow our desires to live in lower-cost warmer climates. [107]

But if we do not compassionately share our fortune with the local Mexicans, we might encounter bitterness and vexation that could threaten our security. That brings us back to our current obsession with defending ourselves, and building up our protective/military strength, too often at exorbitant expense and ill-will. Which is not always the best way to approach the world that surrounds us.

Stephen Lendman's 2007 article is actually entitled 'The Militarization and Annexation of North America.' On page five, under the subchapter heading 'Militarizing a Continent As A First Step' we can read:

> No nation is more militarized today than America. It spends more on national defense and homeland security than all other nations combined. Add to those budgets all others related to defense, still others for intelligence and covert actions, plus the net interest cost attributable to past debt-financed defense outlays and it totals over $1 trillion for FY 2007 according to one analyst's estimate and heading way above that in FY 2008 if current budget proposals pass and become law which is almost certain.
>
> Canada and Mexico are expected to share the load as part of Washington's "war on terrorism" and are doing it. Supporting Washington is central for Canada's Stephen Harper conservative administration. It includes adhering to the 2002 Binational Planning Agreement allowing US military forces to enter Canada on its own discretion, set up shop, and exercise authority over Canadians in their own country. [108]

Should this make us happy in America, as free people who can pursue life, liberty and happiness - - as provided in our enduring revered constitution? Meanwhile, in Mexico, Lendman sites 'a 1994 Pentagon briefing paper declassified under FOIA, [that] hinted at a US invasion if the country became destabilized or the government faced the threat of being overthrown because of "widespread economic and social chaos" that would jeopardize US investments, access to oil, overall trade, and would create great numbers of immigrants heading north.

Plans are in place and are playing out to snuff out trouble before it spirals out of control, and the proposed US immigration bill was to provide funding for it through stepped up militarization. But even with the bill defeated, the money's coming and US forces will follow if needed. Congressional budgeting calls for millions in Mexican military aid and massive new border detention centers for up to 30,000 detainees for starters with two notorious ones discussed' [in his article]. [109]

Included here would be Halliburton, with a 'contingency contract worth up to $385 million to build US-based detention centers.' [110] Supposedly they would be for:

"detention and processing" in case of an "emergency influx of immigrants....or to support the rapid development of new programs (for planned) expansion facilities (able to hold 5000 or more persons)." [111]

Though dissenters to the implementation of a North American Union, coming from north or south of the Rio Grande could end up incarcerated in them.

Since the 2006 Mexican election, in which Felipe Calderon became president amongst cries of fraud, Lendman reports that 30,000 troops are present 'in a third or more of the nation's states.'[112] Kristin Bricker, who currently lives in Mexico, in a June 2008 Indypendent article notes 25,000 of these soldiers were 'immediately deployed [after the election]...to drug-cartel dominated states, militarizing a large part of Mexico without legislative approval.'

Felipe Calderon, Mexican President, utilizing the military and Plan Mexico to increasingly militarize Mexico, in league? with efforts to subvert democracy of Mexico, USA and Canada into an unvoted-upon North American Union for the benefit of corporations and misguided military/political interests.

Ms. Bricker talks about 'Plan Mexico,' funded to the tune of $400 million for fiscal year 2008 by our U.S. congress via an amendment to an Iraq supplemental appropriations bill, that 'will further militarize Mexican society by providing U.S. resources to the Calderon controlled military.' [113] Funny thing is that 'Plan Mexico won't give Mexico any cash, making it yet another gift to the U.S. military industrial complex: the U.S. will pay defense contractors to provide equipment, training, and personnel to Mexico.' Various proposals have included funding moneys for the notorious School of Americas,

411

also known as the 'International Law Enforcement Agency,' a well-known and abhorred (yet still somehow in business) for its training of death squads in El Salvador and other Latin American countries.[114]

She also relates that:

'Plan Mexico is an indispensable component of the Security and Prosperity Partnership (SPP) and the expansion and militarization of NAFTA...In the name of increasing North America's competitiveness in the global economy, the SPP calls for the standardization of the laws, policies and practices of Canada, the United States and Mexico. It seeks to expedite natural resource extraction and integrate the continental energy supply.

The SPP calls for the further militarization of borders against the conflated "threats" of organized crime, international terrorism, and illegal migration. In order to standardize security practices, North American countries are already coordinating cross-border police and military training and increasing cooperation among law enforcement agencies and armies.

The Center for Economic and Political Investigation and Community Action (CIEPAC) in Chiapas opposes the SPP because "the United States is making it possible to force Mexico and Canada to change their laws, rules, and regulations in order to secure the economic ('prosperity') and political ('security') interests of its government and businesses ... in order to appropriate our natural resources for themselves and to increase their profits."' [115]

This complements what you have read in this chapter about NAFTA, the SPP, the North American Union, the travesty in especially Chiapas, and misnomered 'free trade.' Ms. Bricker goes on to inform us that:

> Critics of Plan Mexico say it will funnel resources and U.S. training into the Mexican military and police. They point to the [Mexican] government's use of paramilitaries and death squads, the rape of dozens of female detainees by police in San Salvador Atenco, and the murder of journalist Brad Will in Oaxaca by off-duty police and government officials as just some of many reasons Plan Mexico will decrease, not increase, Mexicans' security. [116]

After initial rejection of "unacceptable and unjust" human rights conditions by Mexican politicians, Plan Mexico was re-vamped and approved by our Democrat-Party-controlled Congress with less stringent human rights conditions. However, if these conditions are not met, funding will only be cut by a mere 15% of the total $400 million. And that is just the beginning, because this Plan Mexico will be a multi-year (at least) affair that may run into the billions of dollars.

Can you imagine that 'Sen. Patrick Leahy and Sen. Barack Obama have both stated that Bush's proposal [for just $1.4 billion for three years] falls short and that much more money over many more years will be necessary to fulfill Plan Mexico's mandate?'[117]   What could they could be thinking and how much do they know about what's really going on in Mexico and the conference rooms of our most misguided powertrippers?

One last paragraph from Kristin Bricker projecting what she expects might transpire as a result of implementing Plan Mexico, from prior experience with Plan Colombia - - which many of you may think only narrowly concerns the U.S. 'War on Drugs' exported to that troubled Latin American country:

> While Plan Mexico specifically targets drug trafficking, the initiative's South American counterpart, Plan Colombia, strongly suggests that drug war equipment and training will be used for counterinsurgency. Mexico has already demonstrated its propensity to use deadly drug war equipment donated by the United States against insurgents and civilians. Following the 1994 Zapatista uprising in the southern state of Chiapas, the Mexican military strafed indigenous communities using U.S. helicopters earmarked for counter-narcotics operations. [George] Bush's leaked Plan Mexico proposal includes eight helicopters and two airplanes for the Mexican military. [118]

Would it surprise you then that Stephen Lendman tells us 'civil rights are suspended and widespread abuses are reported [in Mexico]
because the military got a mandate to "use all necessary force to resolve disturbances and return peace to society." That's just a hint of what's coming

across Mexico and the continent under full implementation of S.P.P. that won't tolerate opposition and
will crack down hard against it. Mexican law now
allows it [cracking down hard, your author clarifies]
after passage of the draconian "International Terrorism Law" criminalizing dissent, calling it terrorism,
and imposing harsh sentences for using "violence
against persons, things, or public services that
spread (enough) alarm or fear in the population....
to threaten national security or pressure authorities
to take certain determinations." [119]

Social protests may be criminalized as well
with resistance movements like the Zapatistas and
Oaxacan Popular Peoples' Assembly (APPO) labeled
terrorist organizations and their leaders subject
to 40 year mandated prison terms if charged and
convicted. And President Calderon wants Mexico's
Congress to pass an amendment giving him consti-
tutional powers to tap phones and search private
residences without first obtaining court-ordered
approval under any conditions he claims is "urgent."' [120]

Doesn't that sound like what President Bush
wants and now has in the grand old USA? What
would the late Senator Sam Ervin say in these
times of warrantless wantonness being propagated
by our own (?formerly?) universally respected
government, inspiring others to behave in similar
undemocratic fashion? [Calderon's desired power
to perform warrantless searches was removed
from Mexico's final version of its so-called 'Ge-

stapo Law' of 2008, which was apparently inspired by the USA Patriot Act.[121]]

Could such a repressive future actually lie ahead for our land of the free and home of the brave, and our continental neighbors? Perhaps it is time to perk up our antennae, get our media involved and 'demand sovereignty and independence' loudly before our long-struggled-for rights could be lost to our financial/political/military wielders of under-reported unrestrained power?

Former North Carolina Senator Sam Ervin, chairman of the Watergate-Pres. Nixon impeachment proceedings. Had great concerns about freedom and civil liberties, and the use of a warrant to enter a citizen's home. That the knock upon the door of a home by a law officer "was the only way the occupant could know it was an officer of the law and not a burglar." Ervin said "I think that the greatest hunger of the human heart is for a place where we can be free, enjoy our privacy, without fear of unwarranted governmental intrusion." [122]

New Hope With A Democratic Majority In US Congress?

You may have been given hope by the recently elected Democratic congressional majority voted into place in 2006. You might expect its influence and commitment to exact changes in America's trade policy. Alas, be wary of that 'deal' Nanci Pelosi and her confederates publicized in May of 2007 to improve small Free Trade Agreements with Peru, Panama, Colombia and South Korea. The important environmental, labor, health, legal and sovereignty issues were only tackled on a 'conceptual level.' 'No union, environmental, consumer or small business group supports the 'deal' while all of Big Business does.' In reality, this 'deal' 'could facilitate more Bush trade deals that contain the worst provisions of NAFTA and CAFTA....pasting new labor and environmental provisions onto a NAFTA-style pact is like putting truly tasty frosting on a poisonous cake,'[123] is how Public Citizen describes it.

You've read the information above. If it troubles you, and the light is shining somewhere between your mind and retinae, and your heart is thumping, why don't YOU start doing something about it!? Call your Congresspeople. 202-224-3121 to call your Senators; 202-225-3121 to call your Representative in the House. Spread the word to your friends and foes. The truth goes a long way; lies and deception have to be repeated over and over to fool the audience of citizens who continue to doubt them, but may wearily surrender to their perpetration and perpetrators if enough of us who know better quittingly remain silent.............................

P.S. There actually is cause for some new hope,

by the way. As a result of a May 30th, 2003 incident in Burma, where approximately 80 people were killed by government thugs, the Congress of America was so enraged both houses overwhelmingly passed a law banning all imports from Burma. (Unocal, however, was grandfathered in, exempting them and their petroleum products from the ban.) No WTO challenge has yet been issued to this 2003 law[124], as of the writing of this book.

P.P.S. With current attempts at furthering the WTO's and corporations' inroads into countries around the globe being frustrated (?temporarily?) at ministerial meetings like the ones in Cancun, Doha, and Seattle, the USA has engaged in various smaller free trade agreements. Most are bilateral. For example, with Chile, Bahrain, and Australia, plus the other ones mentioned in this chapter. According to trade expert Patrick Woodall: "the goal of all the trade agreements is to establish trade pacts that guarantee multinational corporate interests will be protected... The agreements deem many national and local environmental regulations to be illegitimate expropriations of profit. Copyright and trademark protections are enforced with trade sanctions, but violations of labor and environmental law are at best subject to fines and more frequently are totally ignored."[125]

Mr. Woodall sees a two-fold purpose in these pacts:
1) "to isolate some countries who were skeptical of the U.S. trade agenda at the WTO and the FTAA."[126]

2) "to set markers for future trade deals based on what countries acquiesced to in bilateral agreements."[127]

As we and aggressive corporations attempt to intimidate reluctant countries that we can isolate one-on-one, we see millions of our jobs being 'offshored' to nations with lower wages and lesser environmental and labor standards. It's a race to the bottom of the barrel, while waving the battle-flag of profit. And some of these offshored jobs involve 'computer programming that operates some U.S. electricity and water systems,'[128] in addition to airplane maintenance, that could lead to sabotage. OK, I'm paranoid. We're worried about terrorism, but big corporations using overseas workers that aren't subjected to background checks, etc., like they would be in America, are happy. And then there are the offshored medical records computerizations and billings by HMO's and accountants; and what about your offshored credit card transactions? Identity theft, anyone? Breaches in privacy[99] any concern, not protected like the way the ole USA does it?

The next subchapter in this story is still to be written.......

(More information on two specific cases is available in the Appendices -- re the Internet Gambling decision, and the complaint vs. the European Union banning GMO's.)

Chapter Five:
Radioactive Wastes In Your Dump, Air, Water, Utensils, Baby Stroller, Zippers, Anyone?
(Are They Too 'Trivial' To Monitor Anymore, Anyway…?…)

"You may soon need to fear household products you have the most contact with...this is the legacy of an industry gone mad."
–Richard Clapp, Associate Professor, Boston University School of Public Health, concerning the nuclear industry, and 'recycling' radioactive waste into the 'Free Market'

©1990 GULLIVER at the BIG BEND SENTINEL

NUCLEAR ENERGY
SAFE · CLEAN ·
NO GREENHOUSE
GASES

DANGER
RADIOACTIVE
WASTE
KEEP YOUR
DISTANCE

You've all heard that 'nuclear is green' lately. Spoken by politicians and people who are supposed to be practical and patriotic. But what about radioactive waste, the most toxic, long-lived entity known to man?? Can that be 'green?' Have our talking heads on television been cut off from their brains and hearts?? Do they know about radioactivity? That it can adversely affect the body and the cells of man, animal, and plant, and cause death and cancer and disease for tens of centuries??? That this is the most troublesome product of nuclear power?

Sometimes, if you repeat a misled claim often enough, and the Thomas Friedmans, Patrick Moores and Christy Todd Whitmans of the world are not frequently contradicted, the consciousness numbs, and the soul pines away.

For a while.

But, soon the tide comes back in; soon the truth surges over the boulders of smooth suppressive public relations campaigns that seem victorious. That shoo the worry from 'emotional' mothers and fathers, feeling helpless to fight today's powers that be. . . meanwhile harboring anger at what is 'legally' being done to them and their children . . .

Radioactive waste or nuclear waste, what we

are talking about in this chapter, comes mostly from our 104 nuclear power plants, and our production of nuclear weapons. Each nuclear power plant produces over 500 radionuclides or radioactive elements every day as uranium is fissioned or 'split' to produce heat to boil water to produce steam to turn a turbine to produce electricity, as mentioned in Chapter One.

While these various radionuclides can silently invisibly kill us and mutate our descendants via striking our DNA inside our cells, the nuclear waste generators and polluters of our civilization are attempting to <u>deregulate</u> their waste and release it unmonitored into 'the marketplace,' or into your neighborhood dump - - or even SELL it! Oh, and why not import more of it, from other nations, too!? Even though we still do not know how to store it safely for as long as it is necessary to do so.

Isn't that all part of the 'green revolution?'

I wish I could say I was "Just kidding." But this is the era of the Hummer, of the soul insatiable, when we see the polar bears struggling for survival because of the loss of Arctic and Antarctic ice, while we deny that our combustive emissions could be dangerously warming or polluting the planet. Unfortunately, because of the ultimate ultratoxicity of nuclear waste, and the inevitable danger of that next Chernobyl nuclear accident happening somewhere soon, nuclear power is not the answer to our energy production. As energy guru Amory Lovins states "every dollar invested in nuclear expansion will worsen climate change by buying less solution per dollar."[1] Not investing in wind, solar, hydrogen, biomass, energy efficiency and other alternative energies now, will deepen the hole we are digging nuclearly in economic, toxic, and practical terms.

You may not want to know, but nuclear waste-containing soil from Los Alamos National Laboratory has already been incorporated into one of your favorite nearby golf courses within the state of New Mexico. Also, using the excuse that background levels are already high there, so why not add more because it wouldn't matter, mixed waste including nuclear waste 'was shipped across the country to a site not licensed for radioactive disposal next to the Lewiston-Porter schools' in upper New York state.[2]

I'm not saying they do it for the money, but why not be aware that the Nuclear Regulatory Commission (NRC)-licensed Alaron corporation is processing nuclear waste from Kentucky and Ohio over in <u>Wampum</u>, Pennsylvania?

Then we have your Aerojet company in Jonesborough, Tennessee, being licensed for 'oxidizing (incinerating) metallic uranium chips and grinding fines for disposal as dry solids...[Tennessee being] the only state in which the Department of Energy (DoE) is burning radioactive waste.'[3] Though, actually, due to community protests preventing incinerating nuclear waste anywhere else in the country, the sole <u>currently operating</u> radioactive DoE incinerator is located in Oak Ridge, Tennessee.

What's the big deal, incinerating nuclear/radioactive waste? Why can't I have one of those incinerators right next to my house, so I can watch the dying sparks fly over and onto my roof, like beautiful little shooting stars?

Romantic as you might imagine the possibilities, 'heat does not destroy radioactivity. The chemical bonds may break and chemical structures and phases change, but the radioactive isotopes remain just as long as if not heated. So the process has the very large danger of dispersing radioactivity into the air [from which we can inhale it directly into our lungs, then circulate it around our body; or get it into us and our environs as it drops into the food chain and our water] . . . concentrate in ash, filters or other solids remaining after the thermal process and contaminating the incinerator.'[4]

Remember from Chapter One that radioactive elements like strontium and plutonium, just to name two of our most prominent radionuclides, have 'half-lives'[5] of 28 YEARS and 24,000 YEARS, respectively, during which time we have to worry about them. Well, really, you might recall that experts tell us that <u>ten to twenty</u> 'half-lives' are the period over which we really have to worry about exposure to radioactive elements and their radioactive beamings. This 10-20 year period is called an element's 'hazardous life.' So, for strontium that is 280 to 560 years, and for plutonium it is 240,000 to 480,000 years.

That's an awfully long time, isn't it? How could we get ourselves into such a fix? All this waste, that really probably can never be 'cleaned up,' plus the billions upon billions of dollars somebody (think <u>YOU</u>, Dear Taxpayer) has to pay to incompletely get the job done. As noted earlier, I informed you that each nuclear plant produces 400-1000 POUNDS of plutonium each year, and just one microgram is enough to cause a lung cancer to seed itself in your lung. So, twenty pounds theoretically could produce lung cancer in every member of the human race if tiny particles are released in some catastrophic event, like a nuclear plant explosion and fire. However, the particles have to be dispersed in less than four microns[6] size, and be inhaled by each of us, and stay within the lung sac or 'alveolus' long enough to mutate some cell(s)'s DNA so that a cancerous transformation results that multiplies and multiplies and multiplies. . . A lung cancer . . . that you would never be able to detect the origin of . . .

The bottom line to nuclear industry and governmental bureaucrats is that paying to contain and store this nuclear waste produced by nuclear power plants and nuclear weapons is very expensive. If we can pretend that the stuff is 'safe' to release, especially free from monitoring, into the public domain, why, there would go those burdensome limitations on an industry gone mad from one very serious responsibility they are attempting to avoid assuming.

Here is a quote for you on this responsibility, as meted out by one of our federal laws defining 'disposal':

> The term "disposal" means the permanent isolation of low-level radioactive waste pursuant to the requirements established by the Nuclear Regulatory Commission [NRC] under applicable laws, or by an agreement State if such isolation occurs in such Agreement State.[7]

Note the word 'permanent' please.

'Low level' radioactive waste (LLRW) may not sound too serious, but it includes irradiated pipes, control rods, resins, sludges, filters, contaminated concrete (13,000 tons from your typical 1000-megawatt nuclear power plant) and contaminated re-inforcing steel bar (over 1,400 tons from the same kind of plant) as your nuclear electricity-generating dome becomes too radioactive to operate after 20 to 30 years.

Most important to know: even though such waste may be called 'low level,' it still can have any and all of the more than 500-radionuclides present that a nuclear plant produces daily. These may or may not be present in small quantities. But these radionuclides do <u>not</u> have to be individually quantified. And remember that just a microgram – a millionth of a gram - of plutonium, as one example, is enough to cause you lung cancer.

According to a 1998 Government Accounting Office (GAO) document, 'Class C "low level" radioactive waste (which includes heavily loaded resins, and is of higher toxicity than Classes A and B) can give a lethal dose, if unshielded, in less than an hour (20 minutes for doses of 500 rads[8] per hour).'[9] (<u>High level</u> radioactive waste, which might spill out of a train or truck in an accident, to become unshielded on your street or highway, can deliver the same lethal dose in a mere 10 SECONDS of exposure if you are but three feet away from it.)

You may hear about medical waste sounding like a prominent part of the low level radioactive waste total, but it is less than one percent of the amount, emitting minimal quantities of 'curies' (see below) with very short half lives. Gallium-67 has a half-life of 78 hours and a hazardous life of 1-2 months; technetium-99m has a half life of 6 hours and a hazardous life of 2.5 to 5 days. While the 'medical waste from diagnosis and treatment shipped in one year from most states usually gives off a fraction of one curie of radiation...each nuclear reactor generates hundreds and thousands of curies in "low-level" waste every year."[10]

A 'curie' is a measure of radioactive energy being emitted, by radioactive waste, for example. It is named after Marie Curie, who co-discovered radioactivity. She worked with radium. One gram of radium emits one curie per second.[11]

Radium is the element that watch makers once used to have the numbers and hands of their watches appear iridescent. We had a factory here in Sag Harbor, New York, where workers would use tiny paint brushes, dip them in the radium mix, touch the tip of the paint brush to their tongue to draw it to a point, then apply it to the watch.

In this case, a little radiation wasn't good for ya. Many of these watchmakers contracted cancer of the tongue and died. The watch factory was closed long ago.

As a quantitative measurement, a curie equals 37 billion disintegrations or radioactive emissions per second from a radioactive material. Although medical waste may total in the one to two curie range per year, primary components from a nuclear plant "average 1000 to 5000 curies per cubic meter...cartridge filters emit about 20 curies per cubic meter...demineralizer resins emit about 160 curies per cubic meter."[12]

So don't let some actor-doctor mislead you with his shiny teeth and stethoscope, panicking you with the end of medicine as we know it, if we can't dump all those medical radioisotopes somewhere. These are just a dot in the mountain ranges of nuclear waste that exist, which could be doubled and quadrupled in the future, should new nuclear power plants be foisted upon the poorer susceptible locales about our country, instead of building windmills and solar paneled homes everywhere, having hydrogen fuel cells nearby at the ready, while employing newly devised technology to increase energy efficiency.

There used to be six main public low level radioactive waste burial sites within U.S. borders, but only the Barnwell, South Carolina site remains open currently. As of June 30, 2008, Barnwell, too, will close its gates to low level waste from all states, with the exception of that from South Carolina, New Jersey, and Connecticut. Otherwise only a private facility administered by EnergySolutions in Clive, Utah, not too far from Salt Lake City, and a facility in Richmond, Washington, may take lesser classes of radioactive waste in relatively limited amounts. (Though that may change if EnergySolutions manages to import United Kingdom radioactive waste from big nuclear power and weapons sites, as they are attempting to contract to do[13] - - along with 20,000 tons of low-level and middle-level radioactive waste from Italy.)

Since most of the closed sites leaked or were 'otherwise deemed unsafe or undesired,'[14] nuclear waste generators are now in a bind as to how and where to get rid of their wastes in this twenty first century.

Would you like to see those filters and resins and concrete just dumped into your neighborhood dump as if they were of no concern? somehow deemed too 'trivial' to be monitored any longer? using computer modeling to rationalize and legalize and deregulate what should certainly be isolated and tracked throughout their hazardous lifetimes?

Nuclear waste generators, in concert with the NRC and the Department of Energy (DoE) and even the Environmental Protection Agency (EPA), have been attempting to 'engage' the public to accept 'official release levels' or 'clearance levels' of radioactive waste over the past few decades. But the public does not want this. Not yet. Probably not ever.

As quoted in the Nuclear Information and Resource Service (NIRS) report 'Out Of Control - On Purpose': 'Because of public, local, state, other industry, worker and union opposition to radioactive recycling and release in the U.S., there is no legal, allowable release level'[15] for radioactive waste. Yet the Department of Energy (DoE) persists in setting up allowable <u>internal</u> contamination levels that they hope one day will be accepted by all America as too trivial to worry about.

Americans may think once the government has decided to do something, citizens can do nothing about it. However, here we have a perfect example of this <u>not</u> being true. 'Because of the insistence of the metal industry, along with public and local and state governmental concern, DoE has halted . . . deliberate commercial recycling of potentially contaminated metal.'

The battle is ongoing though, for much is at stake, especially money and expense for the nuclear waste generators. 'The latest threat is DoE's request for Expressions of Interest for companies to process contaminated nickel and other metal from uranium enrichment for "restricted" use within DoE or the regulated nuclear industry.'[16]

Nickel-63 - - with a half life of 96 years and, thus,

One day all too soon nuclear waste may be 'recycled' to end up in your zippers, bra-clips, utensils, batteries, frying pans, office buildings, car chassis to possibly cancerize vital parts of human bodies.

a hazardous life of 960 to 1,920 years, in which to cancerize you and your descendants - - may end up in batteries and frying pans in the long run, once the first inch might be given, and not taken back. It can be used to convert carbon steel to the more valuable *stainless* steel. If radioactive nickel can somehow be declared 'safe' for 'clearance,' one day it may also contaminate your zipper or your bra-clips or your office building, or your knife, or fork, or your car chassis. And would you know it?

Sadly, probably not. Remember, you can't taste, smell, see, or feel when something is radioactive. It's not like a stinky bitter pesticide, or a rotten egg that your senses tell you to stay away from.

Oh, you could carry a Geiger counter with you whenever you go to Home Depot or Lowe's to check if the metallic object you fancy is radioactive, or if your food might be, but if you don't have the right probe, or you use the wrong technique you still may miss the radioactivity present in the goods you could errantly buy.[17]

How can this happen? you may wonder. You trust the government to protect you, and do the right thing. Is this really possible, that radioactive waste could just be freely released to wander through our industrial products and food chain, to pollute our water, and flicker through our air?

Please keep in the back of your mind that about 2500 metric tons of irradiated fuel, high level waste (HLRW), are generated every year by our 104 nuclear reactors, with no place yet approved for this to go. Total nuclear reactor-generated high level waste is about 60,000 metric tons, plus up to 15,000 metric tons more of HLRW from the DoE.[18] Low level waste (LLRW) totals, for some reason, are more difficult to obtain. No centrally collected overall total is available, but for the past twenty years, what has been reported as received at the various burial sites totals 9.35 <u>million</u> metric tons. This does not include what is stored at, or exists within, each reactor, nor other various contaminations and accumulations that exist about the country. Nor does that number give the overall amount of radiation that could be emitted by all our LLRW, though 12 million curies has been reported for those 9.35 million metric tons.[19]

This is a beginning point to gather the unfathomablility of all the nuclear waste we have generated so far, and will continue to generate in the future. Especially, since we have no real long-term safe way of storing it, and no one wants to accept it – rightfully – in their neighborhood. But, if we can shift the responsibility somehow, lobby the politicians, then we can figure out a way to shaft whomever ends up with the most toxic waste on Earth. Right here, somewhere in the grand old USA. Where nuclear is 'green' – supremely toxic radioactive 'green,' that is . . . .

Accepting Toxic Radioactive Exposure To Yourself And Your Loved Ones

Well, let's start with an arbitrary 'trivial' dose that has been selected to be one millirem per year exposure, per 'practice.' [with no limit on the number of 'practices,' and without the entity 'practice' being specifically defined. What if those 'practices' unpredictably silently add up on *you*? making *you* the innocent unknowing victim, receiving too many millirems of total exposure, from too many 'practices,' that nobody is actually measuring?] Yeah, heck, just one millirem. Why, that's nothing!

See, you get 100 millirems of radioactive exposure every year from so-called 'background radiation.' This is true for most parts of the country. Though up in the mountains of Colorado, it might be 200 millirems of exposure per year. 'Background radiation' comes from cosmic rays from outer space, and other radiation that might come out of the soil and rocks on Earth, for example. Then lately we have added in the danger from radon in rocky cellars and similar environs, that could add perhaps 150 millirems to some Americans' background radiation exposure.

Funny thing how over the last few decades the total numbers for background radiation exposure seem to be creeping upward, along with the nuclear industry's ingrained influence within government halls and offices. So what's another millirem here and a millirem there then? How could it even matter if we dumped a bit more radioactive waste on ya, just to help with the expense of providing your electricity for your heated pool, plus the air-conditioning for your torrid apartment?

Yes, the stuff could accumulate. Yes, hot spots of concentrations of the various radionuclides could beam into your body and cells and mess up your cellular reproduction so you get a bizarre cell that goes wild and multiplies itself repeatedly, often (but not always) at a more rapid rate than its neighbor cells can reproduce themselves, as it becomes a 'cancer,' a cancerous growth or tumor.

From DoE Order 5400.5, get this, 'if it is calculated that the volumetric [thru and thru, as opposed to just surface] contamination will give "individual doses to the public [that] are less than 25 millirem in a year with a goal of a few millirem," the waste can go to a solid waste landfill, as long as the groundwater is protected to state requirements and the landfill operator and state solid waste regulator agree.'

Would that happen at your dump in Beverly Hills, California, or yours over in Palm Beach, Florida? If you have the wealth and power to reject such a practice, very unlikely. Besides, to accept radioactive waste, any said landfill must be a licensed radioactive disposal site. Unless, somehow, behind security fences and walls, the DoE can circumvent such rules and adopt "authorized limits" permitting radioactive waste to be dumped

Paducah, Kentucky landfill that has no license to receive radioactive waste, yet has "authorized limits" to do so thanks to the Department of Energy. Do unconcerned citizens using this dump have any knowledge of this, and the possible consequences?

at a site like the state-licensed C-746-U solid waste landfill in Paducah, Kentucky that is *not* licensed to accept radioactive waste. . .

Under a November 1995 Memo and a 2002 Draft Guidance, the DoE developed a procedure to assist their various radioactively contaminated sites to clear 'nuclear waste out economically...[so] radioactive materials and property can be released into general commerce...These are not laws or regulations approved through any public process.'

'First, the total exposures have to be estimated to be less than the 100 millirems per year individuals are allowed from all sources above background [radiation levels]. Like the landfill releases, they should have a *goal* of a few millirems, but can each give up to 25 millirems per year!

'These "authorized limits" appear more and more like blank checks to let contaminated materials go because there is no process and no effort made to verify the exposures caused. . . Since an authorized limit can be used over and over for different releases, and no overall assessment is needed, it is impossible to know the total amount of radioactivity released under each authorized limit. In addition, if the authorized limit becomes impractical, *supplemental* limits may be approved to allow more or different radioactive releases.'[20]

There is supposed to be transparent public record-keeping for all such releases, but this is not being performed. Neither are authorized and supplemental limits reported on a central database. And continue to be aware that with all this immoral dumping, individual radionuclide monitoring is not being done to protect the American public.

Is this acceptable to you, dear parents and children? We already know about the contamination of Chinese-made children's toys, and the rather promptly erected controls to prevent entry of these products into our stores and homes. But this radioactivity is far more dangerous, and far more long-lived.

What if you knew that 'background radiation' of 100 millirems is projected by our very own NRC 'to result in one fatal cancer in every 286 people?'[21] Double that allowable amount of millirems, and one in every 143 people will get a fatal cancer. And if we go even higher, to say 400 millirems as allowable radiation levels including background and what the nuclear mavens say is OK to add into the 'trivial' portion of the lode they don't mind dumping on us, that would be one fatal cancer in 71.5 people! Nice 'collateral damage,' no?

Just for your information, 500 'rads' is the normally accepted <u>fatal</u> dose of exposure acutely to radiation, to induce 'radiation sickness,' as experienced most memorably by the atomic bombed victims of Hiroshima and Nagasaki in 1945 at the end of World War II. 500 rads and 500 rems are similar amounts of radiation exposure absorbed, with <u>one</u> rad or <u>one</u> rem equal to approximately 1,000 millirems in most cases.[22]

The death that would come from 'radiation sickness' would usually occur within two weeks, during which time your immune system implodes, and you may bleed from every orifice. A very painful agonizing way to leave this Earthly life.

You also should know that the National Academy of Sciences in its recent BEIR V and BEIR VII reports concluded 'that all ionizing radiation exposures carried risk of biological damage to the recipient.' In other words, there is no safe level of radiation dose.[23] This is because each exposure to a radiation emission can break your DNA or injure it. Much of the time, you may be fortunate that spontaneous adequate repair occurs. But sometimes this does not happen, or the repair is faulty, and you have a mutation, which is then reproduced during cell multiplication, and passed on to subsequent generations of your cells and those of your descendants.

That is one way of seeing that your exposure to whatever kind of radiation your body receives is *cumulative*. The affect of each hit or curie or beam or radioactive particle adds up to possibly deform, and then further deform, your precious DNA again and again, like Russian roulette. This is the DNA that transmits the genetic code from you to your children and grandchildren, or that may lead to a cancerous growth forming amidst the protoplasm of what is your very own self.

Then there is the further mind-boggling affect

of 'free radicals' produced by radiation. Even if the DNA is missed, the free radicals resulting from radioactive beams and rays, chemically and physically interacting adversely with and within various parts of the cell, add another order or two of magnitude to possible damage and mutation of vital cell components that can be passed on during reproduction.

Polluters like to downgrade or minimize the perceived effects of their toxic additions to the environment, so they can go on polluting and producing their profit. They are giving us jobs, there are risks and benefits that must be accepted, otherwise how will society go on? is one of their arguments. We must advance, we must find new means to make energy, we hear them say. And people like Al Gore and George Bush accept the risk of nuclear power, along with the minimization of the risk that nuclear proponents persist in propagating.

According to the late and great Dr. John Gofman, and Egan O'Connor, in their essay on the law of CONCENTRATED BENEFIT over DIFFUSE INJURY, they state it to be as follows:

> A small, determined group, working energetically for its own narrow interests, can almost always impose an injustice upon a vastly larger group, provided that the larger group

443

believes that the injury is "hypothetical," or distant-in-the-future, or real-but-small relative to the real-and-large cost of preventing it.[24]

Gofman and O'Connor claim that the above quoted axiom 'accounts for the current promotion of a "de minimis" policy toward nuclear (and other) pollution. A de minimis policy asserts that society should not concern itself with trivia. . . [which includes asserting that *pollution* and its] poisonous discharges and human exposures below a certain level should be treated as non-existent - - because their consequences are allegedly trivial.

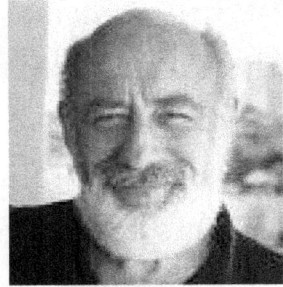

Dr. John Gofman, who first isolated plutonium in adequate quantity for the Manhattan Project, via which the USA built the first atomic bomb. Dr. Gofman later became a passionate - - and perhaps the most knowledgeable - - critic of nuclear power and its ultimate unacceptable threat to human civilization.

Trivial. That is the essence of the axiom. Triumph for each injustice is virtually assured if the advocates succeed in presenting it as trivial.'[25]

Then the "polluters and their agents [can] accuse citizens who oppose them ("activists") of being Chicken Littles and hysterics and ignorant extremists" as they work to attain "the public perception that the injury is trivial,"[26] and that a "calamity" must be proved to ward off their trivial dumping of their waste into the biosphere.

Many of you remember when a disgusted and angry nationwide cry rang out against the 'BRC' or 'Below Regulatory Concern' attempt to deregulate 'low level' nuclear wastes back in the early 1990's. This was an ill-bred effort at trivializing some radioactive dumping anywhere anytime de-monitored across the USA, orchestrated by the NRC and its brothers and sisters in the nuclear industry. There were shows on National Public Radio and the major television networks exposing it and its polluting/cancerizing likely ramifications.

445

In this instance, "trivial" didn't work. Public disapproval was overwhelming, once the media finally did its job reporting properly far and wide on a nuclear issue. BRC as a proposed policy for public consumption and affirmation was summarily withdrawn.

Although many of you may remember this, time allows memory to fade, so that the same old nuclear wine can be brought out again in a brand new faux-green (but still radioactive) bottle. Besides, there are you young pups who were not even born yet by the early nineties, and those of you that were still in your cribs, or riding your tricycles. That's what the nuclear industry banks on. Pretend all is well, that these anti-nuclear hippies and health food elitists have no idea of what they are protesting about, that nuclear is "green," that wind and solar and energy efficiency and other "alternative" sources could never provide our great nation with all the energy it wants to power its businesses and pump the current through our vast electrical cable systems.

What we need is more nuclear power plants, our current leaders tell us, always throwing in that baloney about being independent of foreign oil. [Oil only provides two percent of U.S. electricity.] Yet we are not even independent of foreign uranium. We import 80% of our uranium, in fact.[27] So we want to import other countries' radioactive wastes and reprocess them, meanwhile spewing more new and even worse radioactive waste streams into the communities that would be inflicted with these reprocessing plants - - plus, the dangers of trans-oceanic and whatever other types of transportation, coupled with the possibilities of ambush in this time of terror paranoia, would not project us toward the most sane safe type of world our voracious leaders are fulminating for us.

And, by the way, of course, we would be increasing our production of additional radioactive waste, with no safe technology yet devised for storing the multi-hundred radionuclides' for the duration of their various hazardous lives. On top of all this, if we try to dump these radioactive wastes into our landfills, they will interact in as yet unknown, unstudied ways with the many chemicals and other pollutants that will be accumulating there. To eventually seep into our soil and water.

As Dr. John Gofman repeatedly questioned: "How can we win the war on cancer if we keep adding new carcinogens into the environment?"

Dr. G. Fred Lee, 'landfill groundwater expert,' tells us:

There is no reliable way to properly predict when high density polyethylene liners in a Municipal Solid Waste Landfill or Class C landfill are going to fail. They are going to fail. There is no question they will fail. The issue about that is not if but WHEN and that is unknown. It relates to the fact that there are a whole host of reasons they fail including free radical attack. It can take hundreds of years but that is extrapolating beyond any reasonable approach.[28]

No surprise then that reprocessed irradiated fuel that was 'poured into soil "cribs" and into carbon steel tanks' over in Hanford, Washington, 'is leaking out into the Columbia River watershed and the food-chain.'[29]

But are you Tennesseans and nearby fellow Americans aware that your Oak Ridge Reservation (ORR) 'routinely releases radioactivity into the air and water?' The Tennessee Department of Environment and Conservation (TDEC) DoE-Oversight Division informed us in its 2005 Monitoring report that:

Radioactive contaminants released on the ORR enter local streams where they are transported to the Clinch River, which is used as a source of raw water by local drinking water suppliers...Over 100 miles of surface streams and significant (but unknown) quantities of groundwater in East Tennessee have been contaminated as a consequence of activities on the ORR. Process wastes contribute to this contamination, but the major portion of water pollution on the ORR can be attributed to releases from antiquated and deteriorating waste disposal, transport and storage facilities. Contaminants released from these facilities migrate to groundwater where they are discharged to local streams and are transported to the Clinch River and Watts Bar Reservoir.[30]

According to the NIRS 'Out of Control - On Purpose' publication, 'the downstream Watts Bar Reservoir has hundreds of curies of cesium-137 (half-life 30 years), and mercury contamination located in underwater silt deposits. In the past, marina owners sued the Department of Energy (DoE) for contaminating the reservoir.'[31]

Let me be the one to reveal to you this probably surprising fact: Tennessee 'expressly licenses profit-making companies to import nuclear power and weapons wastes from other states and countries to be re-characterized and released in the state.' This includes allowing 'this "special" waste to go into solid waste landfills [dumps],' and 'streamlining the process of sending nuclear waste to solid waste landfills...creating a systematic way to accelerate the determinations,' [while limiting] the nuclear waste to no more than 5% of each approved landfill.'[32]

'Today there are at least five solid waste landfills in Tennessee [that] have been approved to take deregulated nuclear waste from TDEC-licensed processors.'[33] But 'no requirement appears to have been made to evaluate for the synergistic [combining] effects of radioactivity and hazardous chemicals that could be [happening] in the landfills.'[34]

Of all fifty U.S. states, Tennessee has the most nuclear processors 'and is the most proactive.'[35] 'Since at least the early 1990's, companies in Tennessee have been licensed to make the decision themselves on what is radioactive and what can be considered "clean."'[36]

Does this send shivers down your back, that this really is going on in our country? Sure, it could be sort of kosher, well, theoretically...except, while, 'TDEC inspects licensees and determines they are in compliance. . . the compliance data are not available to the public. [Also] TDEC inspects licensees' programs for release methods and procedures, not the actual releases. TDEC Radiologic Health (as of 2003) did not keep records of what went out. The companies keep the records and they can destroy them when the licenses are closed or terminated. Records of measurements and calculations are maintained until [the] license terminates - then they are destroyed.'[37]

It would be one thing if these were stable local companies that Tennesseans have come to know well and trust, but what has happened over the past two decades is that too many of the few hundred companies involved have been bought up and consolidated under foreign and other-state based entities. When a company changes name or owner, licenses may be terminated, obligations be thrust awry. That is the current way of corporatism and the nuclear industry in the 21st century. Claims of 'national security' may also be sounded as the present nuclear push is expanded, when people want to find out the true story here, but are blocked by information being deemed 'classified.'

Imagine as a Tennesseans (or as a citizen of any sovereign state, U.S. or worldwide), finding out that thanks to your state's unvoted-upon decision, you have been allowed to be 'unknowingly [until this minute], involuntarily exposed to additional man-made radioactivity from nuclear waste. . . one additional millirem a year for every accepted release to state-approved landfills.'[38] If you live near the Middle Point landfill in Murfreesboro (not far from Nashville), or the Chestnut Ridge Landfill and Recycling Center in Heiskell, Anderson County, you could be getting exposed to dozens of streamlined radioactive releases, when you take your diapers and newspapers out to the garbage dump, but, no, no one is physically *measuring* how many millirems of radioactivity your body is absorbing, or anybody else's body is. Nor how many millirems of radioactivity are getting into your groundwater, today or tomorrow, or years and years from now. Nope, it's just 'not practical to verify, enforce or limit the [radioactive] exposures'[39] that you, or your surrounding environs, might innocently be subjected to.

Murfreesboro Landfill near Nashville, Tennessee. One of five Tennessee landfills licensed to accept nuclear waste in USA's 'most proactive nuclear state.' Profit-making nuclear processors are state licensed to import nuclear wastes from other states and countries to be re-characterized and released in Tennessee. How many citizens are aware of the state's nuclear dumpings and policies, and the dangers to them, their families, and their environment?

Yet, for a fee, nuclear waste processors can economically dispose of their 'cleared' waste, while helping the utilities and DoE they serve that have generated nuclear waste for profit, or national defense (when was the last time we dropped an atomic weapon anyway?), be absolved of their liability for said waste by burying it, transformed though it may not quite be, in 'potentially leaking solid waste landfills. The burden is borne by local residents and taxpayers because the added radioactivity potentially increases the risks posed from leakage into ground, surface and drinking water, from use of the landfill gas if radioactive gasses form and from synergistic effects with hazardous or other chemicals in the landfill.'[40]

You would expect your government to protect you, but according to the authors of 'Out Of Control' the DoE 'appears to play more of a role in assisting local sites in preparing defensible data to allow releases than preventing releases or unnecessary public exposures.'[41] By using computer codes instead of actually measuring radioactivity of releases, the false appearance is given that everything is hunky dory, that you are safe, you and your family. Then, when some material is given legalized 'clearance,' via these unvalidated computer codes, it 'is no longer recorded as, labeled or considered radioactive.'[42] Even if the code does not adequately consider different effects of different types of radiation on differently susceptible people.

Babies, young children, the elderly, women, those with compromised immune systems from cancer or multiple sclerosis or odd new types of viral infections, and let us not forget fetuses, and their pregnant mothers, will <u>not</u> react the same way the average seventy kilogram (154 pound) healthy "Standard Reference Man" will to radioactive exposure. Especially if possible concurrent interaction with chemical and other contaminants is ignored, that might happen at your local dumpsite or nuclear facility.

Yet the supposedly credible International Commission on Radiological Protection (ICRP) has inappropriately used Standard Reference Man as its basis for new worldwide recommendations 'that have just been adopted [that] will result in relaxation of many of the [radioactive] dose standards.'[43]

Also, this same ICRP, along with the International Atomic Energy Agency (IAEA) that does the 'inspections' of nuclear facilities from Russia to Rumania to Iran, amongst many of its other pro-nuclear tasks, plus EURATOM (the European Atomic Energy Community) 'on behalf of the European Commission have chosen risk and contamination levels that they consider acceptable and called them "concensus."'[44]

Thank you.

Remember that these three international organizations are 'self-appointing nuclear advocacy groups. Their function is to create recommendations that form the basis for national laws and regulations that allow the government and private industry to engage in nuclear technology. They do not represent those who are exposed, and their committees, processes and reports are exclusive, generally closed from public participation. When public comments are sought, the public's recommendations are regularly ignored, unless they are from the nuclear industry.'[45]

Very much like the World Trade Organization (WTO) and its court's workings on trade disputes, as discussed in Chapter Four.

D'Arrigo and Olson close out the text of their 'Out Of Control - On Purpose' telling us our own NRC uses the recommendations of the above trio of international pro-nuclear groups 'to overcome opposition to unsavory radiation rules.' They warn us to 'watch out for and help challenge the U.S. adoption of the ICRP recommendations, instead demanding greater protection and a goal of preventing unnecessary radiation exposures.'[46]

Before you call your Senator or Congressperson at 202-224-3121, and 202-225-3121, respectively, to voice your opinion and concern about unethical releases of still-radioactive material as if they were not at all of any concern, just merely 'trivial,' you should know one typical story that was a quietly major victory for the nuclear industry, to illustrate what could happen, and what has happened, relative to our theoretically neutral 'protective' agencies vs. the American people.

After the BRC or 'Below Regulatory Concern' battle was lost, in the early 1990's, basically, Mr. Smith went to Geneva and Europe. Actually, it was folks like Don Cool (real name), an executive with the NRC when it attempted to increase allowable radioactivity in U.S. air, soil and water. Mr. Cool and other NRC employees crossed the Atlantic Ocean to help rewrite the international rules on transportation of radioactive goods, and radioactive exposures that could be permitted. Then these international rules could be imposed on the USA, under the guise of 'harmonization.'[47]

Nuclear Regulatory Commission (NRC) Region IV Headquarters. NRC executives like Don Cool dedicatedly labor in such modern surroundings to lessen restrictions for releasing radioactive wastes from monitoring and scrupulous control.

When we talk about 'harmonization' here, it is not musical or spiritual harmony that we should envision, giving us calming joy and unity. No, this is the new millennium's industry and corporate harmonization that must be accepted to deviously <u>lower</u> international standards. As employed by the World Trade Organization (WTO) to force us to accept more pesticides and poisons on our foods; more radioactivity into our environment; lessen restrictions on foreign corporations gobbling up natural resources of other, usually poorer, countries; superseding our own and other nations' democratically enacted laws. That kind of one-standard-for-all (lower) 'harmonization.'

The big agent for this international harmonization relative to things nuclear, shalt be the International Atomic Energy Agency (IAEA) that we just mentioned a few paragraphs ago. It has been called a 'quasi UN agency,'[48] and has been a nuclear promoting agency since its creation. It has agreements with other UN agencies like the World Health Organization (WHO), the Food and Agriculture Organization (FAO), the International Maritime Association (IMO), and the International Civil Aviation Organization (ICAO), that no statements on nuclear levels, issues, etc., will be made unless approved or made by the IAEA first.[49]

Remember, our own Nuclear Regulatory Commission (NRC) was originally named the Atomic Energy Commission (AEC), which initially was formed to promote the use of nuclear energy in the United States. However, when regulation seemed to become more important than blatant nuclear promotion, the AEC was disbanded, and the NRC took its place.

What eventually happened, with diligence and persistence, behind the scenes, the media not reporting any of this to most of you, is that international regulations put forth by the IAEA have quietly been approved by the International Maritime (IMO) and Aviation (ICAO) agencies, and then, by our own Department Of Transportation (DOT), plus the NRC. Suits against weakening exemption levels for a majority of radionuclides were dropped by U.S. courts in late 2006 to finally allow this to transpire.[50]

Specifically, a new chart in an IAEA document called 'TSR-1' added two columns to an older document ('SS6') that exempts radioactive quantities *and* concentrations on an individual radionuclide-by-radionuclide basis, re nuclear transport. In other words, e.g., a separate concentration is proclaimed to be 'safe' for 'clearance' - - meaning its shipment would not have to be monitored - - for plutonium-239, iodine-129, strontium-90, cesium-137, etc., for each of about 400 radionuclides, when each of these are not separate in the radioactive shipment necessarily, or usually. And, it's so phony, the nuclear scientists have not developed a realistic way to measure the radioactivity of each radionuclide separately yet anyway. But one day they will, they promise.

What had been standard protocol in our country was that if radioactive shipments were tested, only those measuring less than 70 becquerels per gram of radioactive material were deemed exempt from transport regulation. (A becquerel equals one disintegration or radioactive emission from a nucleus per second.) If that level was exceeded, the shipment had to comply with transportation regulations (such as labelling and proper containerization). However, in January of 2004 our Department of Transportation and the NRC went on and approved TSR-1.

Now, post-suit-droppage, thanks to Don Cool and his Confederates, and the good ole IAEA, we have an increase in exempt or 'clearance' levels for the majority of radionuclides, to numbers as high as 1,000,000 (one million) becquerels per gram for elements like Argon-37.

A shifty component of TSR-1 is that while some clearance *concentration* levels do go down, *quanti-*

ties allowed of the same radionuclide are relatively high, enabling a shipper to regulatively circle around the concentration level allowable.

What does this mean?

Let's use plutonium-239 as an example, with its radioactive half-life of 24,000 years, and its one microgram lung cancer-causing dose. TSR-1 has lowered its allowable exempting *concentration* level from 70 becquerels per gram to 1 becquerel per gram. So, <u>lesser *concentrations*</u> now require labelling and shielding and protective measures, when compared to before TSR-1. In other words, now any plutonium-239 shipment with a concentration of 70 becquerels per gram down to 1 becquerel per gram is no longer exempt from protective measures, as these concentrations used to be. That portion of TSR-1 for a dangerous radionuclide like plutonium-239 seems good and logical.

However, there is the backdoor trick of the *quantity* allowed, of a 'consignment.' For plutonium, that happens to be 10,000 becquerels total, per 'consignment,' in TSR-1. In other words, you could still ship up to 10,000 grams of plutonium-239 in one deregulated 'consignment,' even if the shipment's *concentration* happened to slightly or greatly exceed one becquerel per gram.

'Consignment?' Let that mean 'Any package or packages or loads of radioactive metal presented by a consigner for transport,' as defined by TSR-1.[51] The problem is, the NRC has kept things very vague; there is no limit set on the number of shipments per 'consignment.' Plus, one truckload could have more than one 'consignment.' So, if you want to ship a lot of plutonium, you could evade the *concentration* limits that have been erected, and just play the 'consignment' game relative to the *quantity* of radioactive goods shipped! Very devious, and very WTO!

There is research going on currently to see how many millirems a consignment of radionuclides like plutonium-239 might give our truck drivers, and just plain folk. It could come out to somewhere between 50-100 millirems per year, which would be rather high, adding quite a bit to our usual 'background radiation' totals, which may really be 100 millirems in most of the USA.

Also, be aware that the exposure to plutonium-239 would just account for exposure to only one of the hundreds of radionuclides that we could be exposed to by this TSR-1 deregulation that could give many more of us and our loved ones cancer.

Although some countries like Germany and England are adopting their own variations of the IAEA 'clearance' standards, the day to day reality of testing for radioactivity is not consistently or faithfully followed on most occasions in most countries.

Unfortunately, the idea of running a Geiger counter over the surface of a container of potentially radioactive recycled cement or soil or asphalt or plastics or metals from a nuclear power plant is facetiously deceiving, because that does not measure what is deeper inside the container, or deep within the surface of the substance being shipped. Any radioactive monitoring techniques also can miss the surface contamination and hotspots, if the employed equipment is not calibrated and used properly by trained professionals whose goal is to find contamination.

According to Diane D'Arrigo, the nuclear industry prefers trying to not find contamination, and has no incentive to have any measurements that are made to be done so in an accurate or honest fashion.

When theoretical doses are calculated by computer model projections, and unlimited numbers of releases of (perhaps) one millirem lodes are allowed in the USA and worldwide, our environment and civilization is threatened by unconscionable radioactive pollution.

Yes, it would be terrific if you could put a dosimeter badge on everyone on planet Earth, as we do with our x-ray technicians in our hospitals and labs, to register each citizen's daily and cumulative radioactivity exposure. But this is not practical or possible.

More nuclear power plants is not the answer to our survival in this world. The radioactive waste generated by this technology has certainly not proven to be 'green,' from an industry that is not economically viable without government subsidization. [Watch the latest bid for federal subsidies of up to $550 BILLION dollars to build new nuclear plants, which the nuclear industry claims is necessary, for otherwise these new

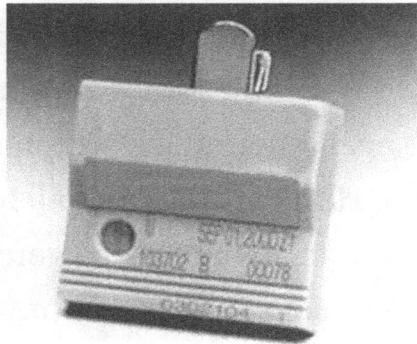

Dosimeter typically worn by our x-ray technicians and those who work with radioactive materials. When exposure adds up to more than 5000 millirems, the worker has to refrain from further exposure. However, it is preferred that worker exposure be kept below 500 millirems total per year.

nuclear plants cannot be built - - because utilities could not entertain the risk of making money from them without such outrageous socialized corporate welfare. These monies will most likely come via the *amendments* to the next version(s) of the 'Climate Security Act,' as sponsored primarily by Sen. Lieberman of Connecticut.]

The future is here now for wind and solar and hydrogen and energy efficiency technologies.

You know from reading this book that countries like Germany are building the near-equivalent of one nuclear plant every two years in wind turbines to generate their electricity safely, without radioactive pollution. Germany and several other European countries are phasing out nuclear power.

Have you heard about Iceland's buses being greater than 50 percent powered by hydrogen fuel cells? Do you know that more than 20% of Denmark's electricity is generated by wind power today?[52]

If you check MIT's *Technology Review* magazine, you can discover that there are new 'printable' cheaper lightweight solar strips being researched by the nanotechnology industry. Within a year or two they will be in your laptops and cell phones, and thereafter in your factories, homes and businesses to help with thermostats, and other devices that utilize sensors to function most efficiently.

Don't be surprised that within the next decade or two, your house, office building, vehicle or that billboard down the road will be spray-painted with photovoltaics that are hooked up to electrodes that connect to the electricity grid.[53]

In other words, despite most of us being blanked from the reality by our delinquent megamedia, our spirit of invention and entrepreneurship is ready to take us past the fossil fuel and nuclear fix to provide our energy. But the nuclear, coal, oil, and gas conglomerates don't want us to realize this. Our corrupt politicians are pushing the wrong agenda, thanks to all those corporate donations and all-expense-paid junkets provided by those whose interests run counter to those of most of us plain old vulnerable American citizens.

And how about this possibility, as suggested by Ace Hoffman of southern California: today there are approximately 200 million cars on the road in the USA. Ford made about 2.5 million vehicles in 2007. Automobile plants and other manufacturing plants have closed and will be closed in the future about our country, putting millions of skilled American workers on the unemployment line.

Why not fund a massive government sponsored nationwide emergency manufacturing of wind turbines at these closed/closing plants to provide our electricity? The two most currently popular larger wind turbines are the 1.5 megawatt General Electric (GE) version and the 2.5 mega-watt Clipper model. The European Wind Energy Association (EWEA) estimates that on the average, a 2 megawatt wind turbine can provide the yearly electricity for 1375 'households.'[54] Our American

If we build 112,222 2.5 megawatt Clipper wind turbines with a similar World War II emergency type of effort today, we could supply the electricity needed for ALL of America's homes! Ford alone in 2007 built more than 2.5 million vehicles, and manufacturing plants continue to close across America – plants that could be used to solve our energy problem without nuclear or fossil fuel pollution.

Wind Energy Association (AWEA) estimates that 270 'homes' can be supplied with their yearly electricity per megawatt of wind energy.[55]

Using the AWEA estimate here, allow us to avidly manufacture the 2.5 megawatt Clipper please for our present population of 303 million people. Estimating four people to the average 'home,' at 675 homes supplied per Clipper, we would then need to manufacture 112,222 wind turbines to provide the electricity for all our homes in the USA![56] What if we quadruple that or quintuple or sextuple that to include our business and government needs! To say one million wind turbines. Why, Ford made 2.5 million vehicles as just one of our national automobile manufacturers in 2007.

And if we think a bit more futuristically, as Mr. Hoffman suggests, and build 10 megawatt wind turbines, which very likely will be developed as supply meets demand and ingenuity, the number of wind turbines needed would then be 28,889 for all of America's homes. Couldn't we make whatever number of those wind turbines would be required with the will we industrious Americans have, when Americans need jobs, jobs we can be proud of, especially in a time of emergency energy need, just as Americans did, for example, during World War II? Our economy

Ace Hoffman, citizen activist and nuclear archivist. Envisions a world with NO nuclear reactors, powered by wind and solar, and other renewable energy resources, possible right NOW! not in ten years - - as described in these surrounding pages.

does depend on us making wise urgent choices to right the financial ship, that seems to be sinking as we errantly continue on our blundering current politically-mottled course.

Donald Rumsfeld required a catastrophe or 'Pearl Harbor' (9-11) to get his 'Project for A New America' jump-started toward weaponizing space (See Chapter Three). We need not have that next catastrophe happen (a Chernobyl nuclear accident in America) here before we get smart and do what Mr. Hoffman pictures we could do urgently to start ourselves on a bright renewable energy road to self-sufficiency without producing eternal radioactive pollution of vast regions of our country.

Then, we could close down all our nuclear plants, as Mr. Hoffman knows we could and can do in the very foreshortened foreseeable future.[57]

What Can You Do?

Yes, you can do many things. Start by calling your Congressperson and both Senators in Washington, D.C., via telephone numbers 202-225-3121 and 202-224-3121, respectively, and tell them how opposed you are to deregulating radioactive waste. Also voice your opposition to importing and dumping radioactive waste from foreign countries within U.S. borders, making America the dumping ground for the world's nuclear waste. Plus, warn them about Senator Lieberman's so-called 'Climate Security' bill that could sneakily authorize up to $550 BILLION in subsidies to build more nuclear plants, via <u>amendments</u> that will be added to the bill – whenever whatever form of the bill gets re-introduced in the Senate all too soon. Say that you vote and you are flatly opposed to such outrageous subsidies.

Then email our delinquent agencies responsible for these travesties by going to www.nrc.gov and www.epa.gov and www.doe.gov to vent your feelings.

Here is a sample resolution that is currently being circulated that your elected representatives and you could support concerning deregulating radioactive waste: (on next page)

Resolution Against Radioactive "Recycling" and Deregulation

Be it resolved that the undersigned entities hereby support a prohibition on deregulation of radioactive wastes and materials.

This includes any and all deregulation of radioactive wastes and materials for "clearance," "release," "recycling," "exemption," listing as "below regulatory concern," or any other legalistic mechanism that could result in the dispersal of nuclear wastes and materials into public commerce, unlicensed disposal, or designation and treatment as non-radioactive.

Such practices pose an indefensible hazard to public health and the environment for current and future generations.

Since there is no safe level of ionizing radiation, nuclear power, weapons and mining wastes should not be forced on an unknowing, unconsenting public.

Radioactive waste should not be treated as an asset or a commodity, and must be contained and isolated from the public and the environment for its entire hazardous life, at the expense of the waste generators.

Contact: **Diane D'Arrigo, Nuclear Information and Resource Service: http://www.nirs.org 301-270-6477**

Also, contact General Electric and Westing-house [now owned by Toshiba], and tell them you don't want them to make any more nuclear power plants. Check your mutual funds and see which ones include these two corporations. Divest, if you discover them as part of the package. Don't buy any products made by any of these companies.

By the way, General Electric is also included in companies researching the new solar stripping technology I mentioned above,[58] and they are the majority owners of the NBC television network, and several other networks.

You don't have to sit back, and be a paralyzed couch potato. Knowledge is power. Go out and use it! Do you think it might be your <u>duty</u> now to share what you have learned from reading this book, with other people who would like to survive this corporate messapotamia our corrupted leaders have dysfunc-tionally erected to ruin our civilization??

Be aware that you are probably feeling akin to many other Americans who have given Congress its lowest approval rating EVER, dropping it into *single digits* for the first time as reported in a July 2008 voter survey.[59]

America is still a democracy, despite the trav-esties we are experiencing that YOU as a rightful citizen can reveal and reverse if you just bound up from your passive role and actively intelligently passionately express your discontent.

Plus, show the positive side of what can and should be done by our leaders, who should be obligated to do what we think is proper and sustainable for our country and our children's heritage.

And, you unfortunate folk from Tennessee. You might ask some very crucial questions to prevent the further radioactive pollution of your nuclear facilitating state - - as suggested by Mary Olson and Diane D'Arrigo:

> Under what authority is nuclear waste removed from control in [your] state? How much radioactive waste is coming in; where is it coming from; what is happening to it; where is the radioactivity going; how much radioactivity is being dispersed into the state's landfills and natural resources? Who guarantees compliance with landfill regulations and quarterly reporting for landfill disposal and other expressed provisions for special (radioactive) waste release and disposal? What efforts are being made to verify the claims about safety? What public education, regulatory and legislative efforts are needed to resume controlling, rather than releasing, radioactive waste in the state?[60]

Further, the authors of 'Out of Control - On Purpose' urge that:

> Follow up is needed in Tennessee to determine whether the public wants to remain a nuclear waste destination for much of the nuclear power and weapons fuel chain and what authority justifies the additional radiation doses to people now and in the future.[61]

So, please do ask the necessary questions. For the people and animals and plants of your state. To help all of us deal with this irascible problem right now to save the sanctity of our future.

Unfortunately, you folks of Utah also are being betrayed by your representatives, and other forces united with the nuclear industry as of mid-2008. As mentioned earlier in this chapter and book, EnergySolutions corporation wants to import 20,000 tons of radioactive waste from Italy; and now, the latest, along with Studsvik corporation, is ready to import large amounts of radioactive waste from the United Kingdom (U.K.). The latter will come from clean-up contracts of nuclear power and weapons sites in the U.K.

Though the Italy waste would enter the country via ships docking in Charleston, South Carolina and ports in Louisiana, and head for Tennessee to be re-classified, whatever is not dumped in the Volunteer state will probably end up in Utah. The same story and routing will probably apply for the U.K. waste, and future radioactive waste from other countries that could arrive in the USA.

Utahns are protesting, but according to a Salt Lake Tribune editorial only Representative Jim Matheson 'has answered the call to action.

Matheson has introduced legislation [HR. 5632] to ban the importation of low-level radioactive waste, and wisely reserve our limited disposal space for domestic waste...But Utah Senators Orrin Hatch and Bob Bennett, and Reps. Chris Cannon and Bob Bishop, all Republicans flush with campaign contributions from EnergySolutions' political action committee and company officials, are witholding their support.'

Foul and unfair as this may sound, the editorial warns that the NRC will not stop this importation, but only 'rule strictly on the technical merits of the request...So, it appears that it will take an act of Congress to stop the plan.'[62]

As you can see, with transportation and encouragement, instead of blockage and denial, it will not only be Utahns and Tennesseans who will be affected by this incredible complication of our nuclear energy push. South Carolineans and Louisianaians, and citizens from all the states in between dockage sites (and what about a transportation accident at sea, along the coastline, for example?) and Utah could be at risk.

Again, a major point to be emphasized in your phonecalls, letters, emails, conversation, etc., dear readers: your condemnation of this bold attempt to make America the dumping ground for the world's nuclear waste. (You might mention that the Governors of Utah and Wyoming <u>have</u> come out against EnergySolutions' plan.[63])

Final Thoughts Concerning Radioactive Waste and Nuclear Power

One last quote from 'Out Of Control' to put the reality of nuclear reactors and other nuclear facilities in perspective, re 'cleaning up' the sites for these:

> Clean-up in the true sense would have a goal of capturing and isolating ALL of the waste and contamination generated by the processes. If this is not technically possible, not reasonable or practical, as most contend, then building nuclear facilities is effectively creating sacrifice zones – labeled or not. Further, the infeasibility of a real clean-up should be admitted *before* any new nuclear facility is opened. This information is rarely, if ever, provided when new nuclear sites are proposed - - in fact, contamination is often denied by proponents.[64]

To really be practical then, in comparison to mankind's relatively short existence on this Earth, think: Radioactive waste is forever. And we must contain it safely, responsibly, for that long, since we made it. Ask yourself and your neighbors if we need to generate seven tons of high level radioactive waste <u>every day</u> in our 104 nuclear reactors, 2,500 tons <u>every year</u> (multiply those amounts by approxi-

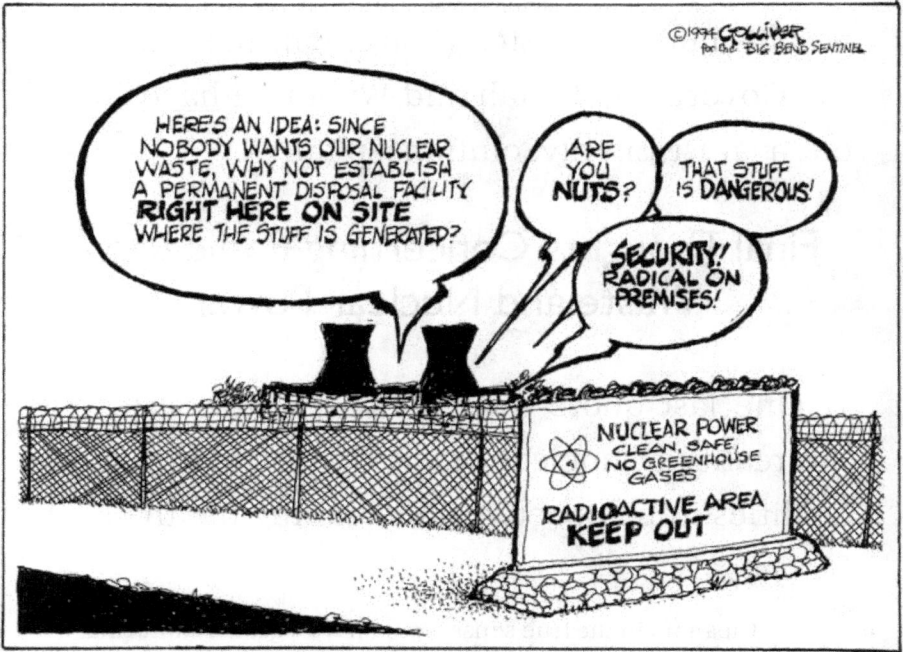

mately five to get worldwide totals)[65], any more to threaten our health and that of future generations? We have viable forms of alternative energy available today that we can develop and implement certainly within a decade to replace the ultimately poisonous nuclear reactor that belongs in the Devil's scrap-heap of obsolete technologies, along with other polluting fossil fuel technologies.

This is our country. We must wake up our politicians to spend our money in the best way possible - - not feed it to an industry that should be closed down - - before we have that next Chernobyl, when it will be too late to pretend that nuclear accidents don't happen. The nuclear reactor belongs with the Edsel and the pet rock in the Museum of Fickle Inventions. We need to slow life down a bit, use a little restraint in consuming the latest fad that scrolls across our screen, before we pollute ourselves into infinity because of bad decisions based on greed, power and fear of not having every last thing we think we must acquire to suc<u>ceed.</u>

Chapter Six:
Preserve Your Dignity, American

As commercialism and the corporate way sweep their web wider and wider, across a world where the individual seems to have less and less control of what is happening to his/her life and the direction of our country, the spirit of the individual, the dignity of the individual, is something we must preserve.

Though the 'War on Terror' may invade our personal space - - frisking and x-raying us on security lines at our airports, or spying on what we may type into our internet communications or say on the telephone - - we still do have our inner *dignus* or worth. As defined in Webster's New

Expanded Dictionary of 1990, dignity is 'nobleness or elevation of mind; honorable place.' Back in 1900 Funk and Wagnalls Standard Dictionary mentioned 'repose or serenity of demeanor,' as part of its first definition of dignity.

Can we hold our head up high, chin out, and walk gracefully ahead, while we accept insults and disinformation (lies) like scaly reptiles, or ducks letting water slip off our teflon-esque backs? We work all our lives, thinking America will maintain its commitment to that dream of justice for all. That what we earn, like our pensions and accumulated benefit time, we will keep. That long hours of labor will be fairly rewarded. That what our bosses and leaders tell us is the truth, will not be subverted, or wiped from the slate, because of 'unavoidable' catastrophes. Their word is their bond. They will behave honorably when called upon to act in our behalf.

Yet, look how we find ourselves today, having to scrape and adapt to 'the marketplace.' Enron and Global Crossing and WorldCom executives plunder the funds of companies that our financial experts have faithfully invested our future into. The companies crumble, our pension funds and stock portfolios are decimated, what we worked for all our lives is halved or worse. Some of us in our 60's and 70's have to go back to work settling for a menial minimal wage job, suffering painful indignities, bossed around by 20-something-year-olds who barely appreciate our experience, or the wisdom we have gained.

Or we have to navigate through the field of 'down-sizing' as our hospital or business falters due to compromised income. A big corporation has bought us out. They have to cut staff, eliminate benefits, decrease our hourly wage when we were told this never would happen.

What can you do? Where can you go? You have three kids. They have to go to college. Or, you thought they had to go. The cost is now $150,000 for four years, on the average. You may have to take another job, or two. . .

How will you maintain your health if you over-extend yourself? Where is the time to find that 'repose or serenity?' So you can maintain your dignity of honorable demeanor? It seems that just about everybody is hassled. Even the bosses are afraid *their* bosses will give them the shaft.

Some 9 year old girl working 16 hours a day in China for a total wage of pennies in our money, may take over the work you've been doing with pride for thirty-three years. And people will go to Walmart or Chainstore Gala-La to marvel at the bargains there: cheaper, usually inferior products made by those foreign slave-labor little girls.

Many of us may not be thinking that shopping in these mega-stores continues and heightens the subversion of our economy, the worth of our own labor, that quite often is a labor of love. When we work, we want to have pride in what we do, and be appreciated. We all need positive re-enforcement to help us continue along our meaningful trail through this Earthly life.

Then you see people like Little Kim tramping around in her underwear on television, selling

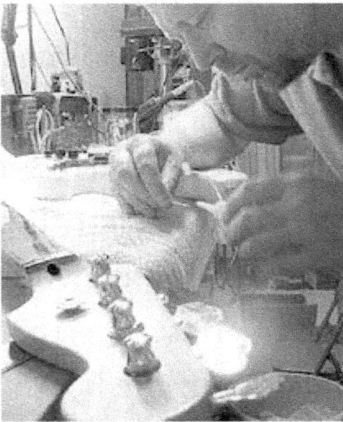

A labor of love. Commercial American Gibson guitar craftsman. Many lost their jobs when Kalamazoo, Michigan factory closed, but company re-organized in late 1980's, making guitars now in Memphis, Tennessee and Bozeman, Montana. Kalamazoo Crew of ex-Gibson guitar makers scrambled to create Heritage Guitars in their city.[1]

her soul. As do so many of our more extroverted desperados compromising their dignity for the dollars and exposure they consider as gain over-all - - especially when those bank tellers docilely receive their ample deposit envelopes. But you blabbered loudly like a fool for A T & T, or put your honor on the line for corn flakes made from a genetically modified crop, laced with biotech-matched herbicides (banned in Denmark[2]) that may lodge in your fiance's breasts for untold years, possibly causing her cancer one unexpected day.

Nah, it didn't matter that you said you used Viagra when you were hitting those 500-plus home runs on the way to the Hall of Fame - - even though you were still in your late thirties. Weren't you embarrassed to confess that you needed that chemical to strengthen your sexual worthiness, for all the world to see? Or are you just lying for the big bucks? And what you say means nothing? Your honor a fleeting trail of moments that mean zero, both in the aggregate, or from one thirty-second span of advertising to the next, when you are vacationing in St. Barth's, or Aspen or Vale? (And your path to the Hall of Fame de-railed....temporarily??...)

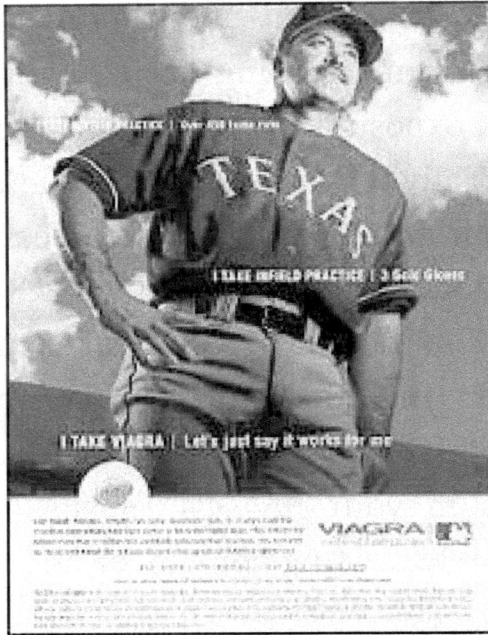

30-something All-Star baseball player Rafael Palmeiro promotes use of Viagra. Apparently lost his shoo-in Hall of
Fame status due to steroid-use controversy and possibly lying
to Congress.

Does our life mean anything? Does God direct
everything? Or Allah? Or Buddha? Can what we
do have any affect on the flow of power in the
halls of Earth's White Houses and palaces? We're
just insignificant cogs in an unimaginably massive wheel that grinds us into impersonalized
chopmeat. We eat the factory farmed cattle-steaks
tinged with hormones and shackled existences.
We support all that with our choice of buying at
McDonald's and Wendy's and I-Hop and Sizzler. If
we ate organic rice, and other organic grains and

foods, and told our children and our friends that they vote for cancer and poisoning of the planet, and destruction of the rainforests, when they spend their hard earned money on inorganically raised meat, would that assist the general survival and sustainability of the world? Or would it just be a silly waste of time and effort?

Does our wisdom have any influence over our children? Do the choices we make in the foods we buy for them and advise them to eat make a difference in the master picture of the planet? Should we maintain our dignity and do what is right, even though others behave ignorantly, and laugh at our Don Quixote behavior?

Well, Germany and Belgium are embracing windmills and windpower. They are abandoning nuclear power, defying the publicized craze that appears to be conventional wisdom. While we Americans have to swallow the insulting lie that 'nuclear is green,' spread like nerve-numbing pesticide over the screens of our national media. And our champion plunderers, our politicians, sacrifice our plight to meet the pay-back for those exorbitant campaign contributions that got them into office and/or keep them there, doing all the wrong things. Impaling our dignity, our intelligence, with their compromising behavior.

When those corporations donate a million $$$, they often profit a billion $$$,$$$ for what really is an investment. 'They' are not even human, accountable entities. Their humans that work within their structure are sacrificeable, easily changeable parts, from the janitor in the 70 story city building to the CEO taking his perks before his reign crashes and burns. Yet the CEO still gets his 'golden parachute,' that he expects can float him beyond polluted seas to that paradise beyond care, filled with golf courses and attractive masseuses. . .

The talking heads, the chiefs, the bosses, seem to speak with dignity and honor when they promise profits and the plumping comfort of wealth inevitable, if you continue to steady the course they have to

The late Ken Lay of Enron, listened-to champion of deregulation - especially by Bush administration. Misled investors and workers countrywide before's Enron's calamitous collapse. Died before society could exact its punishment on him for his ignominious crimes.

follow. Some maintain it - - though their wives divorce them when the sham is over - - claiming whatever they dare to anchor their insecure estates, no matter the sum of the value.

We Americans do not want or like to bow to some Pharaoh or Stalin or Hitler or Saddam Hussein. We are Americans! The strongest and surest *homo sapiens* on modern day Earth! We want to remain Number One, at the top of the pile when Survivor is done.

OK, we may have to undergo certain indignities to reach the pinnacle of fame or progress. But we have always been told that perseverance and dedication will get us to our goal. Whether it be peace in our hearts, or profit amidst the sharks. Life, Liberty, and the Pursuit of Happiness. Yes, that is what it says in our magnificent Constitution. That is definitely a great part of what we want. For ourselves and our children.

We should be able to avoid what is 'coarse or trivial'[3] as we attain that dignity that we deserve as we advance through our lives. We want our country to lead the way to democracy, here at home, and across the rest of the world. Yet some of us are arrested as the war on terrorism drags on, and are unceremoniously placed in jail without the due process and *habeas corpus* that makes America America. We have to rationalize away places like Guantanamo Bay, and the reality of 'extraordinary rendition' allowing torture to be acceptable and preferable for the American consciousness, when we always have championed the cause *against* torture throughout the world.

All this in the dimmed light of our media downplaying real news like the so-called 'Downing Street Memo' wherein 'Richard Dearlove, former head of British intelligence, told [Tony] Blair that [George] Bush "wanted to remove Saddam Hussein through military action, justified by the conjunction of terrorism and weapons of mass destruction. *But the intelligence and facts were being fixed around the policy.*"'[4] (My italics.)

Aren't our leaders supposed to be faithfully speaking for us? Or are they so corrupted, they don't care about our dignity or our integrity, or our democracy that should be a model for 'freed' societies like Iraq or Croatia? That we strive to strengthen each day by day as we make our decisions on what to do, what to say, how to act, what to buy, what to decry.

International justice has always held our nation in great respect, as we stood for the right of the individual, the dignity of mankind, shoring up the foundation of these all-important bases for modern civilization during most of the past two centuries.

Yet, we must ask what has happened to us over the last few decades? Are we beginning to out-Russian the Russians? When Stalin was the revered murderous leader of the USSR, and he entered a room or hall, or gave a speech, everyone was afraid to be the first one who stopped clapping for him, as this might indicate disapproval of the Great Leader, possibly leading to hostile interrogation, exile to Siberia, and/or death.

What about today's younger generation, desensitized by our all-too-common public calls for 'regime change' in other countries, or 'pre-emptive strikes' against our latest enemy of the moment?

Yes, many of us are insulted by the behavior of our fawning or aggressive media front-persons, and our arrogant, militant leaders. They are supposed to be representing us, speaking for us, acting as our watchdogs while behaving honorably to ensure that the world is a better place for all of us, and future generations of Americans, and humans. Is it OK to double the size of Guantanamo Bay, like Mitt Romney frothed to favor? Is it OK to start an arms race in outer space so our military can control the Earth below, for the multi-billion-dollar profits benefiting our biggest industrial export industry, our weapons industry, and the corporations that make these weapons? to protect those un-human corporations that anti-democratically rape and plunder other nations, aided by the World Trade Organization-tainted New World Order, as poorer less powerful nations are unfairly gouged of their natural resources, their labor forces subjugated to the lowest common denominator, the corporations feasting on minimalizing wages, erecting new sweatshops across third world countries that *should* be improving in this modern world of the twenty-first century?

Don't we want peace for our children? Don't we realize that the manner in which we treat others, who actually are our brothers and sisters, and our children and our grandparents, regardless of their nationality or ethnic grouping, affects our own dignity? Because we have to do what we do, ignoring our feelings for what is wrong and right, as we hurt or maim our inner selves just as much as we hurt or maim others with our actions and policies that we have to continually rationalize and re-rationalize.

How long can we suppress our intelligence, our senses? No thanks to our media, who demean leaders and perceptive thinkers from outside the immediate American sphere-of-influence, sophisticated world leaders and thinkers, who have been taken aghast at what we have been doing lately.

Maybe our economy will soon suffer a terrible devastating blow that even the rich do not want to entertain imagining? Witness the banking and subprime mortgage foreclosure crisis threatening us today in 2008.

The subprime mortgage crisis threatens the financial well-being of our nation as the first decade of the new millennium forecloses on too many of us Americans.

Meanwhile, we have to plan ahead, re-arrange our portfolios. Our dignity and wealth must be preserved, if we are to continue living at our current expansive (though frighteningly starting-to-contract) lifestyle. As the rest of us watch the pretty women drive their SUV's, and the handsome movie stars handle the next greatest product made of colorful plastic whose manufacturing process fouls the rivers of beauty or the air of our Texases and Gary, Indiana.

Schizophrenically, we feel proud that we have beaten down our enemies with our immense defense prowess, and its increasingly burdening budget to match. ($\sim$50% of our federal budget now goes for 'defense,' while health and education each get a few meager single-digit percentage-point shares of it.) Too many of us fear we may lose our jobs at any moment, yes. That we might be crawling and fawning when the next World Trade tower falls. But how can we trouble our tempestuous natures, when blind arrogance balloons our biggest buffoons? Which all too pompously, we realize, are us. We are seeing how great we are, as portrayed by our media barons and manipulators and pawns, while our dignity is plastered with ominous dander and dandruff that sticks to us as we stumble on a cracked sidewalk, watching our schools fall apart, as our children ravenously devour synthetic crap we label as food so we can continue filling our fragile checking accounts, the war against cancer failing because more radioactive elements and poisons are being dumped into our biosphere, and our health insurance coverage is abruptly terminated.

Do we want to live so desperately in a world that is deteriorating under us, that we will do just about anything to insure our next day's survival, no matter how ragged and tattered this makes us feel?

The television filters make our skies appear so blue and irrepressibly optimistic . . . How could we dream that there could something amiss in our saunter through our existence?

Is it necessary for us to be careful to create what is good and healthy, rather than sacrifice our souls and environment, Earth and surrounding atmosphere, the water and air we breathe, to ride roughshod across the adversities we deny to satisfy our lust for power and profit? Must we move so fast with our decision-making in order to stay one step ahead of the crowd, the apocalypse, our competitors, death's inevitable claim on our bodies that must wear out? Hopefully, before our minds are Alzheimered, or our souls surrender to what we dreaded in our idealistic youth.

Some of us want to live so badly, we try to evade death, yet ignore the tragic complications that might overtake our human protoplasm. We cross the line of dignity to indignity by having that stroke and getting intubated with a hollow tube shunting air into our lungs from a respirator that we thought would 'save' us. However, if we ever regain our senses, we might gather that we cannot breathe without it anymore, that our family has 'helped' us by having a surgeon insert a feeding tube into our belly, and we have to wear a diaper from now on because we no longer can control our elimination of our wastes.

Or perhaps we cannot remember much anymore, we have to be taken care of by our children, our hands held when we totter across the boulevard, by those precious family members who have to laugh at our demented behavior to keep from crying and being depressed about what has happened to us. Our dignity demeaned by our frantic fear of becoming one with nature again. Death breeding life. Life coming up out of the cells of what has lived on Earth before. We here on this planet for just a short walk. When we want to be happy forever, never subjugated by others, or the woes of life.

If we do the right things, visualizing others as having just as much claim to the dignity of an honorable healthy normal un-artifically un-over-extended existence as we should have, and we treat them as we would want to be treated ourselves, then we can preserve the dignity of all humankind.

Unfortunately, as our current government ignores what made our country great, we much too frequently have to experience not a government of the people, by the people, and for the people, but instead, a government against the people, over the people, and around the people. Witness the unpublicized sneaky maneuverings toward a North American Union that could subjugate our democratic freedoms for the benefit of the corporations and military mightists as described in Chapter Four. When our government is supposed to be granted its existence to represent us honestly and faithfully by the very American citizenry we all are.

Some of us who meditate on our alternatives, our joys and sorrows, gather that each moment indeed is important. That what we do, each one of us, does make a difference. In how the world continues, in what our industrial machines make, in how we protect what is sacred.

Breathe in knowing you are breathing in.

Breathe out knowing that you breathing out.

Then you will be in the moment, experiencing it as deeply as you can. Life is not a conveyer belt to nowhere that you can't get off of, no matter how you try. Each moment is like a footstep that you can take in almost infinite directions, with almost infinite consequences, and rewards.

Others may act like boobs, following the lemmings off the cliff. But YOU can behave differently, because how you live your life is the most important thing you can do, hopefully affecting others around you by the gleamings of your best behavior. As you maintain your dignity and integrity, proceeding proudly by choosing the best alternatives through your time on Earth, you may become fulfilled and happy, knowing that your contribution to man's history and culture has helped preserve the civilization we all want to prosper and develop in an advancing, non-destructive manner.

Speak out and speak up when you can assist those who do not know the facts that you have gathered with intelligence and sensitivity. Spread your good feelings, so others will gain trust in humanity, instead of selfishly stealing and hoarding whatever they can, unsure if their wealth will be protected from what they may perceive to be a government of lying thieving power-hungry demons.

Dignity and honor are two basic traits that we must breed and strengthen to keep man from destroying himself/herself, en route to taking the higher fork in the road toward being the best that he/she can be. And right now, with America being the most powerful country in the world, what each of us Americans does may carry a bit more weight than what a poor humbled person in some totalitarian backwater of this Earth might be able to accomplish. Though we all can set fine examples by the way we live through each day, no matter the martyrdom or pain or glory, on the tenuous trail to righteousness.

My dear American, because our forefathers set up the USA to be a functional brilliant nation, with our laws and checked-and-balanced institutions, YOU still have the power to aide mankind's progress to a wondrous democratic existence. Note the denying of the licenses for the Shoreham nuclear plant on Long Island in New York state, and the Sierra Blanca national radioactive waste dump along the Mexican border in southwest Texas, due to citizen activism. Note the rejection of attempts to quietly introduce irradiated food into school cafeterias in Los Angeles and Washington D.C. and other school districts, and the Congressional response via the Child Nutrition Act that followed in 2004 (see Chapter Three). Note the termination of the attempt by our 'watchdog' Nuclear Regulatory Commission (NRC) to de-monitor nuclear wastes as Below Regulatory Concern (BRC) in the early 1990's, due to media publicity and citizen outrage (see Chapter Five). Certainly what you can do can change the course of history.

Call your representatives, contact and engage your local and/or national media, write to your newspaper (*Letters to the Editor* are good), send out a newsletter via snail mail or the internet, do your own show on radio or TV (public access is a great venue to utilize), participate in video websites like YouTube (I'm there![5]).

Tell those who you reach out to about what should or should not be done concerning labelling genetically altered food or irradiated food, or what's going on under the table with the unpublicized push toward a North American Union or erecting more nuclear power plants, or the guile of proclaiming that 'nuclear is green.' Share with your fellow citizens as you learn what you realize are very important pieces of information and ways of behavior, and preserve your dignity and mine so we all can respect ourselves, our planet, and all the dreams we dream that will help us and our children live this one life we have in a positive wise healthy manner. . .

YouTube video of Dr. Miller surfing wave backside in north-west New Zealand with very hot Drive By Truckers music supporting the ride/images.

Appendices

1: The Electricity Grid and Power Generation (With a Look Back at the Northeast Blackout of August 14, 2003

2: The "Inherently Safe" Pebble Bed Reactor

3: Two Prominent WTO Dispute Cases Involving the USA

4: Concerning The Possible Contamination Of Half of USA's Non-Organic Sugar With GMO's: 2 Letters Between The Organic Consumers Association And Kellogg's Corporation

1

The Electricity Grid and Power Generation

(With a Look Back at the Northeast Blackout of August 14, 2003)

Most of us recall the northeast power blackout of August 14, 2003. Why did it happen, and what does the future hold for our generation of electricity?

Can we admit that most of us Americans want our lights and electric toothbrushes to go on without a thought, at the reflexive flick of a switch? We have heard a lot about natural gas as a promising means of supplying our heat, and perhaps, electricity somehow. But with natural gas supply expected to peak at 2025, and more than 40 percent of the future reserves of natural gas situated in the Middle East,[1] this much-heralded savior of

503

our energy security may not provide the answer we idealistically desire. Whether it be gas, nuclear, or coal at that centralized power plant, sending out the electricity in new superconducting cables that can 'carry up to 25 times the amount of electricity than today's standard copper cables can,'[2] the grid system we have is only as strong as its weakest link. Without federal oversight and regulation of the grid, there will be more and worse blackouts in our all too immediate future.

Here's your numbers for the hodge-podge electric industry that Thomas Alva Edison might still be comfortable with if he came back for a time-travel visit:

There are 3,100 utilities operating 10,000 power plants across the USA. With 157,000 miles of high-voltage transmission lines, plus distribution facilities, the industry is valued at $1 trillion. America pays $247 billion each year as its national electric bill, 'paid by 131 million households and businesses...Industry consultant Eric Hirst estimates that at least $56 billion will have to be spent in the next decade just to maintain "adequacy"'[3] of our transmission lines.

So those quivering cables in Ohio that were blamed for the northeast blackout, and most of the others like it, will need to be upgraded, is what we should know. But with a perhaps over-paranoid terrorists-might-sabotage-something mindset, isn't this an obvious focus of concern for our energy supply future? Decentralized power sources, like those solar cells on your roof, or a neighborhood wind generator, or those hydrogen fuel cells not yet being mentioned enough: would such less sabotage-worthy technology make more sense as we blueprint what we should do to make us more energy secure?

Obviously, those 3,100 utilities would be very reluctant to lose their control of energy genera-tion and the profits they enjoy now. Politicians would be deprived of that special interest money so coveted to pay for their television campaign advertisements, as contributed by all these pow-erful utilities. Similarly, the media would hate to see a dip in their revenues that will exceed one billion dollars for the presidential campaign year cycle 2007-2008.

Can what is right for America's energy future be presented by our media in an impartial man-ner? Can our politicians be relied upon to make the best decision for the citizens they are sup-posed to honorably represent?

On the other hand, if we wisely moved away from polluting, deadly fossil and nuclear fuel generated power to large wind or solar 'farms,' in the midwest and elsewhere, could superconducting cable and other advancing technology efficiently distribute the electricity produced? And how far would be practicable? Would there have to be more 'alternative' type 'farms' throughout the country nearer to users and cities and businesses? Wind and solar power potential actually exists in every state in the union in various degrees.

Although physicist Dr. Richard Rosen, of the Tellus Institute, tells us that "It's only efficient to transmit electricity for a few hundred miles at most,"[4] because of power consumption, resistance, etc., produced in the process, going along with much that we hear in our media, actually, as you've read in Chapter One, DC transmission only loses ½ percent per 100 miles. Going 3000 miles across country would lose 15%, an acceptable figure to most experts. With super-cables and super-conductors transmission losses could be much much less.

According to Grant, Starr and Overbye in their June 26, 2006 *Scientific American* article their DC super-cable 'could transmit about five gigawatts for several hundred kilometers at nearly zero resistance and line loss.'[5] A gigawatt is 1000 megawatts, so five gigawatts would equal the output of 3,333 1.5 megawatt wind turbines, or five average sized but dangerously polluting nuclear power plants. A kilometer is 5/8th of a mile, so 800 kilometers would equal 500 miles.

The above-mentioned trio of co-authors suggest that with a 'Continental SuperGrid evolving gradually alongside the current [nationwide electricity] grid, strengthening its capacity and reliability... Over the course of decades, the Super-Grid would put in place the means to generate and deliver not only plentiful, reliable, inexpensive and "clean" electricity but also hydrogen for energy storage and personal transportation. Engineering studies of the design have concluded that no further fundamental scientific discoveries are needed to realize this vision.'[6]

Grant, Starr and Overbye estimate that the cost might be 'enormous: perhaps $1 trillion in today's dollars and in any case beyond the timescale attractive to private investment.'[7] In other words, the taxpayer, via our government, would have to fund this massive project. However, just to keep the numbers/dollars in perspective, our engineering trio remind us that our current grid infrastructure is valued at 'more than $1 trillion.'[8] Recall also that the present USA federal annual budget is $2.9 trillion.

Grant, Starr and Overbye like isolated clusters of nuclear power plants to generate their hydrogen to aid super-conduction. But recall the very-long-term cancer and mutation concerns described in this book for this source of energy generation, as opposed to the untoxic capability to generate all USA electricity in an 80 mile square area of desert via solar power, or the capability of a 1.5 megawatt wind turbine to power 400 homes. Can these sustainable forms of electricity generation produce enough energy for these engineers to make their coolant super-conductors? I believe that with our inventiveness and ambition, we can develop whatever is necessary on this front without poisoning our fragile biosphere foolishly forever.

With wind turbines, solar systems, and fuel cell technology privately owned all over the country one day very soon, supplying individual homes or energy-wise communities with their electricity, true American entrepreneurship and independence could send excess de-centralized electric current INTO the grid at a profit for so many practical-minded enterprising citizens.

The time is coming quite soon for this to happen. Despite utility companies influencing policies opposing fair remuneration for years, 'net metering' has become fashionable. If you exceed what you use, your electric meter spins backwards, reducing the tally of your net kilowatt use as that extra electricity flows back into the grid. Thirty two states have passed laws facilitating more equable interconnection and compensation for energy transmission into and out of the grid.[9]

Everything will not forever be as it is today, with our fossil and nuclear fueled interests controlling the powering of our country, and the direction in which our politics roils us. It's time for YOU to voice your knowledge and wisdom from the facts that you have gathered in this book to assist your fellow citizens and representatives in making better decisions about our energy future.

If you want to call your Congressperson, it's easy. Just dial 1-202-224-3121 for your Senators. If you don't know their names, you do know what state you come from. Mention your state and the operator answering the phone will get you to one of your Senators' offices. Then call back the same number and get the other Senator. Tell the Senate staffers that you vote, and inform them about the GE 1.5 megawatt wind turbine and its implications, or the toxicity of plutonium and that nuclear power is in no way 'safe and clean,' or 'green,' or whatever you want to say. Change comes from one person taking a stand, doing something, informing someone, maintaining her or his integrity, following one's conscience, sharing the prospects of a sustainable peaceful future with someone else.

Oh, to reach your representative in the House of Representative, same story, my fellow American. Just dial THIS number instead: 1-202-225-3121. You are not a helpless piece of pollen blowing in the breeze. Be human, express your concerns as a free spunky American to help redirect our troubled country from the course our belligerent leaders are charting to apocalypse. What is, does not always have to be. We're only here on this planet for a short walk, let us make each step, each day, each positive thought count on our journey that can produce love and sharing toward a constructive future, instead of greed and hate that only can lead to the destruction and poisoning of our souls and the only Earthly habitat we have.

2
The "Inherently Safe" Pebble Bed Reactor

Ah, yes: "inherently safe." A buzz-phrase to beware of concerning all old and new nuclear reactors.

As one of the recently heralded 'new' generation prototypes of nuclear reactors, let us take a look at the 'Pebble Bed' Modular reactor (PBMR). They say it could be our nuclear saving grace, constructed right in the middle of your neighborhood. An advanced form of nuclear power technology that is so "inherently safe" you wouldn't even have to worry about it. It wouldn't even need to have any containment structure! Nor would it need that absurd ten mile evacuation zone around it (standard in America, and around the world). Half a mile is fine for this reactor.

When Exelon, the nuclear corporation based in Chicago, applied to the Nuclear Regulatory Commission (NRC) to certify the design of the pebble beds here in America, Exelon planned to base such approval on a South African experience.

Unfortunately, that South Africa 'experience' has not happened quite yet, because no reactor has been built there yet. In fact, its demonstration model will not be ready until 2014, at the earliest.[1] And people in South Africa are fighting its erection.

Meanwhile, the latest going on here in America on the pebble bed front is that the NRC asked too many hard questions concerning the PBMR design for Exelon, so that corporation pulled its request for USA certification. However, the actual pusher/maker of the PBMR hoping for worldwide commercial approval, Eskom, the South African government-owned utility, is still trying to get the pebble bed approved here in America. Currently, Eskom is 'in the pre-application stage for design certification from the NRC through an undisclosed second party,' according to Paul Gunter, who heads the Reactor Oversight Project at the Beyond Nuclear organization.[2]

Description of the Pebble Bed Reactor

The pebble bed reactor can be described to be like a gumball machine, with 2 inch diameter lemon-sized fuel balls rolling into the core, fed out of a canister pneumatically. Each ball weighs about as much as a baseball and has uranium-235 imbedded as 10 to 12 thousand 'microspheres' in reactor grade graphite. Theoretically, because graphite (a form of carbon) has such high melting and boiling points above 3500 degrees centigrade, it would be very unlikely to catch fire. Except that when the Chernobyl reactor blew, there were graphite fires that burned and burned for at least ten days after the accident. You always have to worry about Murphy's Law: Anything that can go wrong, will go wrong.

There would be about 320,000 of these fuel balls in the typical 110 megawatt pebble bed reactor (an average-sized commercial nuclear reactor is designed to produce about 1000 megawatts of electricity). When damaged or when they become too radioactive, the fuel balls are then sent off to a waste canister, which is air cooled, the decay heat naturally conducted to the world outside its sealed and welded walls.

The pebble bed works by nuclear fission in the balls producing heat, which heats helium, which eventually turns a turbine and produces electricity.[3]

As with most gumball machines, shouldn't we

worry that the radioactive gumballs will possibly get gummed up one day? This did happen, actually, in Germany in 1986 at the Hamm reactor of the pebble bed type. Operators tried to blow out a fuel ball that got stuck, then used control rods. Somehow the coverings of enough balls were breached, which led to a very serious radiation release.

Because this happened only eight days after the Chernobyl disaster's spewing of toxic radionuclides all about northern Europe especially, the Hamm reactor's owners claimed that the radiation its plant had released to the surrounding area came from Chernobyl. However, it was discovered that nearly 70 % of that radiation <u>was</u> from the Hamm reactor. This led to the reactor being closed down by a very angry government and citizenry for 'a routine inspection.' More problems with damage to unused fuel balls -- or 'pebbles,' as they are sometimes affectionately called -- plus other deficiencies of quality control, led to the reactor's final closure in 1988, after less than four problem-plagued years of semi-function.[4]

Basically, that is your Earthly history of the latest type of commercial electricity-producing pebble bed reactor use. England, Germany and

France have rejected them as commercial fail-ures. They are not so 'advanced' as they are be-ing hyped to be.

Alas, even though they are to be smaller than most commercial nuclear reactors of today at 110 megawatts, the pebble bed reactors will still pro-duce nuclear waste when their fuel balls or peb-bles become too radioactive, or breach their clad-ding/covering. And, remember, all 320,000 of those fuel balls must be manufactured perfectly, or there will be radiation released. Plus, what if the separation or pneumatic feeding of balls to/from whatever canisters are involved, malfunc-tions? Or, what if the air-cooled waste canister gets too hot somehow, and a fire develops inside it? How will it be put out? There does not seem to be any water sprinkling-type extinguishing system engineered into the pebble bed's design. For water and super-hot nuclear waste don't mix, they could produce criticality and an untoward event, like a fire, explosion, etc. Inherently safe? So much so that Exelon wanted legal assuranc-es written in, before they would consider build-ing their first USA reactor, that any accident that might occur would be covered under a limited liability to them under the Price-Anderson Act, at $10 billion per reactor accident. Again, anything above that $10 billion would be paid for, not by Exelon, nor the South African company which actually is manufacturing the PBMR, Eskom, but by you and me, the <u>USA</u> taxpayer.

Perhaps the Bush administration might figure the American public would like these cute little 110 megawatt reactors, or that they'd hardly be noticed, built in proximity to some poverty stricken community without the resources to fight their siting. No matter where they could be constructed, their juice produced would go into the electricity grid, and their waste would be added to the growing pile causing us infinite headaches worrying about the cancer such nuclear garbage could generate............

Although Paul Gunter thought the pebble bed reactor was "dead in the water" a few years ago,[5] we have the very very pro-nuclear Bush administration ever-ready to wreak new nuclear havoc on the country and the planet. And then there's presidential candidate John McCain and his wanting to go 80% nuclear, like the French....

The South African government's Director General of Environmental Affairs and Tourism says that he is satisfied the pebble bed reactor is "safe from an environmental impact point of view."[6] Plus hundreds of millions of dollars are flying around South Africa to fund the pebble bed's development[7] from many different sources.

It could happen just around the corner from your house. Let's hope not though. (But from the activistic info I've given you throughout this book, you know who to call, you know what to do to prevent this. Right?)

P.S. The latest on *containment* for the pebble bed reactors is that they do indeed have containment! Yes, each ball, each pebble is its OWN containment. Just 320,000 pebbles we have to worry about breaching their own individual containment.

But, Oh, yes, those pebbles have to be changed, re-stocked periodically, as does fuel for all nuclear reactors. That means we actually have to worry about millions of radioactive pebbles maintaining the integrity of their 'containment' over the pebble bed's possible lifetime. Isn't that logical, wonderful? Or just a nightmare to even contemplate the challenges of evaluating/regulating/inspecting this miraculous nuclear gizmo....?....[8]

3

Two Prominent WTO Dispute Cases Involving the USA

The Internet Gambling Dispute

Two very prominent WTO (World Trade Organization) dispute panel cases involving the USA recently had interesting outcomes. One is the internet gambling case brought to the secret court by Antigua and Barbuda, two small islands that comprise a country in the Caribbean Sea, versus the USA. After losing the case in which the island nation filed for $3.4 billion, the USA decided to pull a fast one.

The original challenge made in March of 2003 was that remote U.S. gambling laws were "barriers to trade" in "cross border gambling services" under the GATS agreement.[1] GATS stands for General <u>A</u>greement on <u>T</u>rade in <u>S</u>ervices, and is one of 18 side agreements that exist under the WTO umbrella. Arguing that our laws were "'necessary to protect public morals" and therefore were protected by a WTO exception'[2] did not help us evade the inevitable on all counts.

What we, the USA, did, was appeal the guilty verdict that was laid down in November 2004, to the WTO Appellate Body, on the above grounds, adding 'that Internet gambling nurtures gambling addictions and makes it difficult to screen out minors and prosecute fraud.' This argument was accepted surprisingly by the Appellate Body for three challenged laws, but our Interstate Horse Racing Act was ruled to be in violation of the GATS agreement 'because it allowed bets to be placed remotely across state lines, but not from Internet servers based in foreign nations.'[3]

Here's where the hyprocrisy comes in, and the unilateralism that we've become infamous for over the initial decade of the new millennium. Though, of course, you could argue that we want to protect our sovereignty in making our laws; that the WTO should not have the right to override our democratic process, and the needs of our independent people and their legislative freedoms.

What we claimed was that we made a mistake. That we 'should not be liable under WTO rules because no one had envisioned the availability of online gambling, when the Clinton administration signed the trade agreement in 1994. "It never occurred to us that our schedule could be interpreted as including gambling until Antigua-Barbuda brought this case,'"[4] Deputy U.S. Trade Representative (USTR) John Veroneau said.

In other words, we were given the option to <u>not</u> subject gambling to our list of services susceptible to GATS rules and challenges when we signed onto the agreement originally - - in our 'scheduling' of such services. Now we have decided to withdraw our gambling sector from WTO jurisdiction.

Mark Mendel, Antigua-Barbuda's lawyer in the case, discounted the U.S. 'mistake' argument by noting that such an argument was "a horrible thing to say to your trading partners, and what an awful precedent to set. However, I don't see it working on the rest of the world, because the rest of the world knows that 'mistake' is irrelevant under international law."[5]

As Mary Bottari of Public Citizen says '"I find it ironic that the United States is one of the primary backers of the WTO ... and it has pretty much been thumbing its nose at the WTO rules ... They have done everything they could not to follow the rules." Bottari argues the ruling shows the dangers of the GATS - and that the United States should not demand others to adhere to rules from which it seeks to escape.'[6]

At this stage, small Antigua-Barbuda has been awarded $21 million, not billion, in compensation for winning against the mighty USA. Negotiations are going on between our representatives and those of other nations like Japan, Australia, Macao, Canada, 15 Caribbean nations, the European Union, in addition to Antigua-Barbuda, for damages and compensation.[7] Instead of paying money, one alternative being considered is facilitating access to U.S. water resources by our United States Trade Representative (USTR).[8] Meanwhile, WTO arbitrators authorized Antigua-Barbuda 'to impose sanctions by suspending its obligations to respect U.S. patents and copyrights' - - to leverage the U.S. to change our laws.[9]

Critics of the WTO say, although the U.S. withdrew its gambling sector 'from GATS control, the agreement continues to endanger U.S. (as well as other countries') legitimate regulatory policies. "It's good news that the Bush administration finally is listening to the state attorneys general and others who have asked for the U.S. Trade Representative to remove gambling from the WTO jurisdiction and thus eliminating further attacks on U.S. gambling regulation," says Lori Wallach, director of Public Citizen's Global Trade Watch. "The WTO's ruling against the U.S. Internet gambling ban was not some fluke, but rather a preview of coming attractions given how extensively the WTO's service sector rules interfere with non-trade domestic policies regulating the conduct of services operating within our own country."'[10]

USA Challenges European Union Genetically Modified Food Bans

The second case worth mentioning here, that was presented prominently in Chapter Four, is our challenge of European Union (EU) bans and moratoriums concerning genetically modified organisms (GMO's). As discussed in several places in this book, with documentation from many official scientific and citizen sources, the people and nations of the EU effectively have rejected this technology, its products, and its crops.

Although the USA claimed victory in our challenge, the reality is that only two minor claims of the challenge were *not* dismissed. These were that "'product specific marketing bans or moratoriums" were causing "undue delay" in approval procedures for over 20 specific biotech products.' And that these bans and moratoriums did not meet risk assessment requirements under the SPS (Sanitary and Phytosanitary Measures) side agreement.[11]

Although the main thrust of the USA challenge (made along with Canada and Argentina) was to get these bans and moratoriums declared explicitly *per se* illegal, this was not accepted by the WTO dispute panel. In fact, the decision allowed that the moratoriums and bans could become WTO-legal *"if new scientific evidence comes to light which conflicts with available scientific evidence and which is directly relevant to all biotech products subject to a pre-marketing approval requirement."* The panel further said that *"depending on the circumstances [it could] be justifiable to suspend all final approvals pending an appropriate assessment of the new evidence."*[12]

Although the USA 'used the publication of the WTO draft ruling' on this issue 'to threaten the rest of the world *"against following the European lead in throwing up bans or partial bans against genetically modified crops,"'*[13] such behavior has been ill-based relative to the facts of the case. Following the February 2006 stage of the ruling, Hungary declared itself henceforth 'GM-free, and Greece and Austria ... affirmed their total opposition to the crops.'[14]

What Friends of the Earth (FOE) recommends in their briefing on this ruling is that 'since the Biosafety Protocol [the only legally binding international treaty that deals with the transboundary movement of GMOs – to which the USA has *not* signed-on with 130 other member countries] is the most specialized and expert international forum in the field of GMOs, it makes it the most sensible place to deal with disputes related to the functioning of GMO regulatory regimes. Issues related to the safety of GM crops and risk assessment, key elements of the Protocol, are also a natural fit.'[15]

FOE also notes that there is a Compliance Committee that can deal with the 'undue delay' accusation, in conjunction with the Protocol's established provisions that already 'contain a precise timeline in the approval process, as well as describing the elements that can delay such procedures.'[16]

While the USA may claim 'to have won a great victory for "free trade,"' giving our government grounds to warn 'other parts of the world – particularly nations in Africa and Asia – against following the European lead in throwing up bans or partial bans against GMOs,'[17] FOE contends that the WTO is not the right forum to 'deal with conflicts between trade rules and environmental protection, which according to international agreement should be based on the Precautionary Principle.'[18]

Remember, the Precautionary Principle basically is: first *prove* it is safe, then allow it to be released, consumed, used.

Although the USA may badger the rest of the world, and intimidate poorer nations who attempt to do what the EU nations have done concerning GMOs with a challenge in WTO court, justice will most likely not be served properly. The media of the planet may accept that the USA has indeed secured a victory in this case, while even the citizens of our country remain in the dark on the scientific facts of GMO dangers, as presented in this book. The threat of a WTO challenge then may stop less economically endowed nations from best protecting their own people, as the USA claims victory on narrowed yet inhibiting grounds re banning the import of GMOs.

The latest political wranglings in this case are complicated by a new European Commission regime headed by the former Prime Minister of Portugal, Jose Manuel Durao Borrosso, with much more of a pro-George Bush leaning, favoring 'free-traderism.' Compliance is the biggest apparent issue currently: USA compliance with a ruling against our subsidies in the Upland cotton case before the Europeans will comply and hurry up their approval/evaluation of biotech products/crops/seeds; then the argument that using the Biosafety Protocol for a forum in that entity's Compliance Committee is unacceptable because of its insufficient enforcement capability as compared to the WTO's.

Also, various forces plied by unknown fundings are publicizing worry in the EU that "OH! If we don't take in GMO's we won't have enough animal feed for our animals."[19] (For an overwhelming percentage of USA and other pro-GMO nations' animal feed is predominantly full of GMO soybeans and corn, because we can't sell these patented crops otherwise in ample quantities to informed hostile consumers worldwide - - especially in countries requiring labelling of GMO's.)

According to Steve Suppan of the Institute for Agriculture and Trade Policy (IATP), however, with the addition of Poland (which prohibits use of GMO's countrywide, <u>including</u> animal feed[20]) and Bulgaria, adding in to the now 27 countries of the European Union (EU), there is plenty of land, water and technology available in the EU today to generate all the animal feed needed there, without any GMO tainting.[21]

4

Concerning The Possible Contamination Of Half Of USA's Non-Organic Sugar With GMO's: 2 Letters Between The Organic Consumers Association And Kellogg's Corporation

June 12, 2008

A. D. David Mackay
President and Chief Executive Officer
Kellogg Company
One Kellogg Square
Battle Creek, Michigan 49016-3599

Dear Mr. Mackay,

This letter is to notify you that the Organic Consumers Association and other non-profit organizations listed below are planning to call for a consumer boycott of all Kellogg's products because of your plans to use sugar from genetically engineered sugar beets in your products.

In a November 26, 2007 article in The New York Times titled "Next up for U.S. farmers: Genetically modified sugar beets" there was a quote from a spokesperson from Kellogg's. Here is how the newspaper reported on Kellogg's position:

"A Kellogg spokeswoman, Kris Charles, said her company, the top U.S. maker of cereal, 'would not have any issues' purchasing such sugar for products sold in the United States, where she said 'most consumers are not concerned about biotech.' "

In reality, consumers worldwide are concerned about biotech foods. However, in the United States, consumers have been kept in the dark about these risky genetically engineered foods because of a lack of labeling requirements. Genetically engineered foods are required to be labeled in all the European Union nations and in Japan, China, South Korea, Thailand, Australia, New Zealand, and many other countries around the world.

How unaware are American consumers about the presence of genetically engineered foods in their diet? In a November 2004 report funded the United States Department of Agriculture and conducted by Rutgers University, researchers found that only 31 percent of the American public thought that they had ever eaten a genetically engineered product. Yet in reality, since soy and corn are both used in many processed foods, the majority of people are eating these unlabeled biotech products on a regular basis.

When researchers for this study asked consumers if they thought genetically engineered foods should be required to be labeled, a whopping 89 percent thought that they should be required to be labeled and only 10 percent did not think that labels should be required.

CAMPAIGN DIRECTOR
RONNIE CUMMINS

POLICY BOARD
WILL ALLEN
Vermont Farmer

MAUDE BARLOW
Council of Canadians

JAY FELDMAN
National Coalition Against
the Misuse of Pesticides

JIM & REBECCA GOODMAN
Wisconsin Organic Farmers

JEAN HALLORAN
Consumers Union

TIM HERMACH
Native Forest Council

JULIA BUTTERFLY HILL
Forest Activist and Author

ANNIE HOY
Ashland Community
Food Store, Oregon

MIKA IBA
Network for Safe and Secure
Food & Environment Japan

PAT KERRIGAN
Emergency Food Shelf
Network, Minnesota

JOHN KINSMAN
Family Farm Defenders

AL KREBS
Agribusiness Examiner

BRUCE KRUG
New York Dairy Farmer

HOWARD LYMAN
Voice for a Viable Future

VICTOR MENOTTI
International Forum
on Globalization

FRANCES MOORE LAPPE
Author

ROBIN SEYDEL
La Montanita Co-op,
Santa Fe, New Mexico

VANDANA SHIVA
Research Foundation for
Science, Technology, and
Natural Resource Policy, India

JOHN STAUBER
Center for Media
and Democracy

Organic Consumers Association · 6771 S. Silver Hill Drive · Finland, MN 55603
· Phone: 218-226-4164 · Fax: 218-353-7652 · Email: campaigns@organicconsumers.org
www.organicconsumers.org

Kellogg's

David Mackay
President
Chief Executive Officer

June 27, 2008

Mr. Ronnie Cummins
National Director
Organic Consumers Association
6771 South Silver Hill Drive
Finland, MN 55603

Dear Mr. Cummins:

Thank you for your recent letter dated June 12 regarding biotech ingredients.

While we don't currently use sugar from genetically engineered sugar beets, it's important to note that there is worldwide scientific consensus that there are no safety concerns with the currently commercialized genetically modified agricultural products on the market, including sugar. The World Health Organization, the Food and Agriculture Organization of the UN, the U.S. National Academy of Sciences, and the American Medical Association all share this assessment.

Being a global organization, our focus has always been on meeting the needs of our consumers worldwide and being responsive to a variety of consumer preferences. Our decisions on whether or not to use biotech ingredients are made on a market-by-market basis and depend on a variety of factors specific to each market.

Consumer preference is the critical factor Kellogg uses in determining the products being provided in each market, and those preferences are not the same in every country. Ms. Charles was referring to a U.S. study conducted in September 2007 by the International Food Information Council stating that U.S. consumer "concerns about the usage of biotech ingredients in food production are low." Public acceptance of biotechnology in Europe is lower than in the United States. As a result, all Kellogg products sold in Europe are free of any ingredients derived from biotech sources.

All of our products comply with the food labeling requirements in the markets in which they are sold throughout the world.

We are as committed today to protecting and promoting consumer confidence in our products as our founder, W. K. Kellogg, was when he had his name placed on every product as his personal assurance of quality.

Thank you for taking the time to share your organization's point of view and for allowing Kellogg Company to respond in kind.

Best regards,

David Mackay

kc/rkj

Kellogg Company / Corporate Headquarters
One Kellogg Square / P.O. Box 3599 / Battle Creek, Michigan 49076-3599 (269) 961-2000

531

Endnotes

Frontmatter

1 '81% in Poll Say Nation Is Headed On Wrong Track,' by David Leonhardt and Marjorie Connelly, April 4, 2008, New York Times http://www.nytimes.com/2008/04/04/us/04poll.html?em&ex = 12 07540800&en = 691578229c7679c7&ei = 5087%0A

2 Disney (ABC, ESPN, Disney Channels), Viacom (CBS, MTV, VH1, UPN), General Electric (NBC, Bravo, USA Network), News Corporation (Fox, MSG, Speed, Fuel), Time-Warner (CNN, WB Network, HBO). (Those networks listed in parentheses are just a few of many networks, stations, movie studios, etc., owned by these megacorporations.)

3 Karl Grossman, article on Chernobyl Nuclear Disaster, 4 26 07, page A13 Southampton Press.

4 'Poll: Many Won't Buy Genetically Modified Food,' CBS News Poll reported May 11, 2008, http://cbs4.com/national/CBS.News. New.2.721469.html   Also stated in this article: '53 percent of Americans say they won't buy food that has been genetically modified.' and from nutritionist Marion Nestle about the Monsantos and Syngentas 'one basic fear' in the fight over genetically modified foods: "They didn't want it labeled because they were terrified that if were labeled, nobody would buy it."

5 Seeds of Deception by Jeff Smith, page 12, published by YES! BOOKS 2003.

6 "Frankenfoods' Survey of the Week,' '99%: Estimated likelihood that the U.S. sugar supply will start to be sourced from genetically engineered plants this year.' Organic Bytes #135, May 22, 2008, http://www.organicconsumers.org

7 'How Consumers Drove Genetically Engineered Foods off the Shelves in UK, Europe,' The Independent, 9 July 2006, available at

http://www.organicconsumers.org/articles/article_10140.cfm

8 Global Network Space Newsletter 17, Winter 2006, http://www. space4peace.org

9 'Connie Fogal says the Security and Prosperity Partnership (SPP) is the "hostile takeover" of the apparatus of democratic government and an end to the "rule of law". There has been a kind of coup d'état over the government operations of Canada, U.S.A. and Mexico.' published by Global Research. Selected as co-winner for Story #2 'Most Censored Stories' Project Censored awards 2009.

10 See http://www.nirs.org for the latest on this story.

Chapter 1: Bush's Nuclear Push

1 'BBC report: Bush said God told him to invade Iraq,' Agence France Presse, October 6, 2005, 8:08 PM EDT, http://www.newsday.com/news/nationworld/world/ny-wobush1007,0,4758135,print.story

2 'Feds to assess Godley water, ATSDR will act on request from Durbin,' June 22, 2006, Morris Daily Herald / Illinois, By Jo Ann Hustis.

3 Joe Cosgrove, Director, Parks Department, for Godley, Illinois; telephone conversation June 19, 2006.

4 'Why EPA's Tritium Standard For Drinking Water [20,000 picocuries per liter] Is Undoubtedly Way Too Lax, & A Suggested New Standard," Jan 17, 2006 by Russell Ace Hoffman.

5 'Money is the Real Green Power: The Hoax of Eco-Friendly Nuclear Energy,' by Karl Grossman, Extra! Feb. 3, 2008.

6 "Nuclear Relapse Goes Global," WISE/NIRS Nuclear Monitor #642, February 24, 2006, page 1. Mr. Keegan is the chairperson of the Coalition for a Nuclear-Free Great Lakes.

7 http://www.nea.fr/html/ndd/reports/2002/nea3808-kyoto.pdf page 8.

8 Ibid., page 2.

9 See http://www.nirs.org for the ongoing particulars on this boondoggling legislation for the nuclear industry. The first attempt at passing the Climate Security Act, or the Cap and Trade Act or, perhaps, as some say, the Bait and Switch Act, failed in June 2008. The danger for taxpayers and human beings will be in the amendments, which will introduce the nuclear subsidies, probably not to be mentioned at all in the bill itself. To doubters who think that such a scam is impossibly preposterous, be aware that an aide to Senator Lieberman told *E and E Daily*

as published Feb. 8, 2008 "The bill, as reported out of committee, would be the most historic incentive for nuclear in the history of the United States ."

10 The $300 billion cost figure has been a commonly quoted number but is probably low, as according to the following document the expected costs for Byelorus alone are projected to be $235 billion. And the area around Chernobyl is rather small town/rural, as compared to the NYC metropolitan area where the Indian Point nuclear reactor is leaking strontium for starters, and could be the next nuclear disaster to spring out our pooh-poohing eyeballs. www.cleanenergy.org/hottopics/index.cfm?id = 64

11 'Reasonable Doubt,' by Ian Fairlie, *New Scientist*, April 24, 2008.

12 Karl Grossman, article on Chernobyl Nuclear Disaster, 4 26 2007, page A13 Southampton Press Eastern Edition, quoting from Dr. Alexey Yablokov's new book. Mr. Yablokov was president of the Center for Russian Environmental Policy and former environmental advisor to the late President Boris Yeltsin. Also noted was that life expectancy in Russia, which had been the same as that of the United States, has dropped to 59 years for men, and 64 years for women, a fact that Dr. Yablokov attributes principally to Chernobyl: "You see longevity dropping
precipitously right after 1986 and the accident."

13 'Radioactive Isotope Found In Third Well at Indian Point Plant,' by Greg Clary, http://a4nr.org/library/safety/03.22.06-poughkeepsiejournal

14 Telephone conversation with nuclear engineer Arnie Gundersen June 24, 2008.

15 Telephone Interview with Arnie Gundersen, July 19, 2006.

16 Telephone conversation with Arnie Gundersen, June 25, 2008

17 "Nuclear Relapse Goes Global," WISE/NIRS Nuclear Monitor #642, February 24, 2006, page 2.

18 "The Dangers of Reprocessing," WISENIRS Nuclear Monitor #643, March 17, 2006, page 6.

19 "How To Help Terrorists Get The Bomb," by Edwin Lyman and Lisbeth Gronlund, Catalyst Magazine, page 8, Spring 2006, Volume 5, Number 1.

20 Op. Cit., "The Dangers of Reprocessing," WISE/NIRS Nuclear Monitor #643, page 7.

21 'Nuclear Fuel Recycling: More Trouble Than It's Worth,' by Frank N. von Hippel, April 28, 2008, *Scientific American*.

22 Op. Cit., "The Dangers of Reprocessing."

23 Ibid.

24 Ibid.

25 Ibid.

26 Ibid.

27 Ibid., page 9.

28 email attachment sent to me by Matt Bivens, Fellow at the Nation, May 1, 2002.

29 'Move Over, Oil, There's Money In Texas Wind,' by Clifford Krauss, February 23, 2008, New York Times, http://www.nytimes.com/2008/02/23/business/23wind.html?hp

30 Ibid.

31 Ibid.

32 Telephone conversation with Paolo Berrino of the European Wind Energy Association, 6 18 08. See more at http://www.ewea.org

33 'Lessons From Germany's Energy Renaissance,' by Eric Reguly, Saturday's Globe and Mail, March 21, 2008. http://www.theglobe-andmail.com/servlet/story/RTGAM.20080321.wrcover22/BNStory/energy/?page = rss&id = RTGAM.20080321.wrcover22

34 Ibid.

35 Telephone conversation with Paolo Berrino of ewae.org 6 19 08. Numbers in megawatts of windpower's top four nations' wind capacity as of the end of 2007: Germany: 22,247 megawatts of windpower installed, USA: 16,818 megawatts, Spain: 15,145 megawatts, India: 7,845 megawatts.

36 Telephone interview April 9, 2008 with Kathy Belyeu of the American Wind Energy Association. More info available at http://www.awea.org

37 In a conversation with a representative of the cooperative KUIC of Kauai on June 23, 2008, I was told that Kauai consumed 466,896 megawatt-hours of electricity in 2007. Using a *capacity factor* of 33% for a 1.5 megawatt wind turbine over 8760 hours that exist in a year, gives us 4,380 megawatt-hours produced per 1.5 megawatt wind turbine produced per year. This would then indicate that for ALL Kauai's electricity needs, including hotels, businesses, etc., 107 wind turbines would supply adequate power. However, there also is the 2.5 megawatt Clipper wind turbine as the number two wind turbine popularly available. The math there would, with the same capacity factor, require 64 of these 2.5 megawatt wind turbines to provide the 466,896 megawatt-hours of Kauai electricity.

38 Telephone interview March 14, 2008 with Ron Kimmel of the American Wind Energy Association. More info available at http://www.awea.org

39 Telephone conversation with Professor and Physicist and former Vice Provost of the California Institute of Technology, David Goodstein, May14, 2008. Prof Goodstein is the author of *Out Of Gas: The End Of The Age Of Oil.*

40 http://en.wikipedia.org/wiki/Fuel_cell

41 The Hydrogen Economy, by Jeremy Rifkin, pages 194-6, published by Jeremy P. Tarcher/Putnam, a member of Penguin Putnam Inc., 375 Hudson Street, NYC, NY 10014, 2002

42 'Amory Lovins: Energy Efficiency' by Derek Reiber, May 27 2003, www.ecologicinvestor.com

43 Your devoted Author's note. Why would Mr. Lovins not include the radioactive pollution from nuclear power plants?...

44 'Negawatts for Buildings,' by Amory Lovins & William D. Browning, www.rmi.org

45 'A Reliable Renewable Electricity Grid in the United States,' by Arjun Makhijani, Ph. D., 'Science For Democratic Action, Volume 15, No. 2, January 2008, page 9.

46 Op. Cit., Krauss, Feb. 23, 2008, New York Times.

47 'Uranium Prices to Skyrocket,' Nuclear Monitor #642, February 24, 2006, page 15.

48 Telephone conversation with Peter Meisen of geni.org on January 29, 2008.

49 'Colorado Scientists Make A New Form Of Matter,' by Joseph B. Verrengia, AP, http://www.salon.com, January 30, 2004. Also see Appendix for section on the electricity grid.

50 http://www.ewea.org/fileadmin/ewea_documents/mailing/wind-map-08g.pdf

51 'What Price The Atomic State, The Siege of Gorleben,' http://www.nirs.org/mononline/gorleben.htm

52 'Radioactive Wreck: The Unfolding Disasters Of U.S. Irradiated Nuclear Fuel Policies,' Nuclear Monitor No. 643, page 9, March 17, 2006, also available at http://www.nirs.org

53 Telephone conversation of July 3, 2008, with Kevin Kamps, radioactive waste expert with Beyond Nuclear. The fatal 10 second exposure from 3 feet away will occur most likely from high level radioactive waste (HLRW) very recently removed from a nuclear reactor. Older waste, perhaps removed decades ago, might require up to 3 minutes of exposure to produce the fatal radioactive syndrome.

54 Ibid., Nuclear Monitor, Issue No. 643, page 5.

55 'Hot Cargo: Radioactive Waste Transportation,' prepared by Radioactive Waste Management Associates, New York, N.Y., January 1995, NIRS publication, Washington, D.C.

56 Op. Cit., Nuclear Monitor Issue No. 643, page 5.

57 Ibid. Also see http://www.state.nv.us/nucwaste/trans/nucinc01.htm for the Robert Halstead report "Reported Incidents Involving Spent Nuclear Fuel Shipments, 1949 to Present," May 6, 1996.

58 Op. Cit., Nuclear Monitor Issue No. 643, page 5.

59 Ibid.

60 Ibid.

61 'U.S. Nuclear Reactors - Al Qaeda's Original Target,' WISE/NIRS Nuclear Monitor, #573 -- North American Edition, September 13, 2002, page one.

62 'Chernobyl: Two Decades Later," by Cathie Sullivan, Science For Democratic Action, Volume 14, Number 1, April 2006, page 7.

63 'How Much Radiation Was Released By Chernobyl?' Nuclear Monitor 641, January 27, 2006, page 8.

64 Ibid.

65 Ibid., page 6.

66 Op. Cit., 'Chernobyl: Two Decades Later,' page 8.

67 Voices From Chernobyl by Svetlana Alexievich, page 143, published by Picador, 175 Fifth Avenue, New York, New York 10010, in 2005.

68 'Concern Over French Nuclear Leaks,' BBC News, 24 July 2008, http://news.bbc.co.uk/2/hi/europe/7522712.stm

69 Op. Cit. Nuclear Monitor #642, page 15.

70 Ibid..

71 'Interview With Vladimir Chernousenko," Karl Grossman, Enviro-Video 1994.

72 Op. Cit., 'Voices From Chernobyl' page 170.

73 Ibid., page 194.

74 Ibid., page 81.

75 Ibid., page 40.

76 'Chernobyl: Two Decades Later,' by Cathie Sullivan, Science For Democratic Action, Volume 14, Number 1, page 10.

77 Op. Cit. Chernousenko, Enviro-Video 1994.

78 Voices From Chernobyl by Svetlana Alexievich, page 2, published by Picador, 175 Fifth Avenue, New York, New York 10010, in 2005.

79 Op. Cit., Chernousenko, Enviro-Video 1994.

80 Op. Cit., Karl Grossman, article on Chernobyl Nuclear Disaster, 4 26 2007, page A13 Southampton Press.

81 From the US Academy of Sciences, BEIR-5 Report, as annotated in Nuclear Monitor #641, page 7.

82 www.cleanenergy.org/pdf/Chernobyl20Anniv.pdf

83 'Chernobyl: The True Story by Dr. Vladimir Chernousenko,' Synthesis/Regeneration 10 [Spring 1996]

84 Ibid.

85 Op. Cit., Chernousenko, Enviro-Video 1994.

86 Actually "according to the IAEA PRIS database, as of January 1, 2007, 435 nuclear power reactors in operation, 29 under construction, 6 in long term shutdown.' 'Most of 435 reactors are 20-30 years old, only 35 reactors went into operation in the last 10 years, and 100 reactors are over 30 years in operation.'* No nuclear reactor has ever operated for 40 years or longer, yet the Bush administration wants to 'streamline' reactor approval, removing public input for licensure, for 40 years of operation. And they are talking about extending that to 60 years! *Nuclear Monitors #651, page 5.

87 See the Appendix, re Pebble Bed Reactors and their claim to be so "inherently safe" no containment would be necessary, and an evacuation zone of only half a mile would be needed, instead of the standard 10 mile zone.

88 Op. Cit., 'Chernobyl: The True Story."

89 Ibid.

90 'Radioactive Produce still arriving at Moscow's markets,' Nuclear Monitor #641, January 27, 2006, page 10.

91 'Decommissioning The Nuclear Power Industry: Rubble, Rubble, Toil and Trouble,' Nuclear Monitor Special Edition, February 2000, http://www.nirs.org

92 Ibid., page 10.

93 Ibid., page 11.

94 Ibid.

95 Ibid., page 8.

96 'Disarmament Treaties,' http://64.233.161.104/search?q=cache:T2Xb3vKEVLMJ:www.ippnw-students.org/NWIP/factsheets/treaties.html+NPT+article+6&hl=en&gl=us&ct=cln k&cd=3

97 Actually, with North Korea withdrawing from the NPT, this number currently is 187 (countries). Negotiations are going on currently to return North Korea into the fold of co-operating NPT nations.

98 http://www.democracynow.org/article.pl?sid=06/03/02/148233

99 Ibid.

100 Ibid.

101 Ibid.

102 See beginning of the Star Wars chapter, where pages 3 and 10 of the U.S. Space Command's 'Vision For 2020' are posted. These are also available in color at http://www.crestofthewave.com

103 The G-8 countries' representatives behaving enthusiastically in favor of spreading nuclear power and nuclear business world wide, come from the USA, the UK, France, Japan, Canada and Russia. Italy and Germany remain "sceptical" of a nuclear future. As noted in 'More on the G-8 Apocalypse ASAP SNAFU,' By Rob Edwards, Environment Editor, http://www.energyjustice.net/nukenet/.

104 Ibid., 'More on the G-8 Apocalypse ASAP SNAFU.'

105 'U.S., Others Haggle Over Nuclear Agenda,' by Charles J. Hanley, Associates Press/Yahoo, 2005.

106 Op. Cit., 'More On the G-8 Apocalypse....'

107 Op. Cit., Chernousenko 1994 Enviro-Video.

108 Op. Cit., 'More On The G-8 Apocalypse..'

109 'Boycott Toshiba,' email newsletter from Ace Hoffman, December 27, 2007.

110 Telephone conversation with Michael Mariotte, Director of the Nuclear Information and Resource Service, Jan 3, 2008.

111 Op. Cit., 'Boycott Toshiba.'

112 Ibid.

113 'Fury Over 'Hidden Leak' At German Nuclear Reactor,' from Anna Tomforde, 2nd June 1986 The Guardian.

114 'Nuclear power is the problem, not a solution,' by Dr. Helen Caldicott, http://www.ippnw.org 13 April 2005.

115 Ibid.

116 Telephone interview with Paul Gunter of Beyond Nuclear June 23, 2006. A typical two-unit boiling water reactor (BWR) site takes in 2-3 billion gallons of water per day.

117 Telephone conversation with nuclear engineer Arnie Gundersen, May 11, 2008.

118 'Synopsis As Introduced Amends the Environmental Protection Act. Makes a technical change in a Section concerning how the Title regarding water pollution control and public water supplies is construed.' http://www.ilga.gov/legislation/BillStatus_pf.asp?DocNum = 1620&DocTypeID = HB&LegID = 16436&GAID = 8&SessionID = 50&GA = 94

119 'Nuclear Spill Disclosure Bill Clears Senate, Dahl expects Blagojevich will sign tritium-leak response measure into law,' By Jo Ann Hustis, Morris Daily Herald, IL, March 28, 2006, http://www.shundahai.org/3-28-06MorrDlyHrld_Nuclear_Spill_Disclosure_Bill_Clears_Senate.htm

120 Nuclear Release Notice Act of 2006, http://thomas.loc.gov/cgi-bin/query/z?c109:H.R.4825.IH:

121 'Nuclear Leaks And Response Tested Obama,' by Mike McIntire, page one and 17, NY Times, Feb. 3, 2008.

122 As Mr. McCain likes to say, though the latest figure, as divulged earlier in this chapter, is now closer to 75% of France's electricity is generated by nuclear power. 'Concern Over French Nuclear Leaks,' BBC News, 24 July 2008, http://news.bbc.co.uk/2/hi/europe/7522712.stm

123 'France's Nuclear Fix?' by Arjun Makhijani, *Science For Democratic Action*, page 7, Vol. 15, No. 2, January 2008.

124 Ibid., page 6.

125 'Nuclear Power Costs: *High and Higher*' by Arjun Makhijani, *Science For Democratic Action*, page 2, Vol. 15, No. 2, January 2008.

126 Ibid.

127 Ibid., page 3. Also, 'Nuclear Power Plant Electricity: A Simple Costing Model,' by Philip D. Lusk, available at http://www.nirs.org and http://www.crestofthewave.com

128 Op. Cit., Makhijani, 'France's Nuclear Fix?' page 7.

129 According to the European Wind Energy Association (EWEA), as of the end of 2007, Spain had 15,145 megawatts of wind power on line. That ranked them behind...surprise! the USA with 16,818 megawatts; and Germany with 22,247 megawatts at the top of the heap. India ranked number four with 7,845 megawatts. See map in Introduction for European wind megawattage; and also the website http://www.ewea.org

130 'How Green Are *Their* Valleys?' by John McLaughlin, Delta Sky Magazine, March 2008, pages 70-71.

Chapter 2: Your Food: Mutated, Irradiated, or Pragmatically Pure?

1   Copyright 1988 Virgin Records.
2   Telephone Interview With Ronnie Cummins, Director, Organic Consumers Association, December 30, 2006.
3   The Ram's Horn, page 6, Volume 241, September 2006; 'U.S. Organic Food Industry Fears GMO Contamination,' by Carey Gilliam, March 12, 2008, http://www.organicconsumers.org
4   Telephone conversation with Arpad Pusztai, May 2, 2004; 'Farmaggedon, Frankenfoods, and the FDA: The Dangers of Genetically Modified Food,' by Steven Best, http://utminers.utep.edu/best
5   Ibid., Best.
6. 'Cauliflower Mosaic Viral Promoter [CaMV] – A Recipe For Disaster?' by Mae-Wan Ho, Angela Ryan and Joe Cummins, http://www.i-sis.org.uk/camvrecdis.php > > From the same article: "Another factor which affects the safety of transgenic plants containing CaMV promoters and related constructs is that although CaMV itself infects only dicotyledons, its promoter is promiscuous; and functions efficiently in monocotyledons (14), in conifer cell lines (15), green algae (16), yeasts (17) and E. coli (18). The transfer of CaMV promoter to these other species could also give rise to unpredictable effects on gene expression, which may impact on the ecosystem as a whole."
7   'The Food Revolution' by John Robbins, pages 351 and 353, published 2001, Conari Press, Berkeley, California, 94710-2551.
8   Phil Angell, Monsanto's Director of Corporate Communications, New York Times, 1999 as quoted in Robbins, ibid., page 353.
9   '[UK] Government's 10-day public roadshow opens with a whimper,' by Ian Sample; June 3, 2003 * Guardian * http://www.guardian.co.uk/uk_news/story/0,3604,969217,00.html
10 Op. Cit., Robbins, page 342.
11 'ADM Settles Out of Court,' Ram's Horn, July 2004, page 4.
12 'Archer Daniels Midland - Supermarkup To The World,' at http://www.corporatewatch.org.uk/publications/GEBriefings/controlfreaks/adm1.htm
13 Email June 7, 2004, from Patty Lovera, Public Citizen food irradiation expert.
14 'Congress Ignores The Consumer's Right To Know in the Farm Bill [of 2002],' www.citizen.org
15 'FDA Ignoring Evidence That New Chemicals Created in Irradiated Food Could Be Harmful,' Nov. 29, 2001, www.citizen.org >
16 'Human Children Are Actually Guinea Pigs,' Organic Bytes, Issue 15, June 11, 2003, http://www.organicconsumers.org/irradlink.html

17 'L.A. School District Bans Irradiated Lunches,' Public Citizen News, November/December 2003, Page 11.
18 Ibid.
19 Ibid.
20 'Privatizing The Ocean,' by Phil Lansing, Resource Economist and Senior Fellow at the Institute for Agriculture and Trade Policy on http://www.iatp.org October 16, 2002 News. Also http://www.crestofthewave.com
21 'FOE MONSANTO QUITS GM WHEAT,' Tuesday, May 11, 2004, http://www.foe.co.uk/resource/press_releases/monsanto_quits_gm_wheat_11052004.html
22 Ibid.
23 *Seeds Of Deception*, by Jeff M. Smith, page 137.
24 Multinational Monitor, May 2001, http://multinationalmonitor.org/mm2001/01may/may01bushcc.html#ann
25 'Biotechnology Advisory Committee Seen Lacking In Diversity,' by Stephen Clapp and Lucy Ament, Pesticide and Toxic Chemical News April 4, 2003.
26 FDA = Food and Drug Administration.
27 'Likely Allergenic Genetically Engineered Corn Still Spreading,' ORGANIC BYTES #24, Organic & Food News Tidbits with an Edge, Issue #24 12/9/2003, http://www.organicconsumers.org
28 Ibid.
29 Ibid.
30 'How Consumers Drove Genetically Engineered Foods Off The Shelves in UK & Europe,' By Geoffrey Lane, Independent, Sunday, 9 July 2006. http://www.organicconsumers.org/articles/article_1040.cfm
31 Arpad Pusztai, telephone conversation with him May 2, 2004.
32 The Columbia Encyclopedia, Sixth Edition, 2001, Corn.
33 'Unregulated Hazards 'Naked' and 'Free' Nucleic Acids,' Ho, Ryan, Cummins, Traavik, 2000, isis.org.
34 'Scientific Critique of FDA Whitewash of Starlink Corn Allergy Scandal,' www.gefoodalert.org, 2001.
35 'GM Foods: Potential Human Health Effects,' by Pusztai, Bardocz and Ewen, from Food Safety: Contaminants and Toxins, CABI Pub., Cambridge, MA, copyright 2003, page 367.
36 'StarLink Corn Has Contaminated 25% of US Corn Supply' Boston Globe, May 17,2001, by Anthony Shadid.
37 'Banned GE StarLink Corn Still Contaminating 1% of US Corn Crop,' by Paul Jacobs, Knight Ridder Newspapers, 1 December 2003, http://www.centredaily.com/mld/centredaily/news/7386628.htm
38 Op. Cit., 'Scientific Critique of FDA Whitewash of StarLink Corn Allergy Scandal.'
39 Ibid, 'Scientific Critique of FDA Whitewash of StarLink Corn Allergy Scandal.'

40 Op. Cit., Smith, page 38.
41 Ibid.
42 Op. Cit., Choi.
43 Telephone Conversation with Peter Cousins, MSc CBiol MIBiol, Technical Director, Yorktest Laboratories Ltd., May 2nd, 2003.
44 see www.ABCNEWS.com   6/20/01.
45 Wall Street Journal, November 20, 2000, as reported in BioDem News #31, January 2001.
46 Ibid, Wall Street Journal, November 20, 2000.
47 'Greenpeace Report Shows GE Industry In Trouble,' March 28, 2002, http://www.organicconsumers.org/gefood/Green-Peace032802.cfm
48 BioDemocracy News #33 May 10, 2001, page 4.
49 Telephone conversation with Ronnie Cummins, January 1, 2008.
50 'More Evidence on the Health Hazards of GE Food,' www.gaia@gaianet.org   4 30 04
51 'GM Foods: Potential Human Health Effects,' by Pusztai, Bardocz and Ewen, from *Food Safety: Contaminants and Toxins,* CABI Pub. Cambridge, Mass. copyright 2003, page 369.
52 BioDemocracy News #34 July 2001, page 5.
53 Op. Cit., 'Quotable Quotes....' Page 6.
54 Multinational Monitor, December 1997, "Sabotaging Organic Standards."
55 'What are the implications of the Ag Appropriations bill on organic labeling?' Nov 7, 2005, www.organicconsumers.org
56 Center For Food Safety News June 4, 1999; www.purefood.org
57 BioDemocracy News #32, March 2001.
58 'Greenpeace Report Shows GE Industry In Trouble,' March 28, 2002, www.organicconsumers.org
59 Op. Cit., Vint.
60 Ibid., Vint.
61 Ibid., Vint.
62 Ibid., Vint.
63 Ibid., Vint.
64 Ibid., Vint.
65 Op. Cit., 'Unregulated Hazards...." Ho, Ryan, Cummins, Traavik, 2000.
66 http://www.seedsofdeception.com/Corrections.php?menu1_id=3&menu2_id=12
67 Op. Cit., Smith, pages 57-60.
68 Ibid., Smith, pages 178-179.
69 Ibid., Smith, page 60.
70 Ibid.
71 Ibid., Smith, page 48.
72 Ibid.
73 Ibid.

74 Ibid., pages 61-65.

75 Ibid.

76 Ibid., pages 12-13.

77 Ibid., Smith, page 239.

78 Op. Cit., BioDemocracy News #34, July 2001.

79 Telephone conversation with Dr. Michael Hansen of consumer.org on August 3rd 2004.

80 Telephone conversation with Ronnie Cummins, National Director of Organic Consumers Association, January 21, 2008.

81 Ibid..

82 Ibid.

83 http://www.isaaa.org/resources/publications/briefs/35/executive-summary/default.html   By the way, ISAAA stands for International Service For The Acquisition of Agri-Biotech Applications.

84 'How Consumers Drove Genetically Engineered Foods Off the Shelves in UK.' by Geoffrey Lean, Independent, July 9, 2006, http://www.organicconsumers.org/articles/article_1040.cfm

85 Op. Cit., BioDemocracy News #34, July 2001, page 2.

86 'International Scientific Committee Warns Of Serious Risks of Breast and Prostate Cancer from Monsanto's Hormonal Milk,' March 21, 1999, Cancer Prevention Coalition, Univ of Illinois Medical Center School of Public Health.

87 'Milk and the Cancer Connection' by Hans R. Larsen, MSc ChE, 2001, via Joseph M. Mercola, D.O. website.

88 'IGF-1 and Milk: Questions and Answers' Cancer Prevention Coalition, School of Public Health, University of Illinois Medical Center.

89 'IGF-1 and Milk: Questions and Answers' Cancer Prevention Coalition, School of Public Health, University of Illinois Medical Center.

90 Ibid., Univ of Illinois Med Center School of Health Cancer Prevention Coalition.

91 Ibid., Univ of Illinois Med Center School of Health Cancer Prevention Coalition.

92 Mark Worth of Public Citizen Telephone Conversation, 8 20 2001.

93 email from Patty Lovera, Food Irradiation expert at Public Citizen, June 7 2004.

94 'Congress Undermines the Consumer's Right To Know in the Farm Bill,' www.citizen.org

95 'Consumer Groups To USDA: Don't Feed Irradiated Food to School Children,' December 18, 2002, www.centerforfoodsafety.org

96 Telephone conversation with Patty Lovera of Public Citizen, June 4, 2004.

97 Email 6 25 04 from Judy Szela of www.nocobalt-4-food.org

98 'As If Italian Food Wasn't Already Good Enough,' Organic Bytes #36 7/8/04 by Organic Consumers Association, http://www.organicconsumers.org

99 Ibid.

100 Ibid.

101 Will Williams of Surebeam, Telephone Conversation of 8 21 2001.

102 Ibid., Williams, 8 21 2001.

103 'The Dirt On Factory Farms' by Mark Floegel, Multinational Monitor, page 24, July/August 2000.

104 Op. Cit., Robbins, pages 124-134.

105 'How Hazardous Is Your Turkey?' Center for Science in the Public Interest news release, November 19, 1998.

106 Op. Cit., Floegel, Multinational Monitor.

107 Ibid, Floegel, Multinational Monitor; 'Preventing Pathogenic Food Poisoning: Sanitation Not Irradiation,' Samuel S. Epstein and Wenonah Hauter, International Journal of Health Services, Volume 31, Number 1, pages 187-192, 2001.

108 Food and Water Safe Food News page 4, Winter 1993.

109 The Merriam-Webster Dictionary, page 510, 1974 edition.

110 Dorland's Illustrated Medical Dictionary, page 1110, 24th Edition; http://www.usda.gov

111 Michael Hansen of Consumers Union, telephone conversation, 7 25 2001.

112 Op. Cit., Floegel, 'The Dirt on Factory Farms,' page 25.

113 Dean Danilson, Vice President, Technical Services, IBP, May 17, 1999 comment to FDA Docket No. 98N-1038.

114 From Federation of American Societies for Experimental Biology, Evaluation of The Health Aspects of Certain Compounds Found in Irradiated Beef. Report to the U.S. Army Medical Research and Development Command, Bethesda, MD, August 1977; as quoted in 'Preventing Pathogenic Food Poisoning...'.

115 'FDA Ignoring Evidence That New Chemicals Created in Irradiated Food Could Be Harmful,' Press Release, Nov 29, 2001, www.citizen.org; also, 'Congress Undermines the Consumer's Right To Know in the Farm Bill' page 2, www.citizen.org.

116 Ibid., 'FDA Ignoring Evidence That New Chemicals Created In Irradiated Food Might Be Harmful.'

117 From Sun, M. Science 223:1354, 1984 as quoted in Op. Cit.., 'Preventing Pathogenic Food Poisoning,' Epstein & Hauter.

118 From van Gemert, M. Memorandum Re: Final Report of the Task Group for the Review of Toxicology Data on Irradiated Food. April 9, 1982; and van Gemert, M. Letter to New Jersey Assemblyman John Keller, October 19, 1993 quoted in Ibid., 'Preventing Pathogenic......'

119 Ibid., 'Preventing Pathogenic Food Poisoning: Sanitation Not Irradiation.'

120 The Merriam-Webster Dictionary, 1974 edition, page 251.

121 From Federation of American Societies for Experimental Biology, Evaluation of The Health Aspects of Certain Compounds Found in Irradiated Beef. Report to the U.S. Army Medical Research and

Development Command, Bethesda, MD, August 1977, as quoted in Ibid., 'Preventing Pathogenic Food Poisoning…'

122 Op. Cit., Floegel, pages 25-26.

123 Op. Cit., Conversation with Mark Worth of Public Citizen, 8 20 01.

124 Ibid., Worth.

125 'Caesium' from http://www.campusprogram.com/reference/en/wikipedia/c/ca/caesium.html

126 Cesium toxicity, http://www.fortunecity.com/boozers/vines/858/bandazevski/about2.html.

127 email from Judy Szela, June 26, 2004.

128 GRAY*STAR Genesis Irradiator, page one, http://graystarinc.com/genesis.html

129 Op. Cit., Worth Conversation.

130 Ibid., Worth Conversation.

131 Conversation with Will Williams, Surebeam PR, 8 21 2001

132 Op. Cit., Mark Worth Conversation.

133 Op. Cit., Patty Lovera June 4, 2004 telephone conversation.

134 Conversation with Will Williams, Surebeam PR, 8 21 2001.

135 Op. Cit., Mark Worth.

136 Op. Cit., Hauter, Worth, Epstein.

137 'Radioactive Yummy Nummies,' Organic Bytes #17, 7 16 2003, www.organicconsumers.org

138 Ibid.

139 'Filler In Animal Feed Is Open Secret in China,' By David Barboza and Alexei Barrionnuevo, April 30, 2007, NY Times, http://www.nytimes.com/2007/04/30/business/worldbusiness/30food.html?_r = 1&hp = &oref = slog&pagewanted = print

140 'Status of Food Irradiation Around The World,' November 2006, Food and Water Watch.

141 PBS News Hour February 8, 2007, http://www.pbs.org/newshour/bb/science/jan-june07/irradiation_02-08.html

142 Ibid.

143 Patty Lovera telephone interview April 15, 2007.

144 Robert F. Kennedy, Jr. President, Waterkeeper Alliance Letter mid 2001.

145 Updated by me after speaking by telephone with J. Odefey of Waterkeepers Oct 23, 2002.

146 Op. Cit., RFK Jr.

147 Ibid., RFK Jr.

148 Ibid., RFK, Jr.

149 Op. Cit., Floegel, Multimational Monitor.

150 Op. Cit., RFK, Jr.

151 Op. Cit., Floegel, Multinational Monitor.

152 Telephone conversation with Rick Dove of www.neuseriver.com, June 10, 2004.

153 Op. Cit., RFK, Jr.

154 Op. Cit., Floegel, Multinational Monitor.

155 'Largest Recall of Ground Beef Is Ordered,' by Andrew Martin, New York Times, February 18, 2008, available at http://www.nytimes. com/2008/02/18/business/18recall.html?_r = 1&hp&oref = slogin

156 Op. Cit., Floegel, Multinationlal Monitor.

157 Karleff, Ian, "Canadian Scientists Test E. Coli Vaccine on Source," Reuters News Service, August 10, 2000.

158 'Connecticut quarantines 4.7 million chickens to investigate possible avian flu outbreak,' by Donna Tommelleo, Associated Press Writer, www.salon.com March 6 2003.

159 Op. Cit., Robbins, pages 114-115.

160 Ibid., Robbins, page 135.

161 Op. Cit, Andrew Martin, 'Largest Recall of Ground Beef...'

162 Ibid.

163 Op. Cit., Robbins, pages 135-136.

164 Fox, Nicols, *Spoiled: The Dangerous Truth about a Food Chain Gone Haywire* (New York: Basic Books/Harper Collins, 1997), page 10.

165 Op. Cit., Robbins, page 119.

166 WordIQ Dictionary and Encyclopedia, http://www.wordiq.com/ definition/Aquaculture

167 Bill Manci, 'As Expected, Aquaculture Passes A Production Milestone,' www.Aquaculture.com

168 Op. Cit., Robbins, page 296.

169 'Study: Only 10 Percent of big ocean fish remain,' by Marsha Walton, CNN, May 14 2003 http://www.cnn.com/2003/TECH/science/05/14/coolsc.disappearingfish/index.html

170 Op. Cit., Robbins, page 296.

171 'Aquaculture 2002,' FAO report, pages 3-4, http://www.fao

172 'Omega-3 Fatty Acids,' by Dr. Michael T. Murray, ND (Naturopathic Doctor), copyright 2001, Mind Publishing, available via www.naturalfactors.com >

173 email from chemist and environmental consultant, Sergio Paone Ph. D. December 11, 2002. Also, check this web site of our Dept of Agriculture: www.nal.usda.gov/fnic/foodcomp/

174 'Domestic Aquaculture Production Higher And Imports Up,' Aquaculture Magazine, May/June 2004, Vol. 30, No. 3, page 28.

175 The Ram's Horn, page 6, Volume 241, September 2006

176 'Canola Oil May Offer Multiple Benefits To Fish Farmers,' by K.C. Jaehnig, Op. Cit., AquacultureMagazine, Vol. 30, No. 3, page 11-12.

177 Because canola is so susceptible to genetic drift via its pollen contaminating other fields of canola, many experts expect just about ALL of Canada's canola should be anticipated to be considered genetically altered, unless it is somehow organic and sheltered from such contamination/genetic drift.

178 Telephone conversation with Sebastian Belle, January 21, 2003.

179 'Diseases Associated With Salmon Farms, Sea Lice,' http://www. davidsuzuki.org/files/Oceans/SeaLice.jpg >

180 'State Bans Modified Salmon in Preemptive Move' from transgen-icfish@iatp.org> Dec 25, 2002.
181 'Salmon Farming, Summary of Problems,' http://www.davidsuzuki.org/files/Oceans/FFarm.jpg>
182 'Maryland Governor Passes Law Banning The Farming Of Geneti-cally Modified Fish,' AP, quoted in 'Two Articles on Transgenic Fish,' June 21, 2002, transgenicfish@iatp.org>
183 "Frankenfish' Spawns Debate Over Genetically Altered Salmon,' by Jane Kay, SF Chronicle Environmental Writer, available from transgenicfish@iatp.org on April 29, 2002.
184 'Voracious Snakehead Fish Wiped Out In Maryland,' AP story pub-lished in salon.com Sept 18, 2002.
185 'rBGH To Be Fed to Fish Farm Fish in Hawaii,' Industrial FishFarm-ing (industrial-fishfarming@iatp.org> 2/25/03.
186 'Privatizing The Ocean,' by Phil Lansing, October 12, 2002, http://www.iatp.org/iatp/commentaries.cfm?refid=89609; also available http://www.crestofthewave.com
187 'Bush Seeks Expansion Of Offshore Fish Farms,' by Robert McClure, June 8, 2005, Seattle Post-Intelligencer, http://seattlepi.nwsource.com/local/227623_fishfarms08.html]
188 Op. Cit., Lansing.
189 'Offshore Fish Farming - The Selling Of Common Waters,' http://www.pccnaturalmarkets.com/sc/0504/sc0504-fishfarming.html
190 Op. Cit.., Lansing.
191 Ibid., Lansing.
192 Ibid., Lansing.
193 Op. Cit.., Belle telephone conversation.
194 'Cat's Out of the Bag: New Catfish Line Outperforms Others,' by Jim Core, Agricultural Research Service, reported by transgen-icfish@iatp.org> May 8, 2002.
195 "Re: Antimycin A Risk Assessments, Docket No. EPA-HQ-OPP-2006-1002, March 16, 2007, http://www.foodandwaterwatch.org
196 Ibid.
197 'Salmon Farming,' from www.davidsuzuki.org/files/Oceans/Ffarm.jpg>
198 'Salmon Farming,' from www.davidsuzuki.org/files/Oceans/Ffarm.jpg>
199 'Aquaculture,' from www.wikipedia.com
200 Op. Cit., 'Fish Farming' from www.davidsuzuki.org/files/Oceans/Ffarm.jpg>
201 Op. Cit., 'Salmon Farming.'
202 Ibid.
203 Interchangeable terms as per Alex Kirby in 'Prawn Fishing Plunder Seas,' by Alex Kirby, BBC Online News Correspondent, reported Feb 21 2003, seafood-trade@iatp.org>

204 Op. Cit., www.wikipedia.com

205 Op. Cit., www.davidsuzuki.org/shrimpnotes>

206 'Campaigners' Next Target: The Prawn Sandwich' by Angelique Chrisafis, February 19, 2003, http://www.guardian.co.uk/uk_ news/story/0,3604,898255,00.html reported Feb 21, 2003 seafood-trade@iatp.org>

207 Op. Cit., Kirby; Op. Cit., www.davidsuzuki.org/shrimpnotes>

208 Ibid., www.davidsuzuki.org/shrimpnotes>

209 Op. Cit., Kirby.

210 'Campaigners' Next Target: The Prawn Sandwich' by Angelique Chrisafis, February 19, 2003, http://www.guardian.co.uk/uk_ news/story/0,3604,898255,00.html reported Feb 21, 2003 seafood-trade@iatp.org>

211 Op. Cit., Kirby, and Ibid., Chrisafis.

212 Ibid., Chrisafis.

213 'Scientists: Ocean depths being destroyed,' by Matt Crenson, salon.com Feb 15, 2002, AP

214 Anne Mosness of iatp.org in an email of May 23, 2007.

215 'Aquaculture Rapidly Growing,' Feb 26, 2003, on seafood-trade@iatp.org> quoted from FAO 2000 report, figures are for year 2000.

216 NMFS/NOAA Imports and Exports of Fishery Products, Annual Summary 2001 numbers via www.nmfs.noaa.gov> page 5.

217 'Factoid,' page 83, Aquaculture Magazine, May/June 2002.

218 'Responsible Shrimp Farming: A Honduran Example,' by C. Greg Lutz, *Aquaculture Magazine*, Nov-Dec 2001 issue feature article.

219 'Difficult Issues For Aquaculture in the US and Around the World,' by Brenda Jo Narog, feature article May/June issue of *Aquaculture Magazine*, page 44.

220 Ibid., Narog.

221 Ibid., Narog.

222 http://www.smm.org/deadzone/

223 'Against The Grain, A Portrait of Industrial Agriculture as a Malign Force,' American Scholar, Winter 2004, page 15.

f 'On the Increase: Wastelands in the Water – How U.S. farming leads to 'dead zones,' huge marine areas where nothing can grow,' by Kent Garber, U.S. News & World Report, June 16, 2008.

g 'Cornell ecologist's study finds that producing ethanol and biodiesel from corn and other crops is not worth the energy,' By Susan S. Lang, July 5, 2005, Cornell University News Service http://www.news.cornell.edu/stories/july05/ethanol.toocostly.ssl.html

h Ram's Horn, July 2005 issue, pages 2-3.

i Quoted – Annette Hester, G&M, 7/3/07, www.igloo.org/whestern-hemisphere/afreshap in May 2007 issue of Ram's Horn #246 http://64.233.169.104/search?q=cache:JfmuxdrGDIoJ:www.ramshorn.ca/archive2007/246.html+Ram%27s+Horn+ethanol+sugar+cane&hl=en&ct=clnk&cd=1&gl=us

j   July 2007 Ram's Horn issue #248: http://64.233.169.104/ search?q = cache:J5184iRq56YJ:www.ramshorn.ca/ar-chive2007/248.html + Ram%27s + Horn + ethanol + sugar + cane&hl = en&ct = clnk&cd = 2&gl = us

k   http://en.wikipedia.org/wiki/Potash

l   Op. Cit., Ram's Horn, July 2005, pages 2-3.

224 'Factoid,' page 76, May/June issue, *Aquaculture Magazine*.

225 'Solutions for Fish Farming,' www.davidsuzuki.com >

226 Anne Mosness email of May 24, 2007.

227 Mike Skladany, iatp, telephone conversation, March 6, 2003.

228 'Pebble Mine - Bristol Bay, Alaska,' http://www.renewableresourcescoalition.org/pebble_mine.htm

229 Anne Mosness, telephone conversation, May 21, 2007.

230 Op. Cit., 'Pebble Mine - Bristol Bay, Alaska.'

231 Ibid.

232 Anne Mosness, email of May 23, 2007.

233 Ivan Klima, The Spirit of Prague, Pages 106-107, Granta Books, New York, USA.

234 Vandana Shiva, Biopiracy-The Plunder of Nature and Knowledge, pages 1-3, South End Press , Boston, Mass. 1997.

235 Wallach & Sforza, Ibid., Whose Trade Organization? Page 109.

236 Ibid., Wallach & Sforza, page 109, from Letter to U.S. Ambassador to India, April 3, 1998, on file at Public Citizen.

237 Ibid., Wallach & Sforza, p. 105, from WTO, India - Patent Protection for Pharmaceuticals and Agricultural Chemical Products (WT/DS50/R), Report of the Panel, September 5, 1997, at paragraph 7.1.

238 United Nations Development Programme (UNDP), Human Development Report 1999, Geneva (1999)at 68, as referenced in Wallach & Sforza, p. 104.

239 Ibid., 1999 UNDP Report, as referenced in Wallach & Sforza, p. 104.

240 Ibid., 1999 UNDP Report, as referenced in Wallach & Sforza, p. 104.

241 Ibid., Wallach & Sforza, page 111.

242 Gaia Foundation, "WIPO's Mission Impossible," Published by Genetic Resources Action International (GRAIN) in its quarterly newsletter, Seedling, Barcelona, Spain, September 1998, at 10, as referenced in Wallach & Sforza, p. 111.

243 http://www.grain.org/articles/?id = 6; pre-USA/Coalition invasion of Iraq, 97% of Iraqis saved and exchanged their seeds. Private ownership of biological resources had been prohibited historically by the Iraqi constitution, before the 'USA-imposed patent law introduced a system of monopoly rights over seeds' and their ownership.

244 Rural Advancement Foundation Internation, "Basmati Rice Patent," *Geno-Type*, April 1, 1998, as referenced in Wallach & Sforza, p. 108.

245 'Battle Royale of 21st Century Agriculture: Biotech and Patents Policy,' by Kristin Dawkins, Institute for Agriculture and Trade Policy, Minneapolis, Minnesota, 2000, page 9.

246 Ibid.
247 Op. Cit., Shiva, p. 3.
248 'Human Genome Project Opens The Door to Ethnically Specific Bioweapons,' ProjectCensored.org Story #16 2001.
249 Op. Cit., Shiva page 4.
250 BioDemocracy News #34 as quoted from Cropchoice.com 5 21 2001.
251 Ibid., BioDemocracy News #34 from CBC Canada radio broadcast of 6 2 2001.
252 Ibid., BioDemocracy News #34, from Cropchoice.com 5 21 2001.
253 New York Times 6 13 2001, as quoted in BioDemocracy News #34.
254 'Story Time: How Magic GE Beans Make Lots Of Money (Brought To You By Your Friends At Monsanto),' Organic Bytes #32, 5/11/04, http://www.organicconsumers.org
255 Telephone conversation with Brewster Kneen, 7 19 2004.
256 'Item In Farm Bill Takes Away State & Local Rights to Regulate GMOs & Food Safety,' May 25, 2007, http://www.organicconsumers.org/articles/article_5369.cfm
257 Op. Cit., Telephone conversation with Ronnie Cummins, Jan 1, 2008.
258 Op. Cit., 'Item In Farm Bill...' May 25, 2007, http://www.organicconsumers.org/articles/article_5369.cfm.
259 'Victory! USDA Backs Off From Degrading Organic Standards,' Organic Bytes #33, May 28, 2004, http://www.organicconsumers.org
260 Ibid.
261 http://www.organicconsumers.org/sos.cfm
262 Op. Cit., BioDemocracy News #34.
263 'Agricultural Biotech Companies Are Pesticide Companies,' 2001, www.panna.org >
264 'Monsanto's Big Lie Exposed: Roundup Ready Soybeans Use 2-5 times More Herbicides Than Non-GE Varieties,' by Dr. Charles Benbrook, May 26, 2004, www.gealert.org
265 'GM Soya 'Miracle' Turns Sour In Argentina,' The Guardian, by Paul Brown, April 16, 2004, www.gealert.org
266 'Too Much Of A Bad Thing,' Organic Bytes, Issue #20, September 17, 2003, www.organicconsumers.org
267 See http://www.organicconsumers.org/monlink.html
268 BioDemocracy News #35 Bt Action Alert August 17 2001.
269 'Mutation Could Make Organic Pesticide Useless' 6 August 2001 The Independent UK as quoted on organicconsumers.org.
270 CFS News #18 April 16, 1999 reporting on International Meeting of Entomologists in Basel Switzerland, March 1999.
271 Op. Cit., BioDemocracy News #35 Bt alert.
272 Irwin Greenblatt, Professor Emeritus of Molecular and Cell Biology, Univ. of Conn. @ Storrs, Telephone Conversation 7 22 01.
273 Op. Cit., 'Mutation Could Make...'

274 Op. Cit., BioDemocracy News #35 Bt Alert.

275 'Pending Diversity Initiative: Enhancing Economic and Environmental Stability of Rural Landscapes" Univ of Minnesota, Minnesota Institute for Sustainable Agriculture, as quoted in 'The Biodiversity Problem In The Red River Valley,' IATP, August 11, 2000.

276 'Slow Food - An Italian Answer To Globalization' The Nation, August 20/27, 2001, page 11.

277 'U.S. Imposes Standards for Organic-Food Labeling,' by Marian Burros, www.nytimes.com December 21, 2000.

278 'Organic Certifier Denounces New USDA Organic Standards' by Anne Mendenhall, www.oca.org

279 'Victory! USDA Backs Off From Degrading Organic Standards,' Organic Bytes #33, May 28, 2004, http://www.organicconsumers.org

280 Telephone conversation with Ronnie Cummins, National Director of the Organic Consumers Association, July 16, 2004.

281 'USDA Again Undermining Organic Integrity,' Organic Bytes #35, June 23 2004, http://www.organicconsumers.org

282 http://www.organicconsumers.org/sos.cfm

283 Ibid.

284 Ibid.

285 Op. Cit., 'Slow Food,' Alexander Stille.

286 From telephone conversation with Serena Diliberto of SlowFood USA June 24, 2004. And http://www.slowfoodusa.org/contact/index.html

287 2003 Slow Food U.S.A. flyer, www.slowfoodusa.org

288 From 'The Slow Food International Manifesto' endorsed and approved in 1989, 2003 Slow Food U.S.A. flyer, www.slowfoodusa.org

289 Ibid.

290 Sergio Capaldo, local veterinarian heading Slow Food's efforts on behalf of the Piedmontese cow, as quoted in Ibid., 'Slow Food,' Stille.

291 Ibid., 'Slow Food,' Stille.

292 Op. Cit., Jeff Smith, page 250.

293 Ronnie Cummins, Telephone Conversation of 8 20 2001.

294 Op. Cit., Robbins, *The Food Revolution*, page 293.

295 Ibid., page 256.

296 'Brazil's Beef Trade Wrecks Rainforest,' page 7, *The Ram's Horn*, Volume 221, June 2004.

297 Ibid.

298 'The Price of Beef,' *WorldWatch*, July/August 1994, page 39.

299 Op. Cit., Robbins, page 240.

300 Ibid., page 236.

301 Ibid., page 241.

302 *You Can't Eat GNP*,' by Eric A. Davidson, Perseus Books Group, 11 Cambridge Center, Cambridge, Mass. 02142, copyright 2000, page 24.

303 Ibid., page 139.

304 Ibid., page 24.

305 Ibid., page 292.

306 Op. Cit., Cummins.

307 Op. Cit., Dawkins, page 10.

308 'Low Tech Beats GM, Says Prince {Charles of England},' by John Vidal, The Guardian (UK), January 16, 2001.

309 Ibid., Vidal.

310 Ibid., Vidal.

311 Taken from *The Assignation*, by Joyce Carol Oates, page after the Contents.

312 'Background Info [on cotton],' http://www.organicconsumers.org

313 Ibid.

314 Ibid.

315 Ibid.

316 Ibid.

317 Ibid.

318 Ibid.

319 Op. Cit., Robbins, page 219.

320 Op. Cit., Jeff Smith, page 267.

321 Ibid., pages 267-268.

322 Ibid.

323 '*Diet For A New America*, by John Robbins, pages 172-173, Stillpoint Publishing, Box 640, Walpole, NH 03608, copyright 1987.

324 Ibid.

325 Ibid., page 176.

326 'Corrections,' www.seedsofdeception.com

327 Op. Cit., Burros, www.nytimes.com

Chapter 3: Star Wars and Space Dominance

1   'Vision For 2020,' page 3, http://www.crestofthewave.com/show-case/docs/vision_2020.pdf

2   Ibid., page 16.

3   'Bush Crosses The Cosmic Rubicon,' by Chris Floyd, pages 48-49, *The Ecologist*, July/August 2004.

4   Ibid.

5   'The Ultimate Toy Story BUSH LIGHTYEAR To Infinity And Beyond – How The bush regime is planning to use nuclear powered spacecraft and weapons to secure US control of outer space,' by Karl Grossman, *The Ecologist*, pages 46-51.

6   'Vision For 2020,' page 6, also viewable at http://www.gsinstitute. org

7   Craig Eisendrath, former State Dept officer, quoted in 'Space Corps, The dangerous business of making the heavens a war zone," by Karl Grossman, CovertAction April-June 2001, p. 32.

8   Op. Cit., 'Vision For 2020,' page 10.

9   Op. Cit.., *The Ecologist*, Grossman, page 51.

10 Op. Cit., Floyd, page 48.

11 'US Satellite To Test Missile Defence Technologies,' by David Shiga, 24 April 2007, http://www.newscientist.com news service

12 'New Bush Space Policy Calls For Weapons In Space,' 2007 Winter Newsletter, page one, http://www.space4peace.org

13 'Space Defense Program Gets Extra Funding,' by Walter Pincus, Washington Post, November 12, 2007; http://www.washington-post.com/wp-dyn/content/article/2007/11/11/AR2007111101173_pf.html

14 email from Bruce Gagnon, January 21, 2008.

15 Op. Cit., 'New Bush Space Policy Calls For Weapons In Space.'

16 'China Tests Anti-Satellite Weapon, Unnerving U.S.' by William J. Broad and David Sanger, January 18, 2007   NYTimes, http://www.nytimes.com/2007/01/18/world/asia/18cnd-china.html?ei=5094&en=1cccc5a239b55188&hp=&ex=1169182800&partner=homepage&pagewanted=print

17 Op. Cit., 'New Bush Space Policy Calls For Weapons In Space.'

18 Ibid.

19 'The #1 U.S. Industrial Export,' Mary Beth Sullivan, 2007 Winter Newsletter, page 8, http://www.space4peace.org

20 Ibid.

21 'Commonplace Book,' page 141, American Scholar, Spring 2007, quoted from 'The March of Folly: From Troy to Vietnam, 1984.

22 'Some Bush proposals alarm conservatives,' By Tom Raum, Jan. 10, 2004, AP, salon.com

23 Op. Cit., Floyd, page 49.

24 Ibid.

25 Ibid.

26 'Rods from God,' by Joel Bleifuss, 9 3 2003, http://www.inthese-times.com

27 Op. Cit., Grossman, page 49.

28 Op. Cit., Project Prometheus UPDATE FEBRUARY 2004 - http://spacescience.nasa/gov/missions/prometheus.htm

29 Op. Cit., Grossman, page 50.

30 Ibid.

31 Ibid.

32 Ibid.

33 Ibid.

34 Ibid.

35 'Scientists: U.S. power at stake in space race - Nation's success in moon project could prevent wars, earn right to lucrative helium mining,' by Sue Vorenberg, The New Mexican, February 12, 2008, http://www.santafenewmexican.com/SantaFeNorthernNM/Space-Technology-and-Applications-International-Forum-Scientist

36 'NASA Plans Moon Base To Control Pathway To Space,' by Bruce Gagnon, http://space4peace.org

37 Op. Cit., *Weapons In Space*, page 53; and 'Disgrace Into Space,' by Karl Grossman, *The Ecologist*, Volume 31, No. 2, March 2001, page 34.

38 Ibid., 'Disgrace Into Space.'

39 Op. Cit., Gagnon, 'NASA Plans Moon Base To Control Pathway To Space.'

40 Ibid.

41 Ibid.

42 Ibid.

43 "Space Nuke Boffin: NASA Moonbase Needs Nuclear Rockets," by Lewis Page, June 30, 2007, http://www.theregister.co.uk/2007/06/30/space_nuke_boffin_wants_space_nukes/print.html

44 "Space Nuclear Power Complications Mount," Winter 2006 Newsletter, page 3, http://www.space4peace.org

45 Op. Cit.., 'Bush LIGHTYEAR,' Grossman, page 46.

46 Ibid.

47 Op. Cit., Gagnon, 'NASA Plans Moon Base To Control Pathway To Space.'

48 Op. Cit., Grossman, page 47.

49 Ibid.

50 Karl Grossman, *Weapons In Space,* The Open Media Pamphlet Series, Seven Stories Press, New York, 2001, pages 38 and 50.

51 Ibid. Grossman, pages 22 and 38.

52 'The Medical Implications of Nuclear Power,' by Helen Caldicott, MD, distributed by the SHAD Alliance Long Island Region, PO Box 972, Smithtown, NY 11787 phone: 631-360-0045.

53 Op. Cit., Grossman, *The Ecologist*, pages 47-48.

54 Ibid., page 48.

55 www.epa.gov/radiation/radionuclides/plutonium.htm

56 Op. Cit., Caldicott.

57 Op. Cit, Grossman, *The Ecologist,* pages 46-51.

58 'Triggering A New Arms Race?' by Julian Borger, http://www.salon.com/news/feature/2005/05/19/space_weapons/print.html

59 Ibid.

60 'The Space Age At 50,' by Laura Grego, Catalyst Magazine [published by the Union Of Concerned Scientists], Volume 6, No. 2, Fall 2007, pages 5-7, and 12.

61 'Missile Obsession Distorted Threat Priorities,' Carnegie Endowment for International Peace, by Project Director Joseph Cirincione, April 6, 2004, www.clw.org

62 Op. Cit., Floyd, page 49.
63 Ibid.
64 Ibid.
65 Op. Cit., Bleifuss.
66 Op. Cit., Vision For 2020, page 6.
67 Op. Cit., *Weapons in Space*, by Karl Grossman, page 22.
68 email from Bruce Gagnon, January 21, 2008.
69 http://en.wikipedia.org/wiki/NASA_Budget
70 'National Missile Defense: Not Ready For Prime Time,' Carnegie Endowment for International Peace, Special Report, March 25, 2004, www.clw.org
71 Ibid.
72 Dennis Kucinich, "Keynote Address to the Global Network against Weapons and Nuclear Power in Space," April 15, 2000. Videotape available from the Global Network at P.O. Box 90083, Gainesville, FL. 32607.
73 U.S. Air Force Advisory Board, *New World Vistas: Air and Space Power For The 21st Century* (New York: Crown Publishers, 1996), p.420.
74 Op. Cit., Grossman, *Weapons In Space*, pages 58-59.
75 Ibid., Grossman, page 36.
76 Ibid.
77 'U.S. Nuclear Reactors - Al Qaeda's Original Target,' NIRS *Nuclear Monitor*, September 13, 2002, front page story.
78 'Countermeasures, A Technical Evaluation of the Operational Effectiveness of the Planned US National Missile Defense System', April 2000, page 25, Union of Concerned Scientists, MIT Security Studies Program.
79 Telephone conversation with Lisbeth Gronslund, physicist, Union of Concerned Scientists, October 18, 2001.
80 'Odds and Ends,' Winter 2007 Newsletter, page 11, http://www.space4peace.org
81 Op. Cit.,Countermeasures, pages 44-45.
82 'Senator Carl Levin's Opening Statements on Missile Defense,' at the Senate Armed Services Committee Hearing on Missile Defense - March 11, 2004, www.clw.org
83 Op. Cit., 'National Missile Defense: Not Ready For Prime Time.'
84 email from Lisbeth Gronlund Ph D, Senior Scientist and Co-Director of Global Security Program, Union of Concerned Scientists, 5 28 2003.
85 email 2-10-03 from Lisbeth Gronlund Ph D, Senior Scientist and Co-Director of Global Security Program, Union of Concerned Scientists.
86 Telephone conversation with George Lewis, radar expert, January 25, 2008.
87 'AEGIS DESTROYER FITTED WITH "MISSILE DEFENSE,"' Winter 2007 Newsletter, http://www.space4peace.org

88 Op. Cit., telephone conversation with George Lewis.
89 Telephone conversation with George Lewis May 10, 2008. The missile that struck the dying satellite was actually a 'single modified tactical Standard Missile-3' that was launched from the US Navy Aegis ship. According to George Lewis, these are considered 'interceptors,' but not the ones that have/will have the 1000 kilometer range to be placed on the Aegis. http://www.cnn.com/2008/TECH/space/02/20/satellite.shootdown/index.html
90 'U.S. Missile Defenses in Europe: The Putin Alternative,' by Philip E. Coyle, III, Senior Advisor to the Center For Defense Information [CDI], The Defense Monitor, Volume XXXVI, No. 4, July/August 2007, page 2.
91 Op. Cit., 'National Missile Defense: Not Ready For Prime Time.'
92 "U.S. Ignores Failure Data At Outset Of Flights" New York Times - December 18, 2002 - By William J. Broad.
93 Telephone conversation with Philip Coyle III, June 27, 2007.
94 "List of States with Nuclear Weapons," http://www.en.wikipedia.org/wiki/List_of_states_with_nuclear_weapons
95 'Still On Catastrophe's Edge,' by Robert McNamara and Dr. Helen Caldicott, Los Angeles Times, April 26, 2004.
96 Op. Cit., wikipedia list of states with nuclear weapons.
97 Op. Cit., telephone conversation with Philip Coyle III.
98 Ibid., also Op. Cit., McNamara and Caldicott.
99 "A Soviet Glitch (September 26, 1983)," The Defense Monitor, May/June 2007, page 4. Published by CDI.
100 Op. Cit., 'Still On Catastrophe's Edge,' McNamara and Caldicott.
101 'Our Hidden WMD Program,' by Fred Kaplan, April 23, 2004, www.truthout.org
102 'Federal Budget Deficit Hits 395.8 Billion,' by Martin Crutsinger, Associated Press Economics Writer, Yahoo News August 11, 2004
103 'Federal Deficit Hits Record $374.2 billion,' by Sue Kirchoff, USA Today, 10/20/03.
104 'White House warns on national debt limit,' by Martin Crutsinger, May 19, 2003 (AP), salon.com
105 'White House Releases $2.9 Trillion Budget Plan,' by Richard Wolf and David Jackson, USA Today, Feb 5, 2007.
106 Robert D. Walpole, testimony to Senate Subcommittee on International Security, Proliferation, and Federal Services, Feb. 9 2000. {footnote f, p. 12 'Countermeasures.'}
107 'Rumsfeld, Nutrasweet, and Star Wars,' by Linda Bonvie and Bill Bonvie, Food and Water Journal, p. 34, Spring 2001.
108 Op. Cit., Grossman, BUSH LIGHTYEAR, page 49.
109 Op. Cit., Countermeasures, pages 10-13.
110 Op. Cit., Garwin testimony on NMD to Senate Foreign Relations Committee on May 4 1999.
111 'A New Generation of CB (chemical/biological) Munitions,' Jane's Defense Weekly, 3 April 1988, p. 852, as quoted from 'Countermeasures,' p. 52.

112 Op. Cit., *Countermeasures*, page 15.

113 Ibid., page 54.

114 David Wright, UCS Global Security Program co-director, telephone conversation April 28 2004.

115 'Army Selects ERINT Pending Pentagon Review,' by David Hughes, Aviation Week and Space Technology, Feb. 21 1994, p. 93.

116 Op. Cit., Footnote 4, page 50, 'Countermeasures.'

117 Ibid., pages 50-52.

118 'All Exposure Poses a Risk - - Doc questions anthrax theory,' by Laurie Garrett, Newsday, October 31, 2001, page A4.

119 Op. Cit., *Countermeasures*, page 53.

120 'Generals Defend Military's Role In Space,' by James Hannah, March 4, 2004, salon.com

121 email of 7-16-04 from Robert D. Steele of oss.net.

122 "Space, the Cluttered Frontier," Harper's Magazine, December 1999.

123 'Disgrace Into Space,' by Karl Grossman, *The Ecologist*, Volume 31, No. 2, March 2001, page 34.

124 National Research Council, 'Protecting the Space Shuttle from Meteroids and Orbital Debris,' p. 5; Wash D.C.: National Academy Press, 1997.

125 "Space Junk's Out of Hand," Newsday, March 31, 1999.

126 Op. Cit., telephone conversation with George Lewis.

127 Op. Cit., 'Protecting The Space Shuttle...'

128 Op, Cit., 'Weapons In Space,' p. 44.

129 Op. Cit., 'Protecting The Space Shuttle...' page 4.

130 Op. Cit., Newsday, "Space Junk's Out Of Hand."

131 Anything that can happen, will happen.

132 Op. Cit., Grossman, BUSH LIGHTYEAR, page 48.

133 'Emergency Preparedness for Nuclear-Powered Satellites,' p. 24.

134 'Eyewitnesses in Chile Shed Light on Russian Probe's Spectacular Fall,' by David L. Chandler, Boston Globe, Dec. 5 1996.

135 Op. Cit., 'Weapons In Space,' p. 38.

136 Ibid., 'Weapons In Space,' Page 42, and also on page 84, footnote 65.

137 Ibid., page 39.

138 Ibid., page 38.

139 Ibid., page 36.

140 'Rosetta probe heads for comet,' BBC News, March 2, 2004, http://news.bbc.co.uk

141 'ESA Unveils Its New Comet Chaser,' European Space Agency "Information Note." 09-99, Paris, July 1, 1999, p. 2.

142 Op. Cit., Weapons In Space, p. 41.

143 'Space Nuclear Power Complications Mount,' pages 2-3, Winter 2006 Newsletter, http://www.space4peace.org

144 'Nebraska's StratCom And World Domination,' Winter Newsletter 2007, pages 12-13, http://www.space4peace.org

145 Ibid., page 13.
146 'Real-Life Star Wars: The Militarization of Space,' By Stan Cox, AlterNet, received as email Nov. 15, 2007, http://www.alternet.org/audits/67699/?page=4
147 'Arming The Heavens: The Hidden Military Agenda For Space, 1945-1995,' by Jack Manno, NY: Dodd, Mead, 1984, p.11.
148 Ibid., 'Heavens...' Manno, p. 11.
149 Biography of Dr. Wernher Von Braun, http://www.redstone.army.mil/history/vonbraun/bio.html.
150 Ibid.
151 Op. Cit., Weapons In Space, p. 65.
152 Op. Cit., Manno, 'Heavens...' p. 13.
153 Ibid., Manno, 'Heavens...' p. 13.
154 Ibid., Manno, 'Heavens...' p. 5.
155 Ibid., Manno, p. 5.
156 Jack Manno, Interview with Karl Grossman, July 2000.
157 Op. Cit.., Manno, 'Arming The Heavens...' p. 5.
158 Op.Cit., Manno, Grossman Interview, July 2000.
159 Ibid, Manno, Grossman Interview, July 2000.
160 Op. Cit., 'China Tests Anti-Satellite Weapon, Unnerving U.S.'
161 John M. Collins, Military Space Forces: The Next Fifty Years [Washington, D.C.: Pergamon-Brasseys, 1989], p. 23.
162 Ibid, Collins, Back Cover.
163 Ibid., Collins, page 24.
164 Op. Cit., 'Rumsfeld, Nutrasweet, and Star Wars, p. 36-38.
165 Op. Cit., Bleifuss.
166 'Russian Halts Participation In Arms Pact For Europe, Suspension Seen as Response to U.S. Missile Defense Plan,' by Peter Finn, Washington Post Foreign Service, July 15, 2007.
167 'New Project to Eliminate Nuclear Weapons,' by Bruce G. Blair, 'The Defense Monitor' May/June 2007, page 8.
168 Op. Cit., 'Countermeasures,' p. 105-106.
169 Ibid., 'Countermeasures,' p. 100.
170 'Report of the Panel on Reducing Risk in BMD Flight Test Programs,' General L. Welch (ret) et al., Feb 1998 & November 1999.
171 Op. Cit., 'Countermeasures,' p. 101.
172 Ibid., p. 101.
173 Ibid., 'Countermeasures,' Executive Summary, page xx.
174 Report of the Commission to Assess United States National Security Space Management and Organization, (executive summary), p. 16, Washington, D.C., January 11, 2001, pages 11-12.
175 Op. Cit., 'Weapons in Space," p. 62.
176 Frank G. Klotz, Space, Commerce, and National Security (New York: Council on Foreign Relations, 1998), p. 56.
177 Op. Cit., 'Weapons In Space,' pages 78-79.

'178Kofi Annan, "Address on the Opening of the Third United Nations Conference on the Exploration and Peaceful Uses of Outer Space (UNISPACE III), July 19, 1999" (http://www.un.org/events/unispace2/pressrel/e19am.htm).

Chapter 4: Surrendering U.S. Sovereignty To The World Trade Organization

1   U.S. President Abraham Lincoln, Nov. 21, 1864, letter to Col. William F. Elkins, _The Lincoln Encyclopedia_, Archer H. Shaw, MacMillan, 1950, NY, available also at http://www.ratical.org/corporations/Lincoln.html

2   Whose Trade Organization? 2004 Edition, Lori Wallach and Patrick Woodall/Public Citizen, page 110, The New Press, 38 Greene Street, 4th floor, New York, New York, 10013.

3   Ibid., Wallach and Woodall.

4   Ibid., page 4.

5   Ibid., page 111.

6   Ibid., page 115.

7   Ibid., page 114.

8   'GATS: Service Economy Gets The WTO Treatment,' by Ruth Caplan, April 2001, http://www.thirdworldtraveler.com/

9   Ibid.

10 Ibid.

11 Op.Cit., Wallach and Woodall, page 126.

12 'Water for All Campaign Makes Waves for Profiteers,' Public Citizen News, Volume 24, No. 2, March/April 2004, 2003 Annual Report, page 11.

13 Ibid.

14 Ibid.

15 'Asian Longhorned Beetle, Questions and Answers,' http://www.usembassy-china.org.cn/

16 Ibid., 'Asian Longhorned Beetle, Questions and Answers,' http://www.usembassy-china.org.cn/

17 Op.Cit., Wallach and Sforza, and 'Asian Longhorned Beetle, Questions and Answers,' http://www.usembassy-china.org.cn/

18 _Whose Trade Organization?_ by Wallach and Woodall, pages 37-39.

19 Op. Cit., Wallach and Sforza.

20 Ibid.

21 Op. Cit., 'Asian Longhorned Beetle, Questions and Answers,' http://www.usembassy-china.org.cn/

22 'Impacts on Unionids,' http://www.wes.army.mil/el/zebra/zmis/zmishelp/impacts_on_unionids.htm

23 Ibid., http://www.dnr.state.wi.us/org/caer/ce/eek/critter/insect/moth.htm

24 Op. Cit., Wallach and Sforza, page 39.

25 http://www.dnr.state.wi.us/org/caer/ce/eek/critter/insect/moth.htm

26 Op. Cit., Wallach and Sforza, page 38.

27 Telephone conversation with Elizabeth McPartlan, Senior Staff Officer of APHIS, Plant Protection & Quarantine, January 28, 2008.

28 Op. Cit., Wallach and Sforza, page 22.

29 Ibid., page 27.

30 Ibid.

31 Ibid., pages 22-23.

32 Ibid., page 27.

33 'The WTO's Controversial Dispute Settlement Procedure,' #4 Excerpt from the updated version of *Whose Trade Organization*, http://www.citizen.org

34 Ibid.

35 Op. Cit., Wallach and Sforza, page 199.

36 Ibid.

37 Ibid.

38 Ibid., Wallach and Sforza, pages 27-29.

39 Ibid., page 40.

40 Ibid., pages 22-26.

41 Ibid., page X.

42 Ibid., pages 22-26.

43 Robert Evans, "Green Push Could Damage Trade Body – WTO Chief," Reuters, May 15, 1998.

44 Op. Cit., Wallach and Sforza, pages 25-26.

45 Ibid.

46 Op. Cit., Wallach and Woodall, page 31.

47 12/11/96, Ibid., at Article VI. 3.

48 Op. Cit., Wallach and Sforza, page 198.

49 "The United States, Europe and the World Trading System" by Robert B. Zoellick, US Trade Representative, May 15, 2001, Strasbourg, France.

50 WTO, Understanding on Rules and Procedures Governing the Settlement of Disputes (DSU) at Article 14 and Appendix 3, Paras. 2 and 3. [Footnote 7 on page 210 of ' WTO?']

51 Op. Cit., Wallach and Sforza, Page 197.

52 U.S. Trade Representative Michael Kantor, Testimony to the Senate Commerce Committee, June 16, 1994.

53 John Maggs, "Congress Frowns On Clinton Plan to Expand NAFTA," *Journal Of Commerce*, April 5, 1995 (emphasis added).

54 From telephone conversation with Margrete Strand of Public Citizen Trade Watch, June 30, 2003.

55 Op. Cit., Wallach and Woodall, page 84.
56 Op. Cit., Wallach and Sforza, page 187.
57 'Defending the Massachusetts Burma Law,' Harrison Institute For Public Law, Georgetown University Law Center, page 2.
58 Op. Cit., Wallach and Woodall, page 236.
59 Ibid., page 3.
60 Op. Cit.., Wallach and Sforza, page 186.
61 Ibid., Wallach and Woodall, page 236.
62 Ibid., page 188.
63 Op. Cit., 'Defending the Massachusetts Burma Law,' page 2.
64 Op. Cit., Page 192. Footnote number 108 of Whose Trade Organization?
65 Ibid.
66 'Profits at Gunpoint: Unocal's Pipeline In Burma Becomes A Test Case In Corporate Accountability,' by Daphne Eviatar, page 16, The Nation, June 30, 2003.
67 'New Corporate Development from Southeastern Mexico to Panama: Plan Puebla Panama,' pages 1-2, http:www.acerca.org
68 Ibid., page 2.
69 Ibid.
70 Ibid., page 3.
71 Ibid.
72 'What Is In The FTAA Agreement?' page 3, http://www.acerca.org/ftaa/whatisftaa.htm >
73 Ibid., page 4.
74 Ibid.
75 Ibid., page 3
76 'Public Citizen Joins Challenge to Bush Administration: Bring CAFTA To A Vote...' Statement of Lori Wallach, Director of Public Citizen's Global Trade Watch, May 26, 2005, http://www.citizen.org
77 'The Threat to the Environment from CAFTA,' http://www.nrdc.org
78 'CAFTA by the Numbers: What Everyone Needs To Know,' Public Citizen's Global Trade Watch, July, 2004, http://www.tradewatch.org
79 Op. Cit., 'The Threat to the Environment from CAFTA,' http://www.nrdc.org
80 Our Word Is Our Weapon, by Subcomandante Insurgente Marcos, pages 22-23.
81 Op. Cit., page x, Sforza and Wallach.
82 See more on the internet gambling challenge in the Appendix.
83 'Billions In Sanctions Authorized for March 1 Unless Congress Implements WTO-Ordered Change to U.S. Tax Policy; Retaliation Looms in Other WTO Rulings Against U.S. Laws,' http://www.citizen.org
84 'W.T.O. Rules Against U.S. on Cotton Subsidies,' by Elizabeth Becker, The New York Times April 27, 2004.

85 'Step 2 Cotton Subsidies for Corporations,' http://www.ewg.org/farm/step2index.php

86 'African Farmers Hope They Will Be Taken Seriously This Time,' Brahima Ouedraogo, IPS 9-10-03, http://twnafrica.org

87 'More on Brazil's WTO Cotton Challenge,' Food Routes, Issue #96, April 30, 2004, http://organicconsumers.org

88 See more on the Frankenfoods European Union challenge by the USA in the Appendix.

89 "WTO Case on GMO's," 17th June 2003, http://europa.eu.int/comm/trade/goods/agri/pr170603_en.htm

90 Ibid.

91 February 13, 2007 Evaluation 'Re Sunset of Fast Track could Avoid Increase in World Poverty: reports from World Bank, Tufts University, and the Carnegie Endowment for International Peace Point to Net Losses for Poor Countries from Conclusion of Doha Round WTO Escalation,' Public Citizen, http://www.citizen.org

92 'Jose Can You See? Bush's Trojan Taco,' by Greg Palast Monday April 21, 2008, http://www.TomPaine.com

93 'The Security and Prosperity Partnership Agreement: NAFTA Plus Homeland Security,' By Harsha Walia and Cynthia Oka Published on: March 31, 2008 http://www.leftturn.org/?q=node/1160 Also email Kristin Bricker 8 14 08

94 Op. Cit., 'Jose Can You See? Bush's Trojan Taco,' by Greg Palast Monday April 21, 2008.

95 All quotes, statements re Ms Fogal's document come from 'North American Union: The SPP is a "hostile takeover" of democratic government and an end to the Rule of Law,' by Connie Fogal, http://www.globalresearch.ca/index.php?context=va&aid=6456

96 'Amero' proposed by Independent Task Force on North America Vice-Chairman Robert Pastor as the name for NAU currency. Noted in Wikipedia entry 'Independent Task Force on North America,' http://en.wikipedia.org/wiki/Independent_Task_Force_on_North_America

97 Wikipedia describes the SPP at http://en.wikipedia.org/wiki/Security_and_Prosperity_Partnership_of_North_America

98 Also similarly described in the Wikipedia entry 'Independent Task Force on North America,' http://en.wikipedia.org/wiki/Independent_Task_Force_on_North_America

99 'The Militarization and Annexation of North America – The Security and Prosperity Partnership (SPP) unmasked,' by Stephen Lendman, Global Research, July 19, 2007. http://globalresearch.ca/index.php?context=va&aid=6359

100 'U.S.: No funds to keep up pesticide survey-Farmers and environmental groups agree it should be maintained,' http://www.msnbc.msn.com/id/24775125/wid=18298287

101 Op. Cit., Lendman.

102 In 'Building a North American Community,' page 15, May 2005, Council on Foreign Relations, http://www.cfr.org/publication/8102/building_a_north_american_community.html English version, it is stated: "Canada's vast oilsands, once a high-cost experimental means of extracting oil, now provide a viable new source of energy that is attracting a steady stream of multibillion dollar investments and interest from countries such as China, and they have catapulted Canada into second place in the world in terms of proved oil reserves. Production from oil sands fields is projected to reach 2 million barrels per day by 2010. The most serious constraints on additional growth are the limited supply of skilled people and the shortage of infrastructure, including housing, transportation links, and pipeline capacity. Another constraint is regulatory approval processes that can slow down both resource and infrastructure development significantly." Which is typical of the Council on Foreign Relations (CFR) for their utilitarian view of a prospect/oil extraction technology that is very energy intensive and destructive to an environment that will be raped and pillaged far from available scrutiny in the wilds of upper Canada.

In a July 20, 2008 letter to members the Rainforest Action Network (RAN) headlined its call to action on this issue 'The Alberta Tar Sands – The Most Destructive Project On Earth.' The letter itself stated that 'producing one barrel of oil from the tar sands creates a whopping three times as much greenhouse gas as producing a barrel of conventional crude. The heavy tar trapped in sand and clay is so thick that a huge amount of energy is required to heat and separate the tar from the sand and upgrade it to usable oil...mining the tar sands is already turning critical wetlands and once-pristine stretches of Canada's vast boreal forest into desolate landscapes marred by sprawling open-pit mines. Tailing ponds holding more toxic sludge than the capacity of China's Three Gorges Dam – the world's largest dam – can be seen from space.' If not stopped, RAN claims that an area the size of Florida, 54,000 square miles, could be destroyed by clearcutting and strip-mining. Indigenous First Nations peoples are fighting against this threat that could devastate their culture and habitat, which also will include the building of '16 new refineries just to process' tar sands oil at the cost of '$100 billion [for] refineries and pipelines...Dubbed "The Saudi Arabia of the North," the Canadian tar sands reserves hold more than two trillion barrels of oil and are poised to become one of America's primary sources of oil.' RAN states. Note that Mr. Lendman's number of three million barrels produced daily is projected for 2015, as compared to the CFR 2 million barrels projected to be produced daily by 2010.

103 Op. Cit., Lendman.

104 Ibid.

105 Op. Cit., Walia and Oka.

106 Ibid.

107 'Border Invaders: The Perfect Swarm Heads South,' by Mike Davis, #22 Top Story Project Censored 2008 Award. See http://www.tom-dispatch.com/index.mhtml?pid = 122537

108 Op. Cit., Lendman.

109 Ibid.

110 Ibid., also Op. Cit., Walia and Oka.

111 Op. Cit., Lendman.

112 Ibid.

113 'Congress Approves Plan Mexico,' by Kristin Bricker, June 6, 2008, http://www.indypendent.org/2008/06/06/congress-approves-plan-mexico

114 'Plan Mexico: Bush and Calderon Join Forces to Fight Another Endless War,' by Kristin Bricker, to be published in Left Turn magazine in October, 2008. email from Kristin Bricker August 14, 2008.

115 Op. Cit., 'Congress Approves Plan Mexico,' by Kristin Bricker, June 6, 2008.

116 Ibid.

117 Op. Cit., 'Plan Mexico: Bush and Calderon Join Forces to Fight Another Endless War,' by Kristin Bricker.

118 Op. Cit., "Congress Approves Plan Mexico," by Kristin Bricker, June 6, 2008.

119 Perhaps what Mr. Lendman is referring to here may be that part of the Mexican Criminal Code, Crimes Against the Security of the Nation, Terrorism, Art. 139 which states (in translation from a representative of the Mexican consulate in emails to me on 8 20 2008) : 'the penalty for acts against people, objects or public services, that produce alarm, fear or terror in the population or a group, in order to attempt an attack against national security or to pressure the authorities to take a decision, will be punished with prison between 6 and 40 years and a monetary penalty of 1200 days of minimum wage salary.' Also noted by the consulate representative: 'the changes in the Criminal Code (art.139) were effective since June 28, 2008.'

Here are the components of Article 139 as sent to me in Spanish:

Artículo 139.- Se impondrá pena de prisión de seis a cuarenta años y hasta mil doscientos días multa, sin perjuicio de las penas que correspondan por los delitos que resulten, al que utilizando sustancias tóxicas, armas químicas, biológicas o similares, material radioactivo o instrumentos que emitan radiaciones, explosivos o armas de fuego, o por incendio, inundación o por cualquier otro medio violento, realice actos en contra de las personas, las cosas o servicios públicos, que produzcan alarma, temor o terror en la población o en un grupo o sector de ella, para atentar contra la seguridad nacional o presionar a la autoridad para que tome una determinación.

La misma sanción se impondrá al que directa o indirectamente financie, aporte o recaude fondos económicos o recursos de cualquier naturaleza, con conocimiento de que serán utilizados, en todo o en parte, en apoyo de personas u organizaciones que operen o cometan actos terroristas en el territorio nacional. (DR)IJ

Artículo 139 Bis.- Se aplicará pena de uno a nueve años de prisión y de cien a trescientos días multa, a quien encubra a un terrorista, teniendo conocimiento de sus actividades o de su identidad.

Artículo 139 Ter.- Se aplicará pena de cinco a quince años de prisión y de doscientos a seiscientos días multa al que amenace con cometer el delito de terrorismo a que se refiere el párrafo primero del artículo 139.

120 All information attributed to Stephen Lendman quoted by permission from 'The Militarization and Annexation of North America – The Security and Prosperity Partnership (SPP) unmasked,' by Stephen Lendman, Global Research, July 19, 2007. http://globalresearch.ca/index.php?context=va&aid=6359

121 Mexico's 2008 'Gestapo Law,' in addition to allowing wiretapping, allows 'the Mexican government to hold detainees incommunicado for up to 80 days. This...particularly terrifies Mexicans who, in an attempt to stem the disappearances and torture that were so widespread in previous decades, fought for a slightly more transparent judicial process in which detainees must be brought before a judge and presented to the public within 72 hours of their arrest. Rosario Ibarra, a PRD senator and the mother of a disappeared son, says doing away with these protections promotes torture.' See http://mywordismyweapon.blogspot.com/2008/02/mexicos-gestapo-law.html as composed by Kristen Bricker for more on this law.

122 From the album Senator Sam At Home, 'The Greatest Hunger Of The Human Heart,' 1973 CBS vinyl record KC 32756.

123 'Initial Backgrounder and Talking Points: Trade "Deal" Announced May 10, 2007, Public Citizen, http://www.citizen.org

124 Conversation by telephone with Simon Billonness, May 16, 2004.

125 'Small Steps for Corporate Trade Pacts,' by Patrick Woodall, _Multinational Monitor_, May/June 2004, page 19.

126 Ibid.

127 Ibid.

128 Public Citizen letter to members, May 28, 2004, page one.

129 Ibid.

Chapter 5: Radioactive Wastes In Your Dump, Air, Water, Utensils, Baby Stroller, Zippers, Anyone?

1   From www.nirs.org statement for climate change and nuclear power.

2   'Out of Control - On Purpose: DoE's Dispersal of Radioactive Waste into Landfills and Consumer Products,' May 14, 2007, page 64; published by the Nuclear Information and Resource Service, 6930 Carroll Avenue, #340, Takoma Park, MD, 20912; also available on the http://www.nirs.org website

3   Ibid., pages 58-59.

4   Ibid., page 59.

5   A radioactive element's 'half life' equals the time that half of its radioactivity degenerates away. After two half lives, three quarters of the radioactivity is gone. But our experts tell us that it takes 10-20 half lives, which make up a radioactive element's 'hazardous life,' until we don't have to worry about the specific element's radioactive effects on us and our descendants.

6   A micron is one millionth of one meter. A meter measures just a bit more than three feet, or 'about 39.37 inches. One foot equals approximately 0.3048 meter. There are about 1609 meters in a statute mile.' From http://searchcio-midmarket.techtarget.com/sDefinition/0,,sid183_gci523639,00.html

7   'Legal and Regulatory Background,' quoted by Judith Johnsrud, Ph. D., from the federal Low-Level Radioactive Waste Policy of 1980 in 'Out of Control - On Purpose: DoE's Dispersal of Radioactive Waste into Landfills and Consumer Products.'

8   See more detailed information on rads and rems and millirems on pages 14-15 ahead in this very chapter.

9   Op. Cit., 'Out Of Control,' page 58.

10   "Low-Level Radioactive Waste," available at http://www.nirs.org

11   Op. Cit., 'Out of Control,' page 10.

12   Ibid.

13   June 7, 2008 email, Diane D'Arrigo of Nuclear Information & Resource Service (NIRS)

14   Op. Cit., Johnsrud, page 102, 'Out Of Control.'

15   Ibid., 'Out Of Control - On Purpose,' page 45.

16   Ibid.

17   Ibid., page 16.

18   email from Kevin Kamps of Beyond Nuclear, Sept 21, 2007. One metric ton is a larger amount of weight than one English ton. Using the conversion factor of 0.9072 then we get 2103 pounds in one metric ton.

19 hpps//mims.apps.em.doe.gov reports 47,182,270.92 cubic feet of
    LLRW was reported total, between Jan 2 1986 and June 30 2007; con-
    version factors to get this to metric tons are: 0.2185 to get cubic feet
    multiplied down to English tons; and then multiply that number by
    0.9072 to get how many metric tons that number converts to.
20 Op. Cit., "Out of Control" — On Purpose," page 41.
21 US NRC Expanded Below Regulatory Concern Policy of 1990, ex-
    cerpt in Appendix J of Ibid., 'Out Of Control - On Purpose,' as quoted
    in main text, pages 30 and 41.
22 One rad, an absorbed dose of radiation, equals 0.01 joules of energy
    absorbed per kilogram of matter; the letters of rad stand for 'radia-
    tion absorbed dose;' a rem is 'a calculated unit expressing the amount
    of biological damage to tissue from absorbing ionizing radiation.'
    Depending on the type of radiation, the amount of rems will vary,
    depending on the type of radiation inflicted, which will actually have
    a factor/number for its 'biological efficiency.' So 1 rem = 1 rad X bio-
    logical efficiency of whichever radiation is involved. 'Alpha particles
    do 5 to 20 times more damage than gamma rays to tissues they hit,
    so they give higher doses in rems than gamma' rays do. x-rays are
    a type of gamma ray; plutonium gives off alpha rays. - - see page 11
    of 'Out Of Control' as footnoted above and below this very footnote
    itself. To get maybe too electrical: one joule is the 'work done per
    second by a current of one ampere flowing through a resistance of
    one ohm.' --noted on page 436 of the Australian Pocket Oxford Dic-
    tionary published in 1976. For further electrical lingo definitions,
    please consult your favorite reference text on such things, including
    dictionary or encyclopedia or electrician's textbook/manual.
23 Op. Cit., Out of Control, Johnsrud, page 103.
24 Ibid., Appendix M, page 93.
25 Ibid., page 94.
26 Ibid.
27 'Uranium Prices To Skyrocket,' Nuclear Monitor #642, February 24,
    2006, page 15.
28 Op. Cit., 'Out Of Control – On Purpose,' page 57.
29 Ibid., page 39.
30 Ibid.
31 Ibid.
32 Ibid., pages 45 and 54.
33 Ibid., page 54.
34 Ibid; synergistic effects result from the combination and often unex-
    pected interaction of different factors, e.g., radioactivity and hazard-
    ous chemicals.
35 Ibid.
36 Ibid., page 55.
37 Ibid., page 54.
38 Ibid., page 56.

39 Ibid., page 56.

40 Ibid.

41 Ibid., page 63.

42 Ibid., page 43.

43 Ibid., page 104.

44 Ibid., page 20.

45 Ibid., page 21.

46 Ibid., page 67.

47 email from Diane D'Arrigo of the Nuclear Information & Resource Service (NIRS) to me, March 16, 2002.

48 As described by Diane D'Arrigo of NIRS in telephone conversation of June 15, 2001.

49 Ibid.

50 Op. Cit., 'Out Of Control,' page 58.

51 As discussed in telephone conversation with Diane D'Arrigo, August 4, 2004.

52 'A Reliable Renewable Electricity Grid in the United States,' by Arjun Makhijani, Science For Democratic Action, Volume 15 Number 3, Janauary 2008, page 9.

53 'Solar-Cell Rollout,' Technology Review, July/August 2004, pages 35-40.

54 Telephone conversation with representative of EWEA, June 19, 2008.

55 Telephone converations with Kathy Belyeu of AWEA, April 9, 2008.

56 However, as of end of 2007 we have 16,818 megawatts of wind energy already present in the USA. For a 2.5 megawatt wind turbine then, 18,163,440 people could be supplied with electricity NOW, leaving 284,836,560 million in theoretical need of electricity. Doing the calculations, without considering business or government being supplied by these 16,818 megawatts, then the total number of 2.5 megawatt turbines needed would be down to 105,498 to be constructed and assembled to supply electricity to America's homes. [my calculations]

57 Multiple conversations and emails with/from Ace Hoffman June/July 2008.

58 Op. Cit., 'Solar-Cell Rollout.'

59 'Congressional Approval Falls to Single Digits for For First Time Ever,' July 8, 2008, http://news.yahoo.com/s/rasmussen/20080708/pl_rasmussen/ratecongress20080708

60 Op. Cit., 'Out of Control,' page 66.

61 Ibid.

62 'Calling Congress: Washington, We've Got A Nuclear Waste Problem,' Salt Lake Tribune editorial, May 30, 2008 update. Via email June 7, 2008 from Diane D'Arrigo of NIRS.

63 'Should the U.S. be a dump for foreign nuclear waste?' http://www.nirs.org/radwaste/llw/italianwastefactsheeteal.pdf

64 Op. Cit., 'Out Of Control,' page 20.
65 Telephone conversation July 3 2008 with Kevin Kamps, radioactive waste expert with Beyond Nuclear; their website: http://www.be-yondnuclear.org

Chapter 6: Preserve Your Dignity, American

1 See http://en.wikipedia.org/wiki/Kalamazoo,_Michigan
2 Glyphosphate or 'Round Up' has been banned in Denmark for toxicity and health reasons. See Chapter 3.
3 Definition of Dignity #4 in Funk & Wagnalls Standard Dictionary, 1900.
4 'Lap Dogs of the Press,' by Helen Thomas, *The Nation*, March 27, 2006, page 20.
5 Look for video 'Is Nuclear Power Green??' on http://www.YouTube.com Also is footage of Dr. Miller surfing a wave in northwestern New Zealand circa 2000 with the Drive By Truckers' music scooting him along in the background.

Appendix I

1 The Hydrogen Economy, by Jeremy Rifkin, page 125, published by Jeremy P. Tarcher/Putnam, a member of Penguin Putnam Inc., 375 Hudson Street, NYC, NY 10014, 2002.
2 'How To Fix U.S. power grids? Do something. – Ideas surge, but action is often short-circuited,' by Patrick O'Driscoll, Fred Bayles, Jim Cox and Del Jones, USA Today, Page one, August 18, 2003.
3 Ibid., O,Driscoll et al, page 2. Also see footnote 5.
4 'An Industry Trapped By A Theory,' by Robert Kuttner, New York Times/Opinion, August 16, 2003.
5 'A Power Grid for the Hydrogen Economy,' by Paul M. Grant, Chauncey Starr and Thomas J. Overbye, June 26, 2006, *Scientific American*, http://www.sciam.com/article.cfm?id=a-power-grid-for-the-hydr-2006-07&print=true
6 Ibid.
7 Ibid.
8 Ibid.
9 Scott Sklar, telephone conversation 9/24/2001

Appendix 2

1. 'PBMR Is Not The Answer,' Nuclear Monitor #642, February 24, 206, page 10.
2. email to me July 14, 2008 from Paul Gunter of Beyond Nuclear.
3. Andrew Kadak, nuclear engineer, MIT, telephone conversation 4 4 2001.
4. "Fury over 'hidden leak' at German nuclear reactor,' from Anna Tomforde, 2nd June 1986 The Guardian.
5. Telephone conversation with Paul Gunter, August 25, 2003. Now working with http://www.beyondnuclear.org
6. 'Pocket Nuke Project Rumbles On Johannesburg,' by John Yeld, 18 August 2003, The Cape Argus.
7. Ibid.
8. Telephone conversation with Paul Gunter, July 14, 2008.

Appendix 3

1. 'Snake Eyes for US at the WTO,' by Jennifer Wedekind, Multinational Monitor, May/June 2007, page 8.
2. Ibid.
3. Ibid.
4. Ibid.
5. Ibid., pages 8-9.
6. Ibid., page 9.
7. Ibid.
8. Telephone conversation with Mary Bottari of Public Citizen, February 8, 2008.
9. Op. Cit., page 9.
10. Ibid.
11. 'Looking behind the US spin: WTO ruling does not prevent countries from restricting or banning GMOs,' February 2006, page one, http://www.foeeurope.org/publications/2006/WTO_briefing.pdf
12. Ibid.
13. Ibid.
14. Ibid., page 2.
15. Ibid., page 9.
16. Ibid., page 10.
17. Ibid., page 9.
18. Ibid., page 11.
19. Telephone conversation with Steve Suppan of the Institute for Agriculture and Trade Policy (IATP), February 26, 2008
20. 'The Irish and Polish Governments have both decided to prohibit the use of GMOs across the entire country,' April 20, 2008, filed under Eco-News, http://transitioniow.org/2008/04/20/ireland-poland-ban-gmo/

21 Telephone conversation with Steve Suppan of the Institute for Agriculture and Trade Policy (IATP), June 19, 2008.

Physician challenges choices of environmental stewardship:
Nuclear Power, Geneticaly Modified Foods, Arms Race in Space

Conrad Miller, M.D. has taken care of approximately 100,00 patients, mostly as an Emergency Physician. As a world traveler, surfer, humanitarian and political observer, he has become concerned with the uses and abuses of scientific knowledge. In this modern world, technology can help mankind if deployed wisely, or hurt mankind if it is not.

Dr. Miller has written a book focusing on The Most Important Issues Americans THINK They Know Enough About, including nuclear power and radioactive wastes, the corporate influence on our legal system, how we deal with our natural resources and technology, our food supply, and our energy systems. He has also been involved with radio and TV hosting, production, and writing about these issues during the past three decades, and wants to encourage others to be mindful of the choices they make each day.

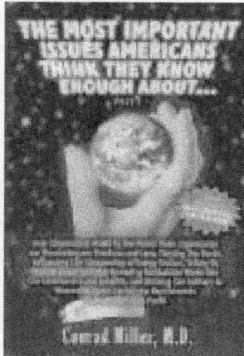

Contact: Conrad Miller MD
631-283-8786
PO Box 180 ,
Watermill, NY 11976
darnoc@crestofthewave.com
http://www.crestofthewave.com

Story Ideas

➢ Nuclear energy is not *green*. Why America should invest in safe ready-to-use alternative energies instead of nuclear power to supply ALL USA homes with electricity.

➢ Do you want nuclear waste thrown in your local dump? Italy wants to ship 20,000 tons of its waste to USA for disposal. How nuclear waste and its contaminating radioactivity can affect our bodies.

➢ Is it safe to eat genetically-modified foods? Have you heard about the study showing rats that consumed them had smaller brains, testicles and livers than control rats? 50% of non-organic USA sugar could be GMO by end of 2008, but this can be prevented by labelling and/or citizen activism.

➢ Fish farming is the fastest growing segment of agriculture, but will it harm our waters and wild fish population? How healthy is it to consume farm raised fish?

➢ Is wind power the answer to solving America's energy crisis? T. Boone Pickens must think so. He's investing $10 billion of his own money in the world's biggest windfarm in Texas.

➢ U.S. Missile Defense is really Offense - - American military and weapons corporations are threatening to start an arms race in space. Fact: weapons are now USA's #1 Industrial Export.

➢ NY State Grand Masters Surf Champion will teach your viewers how to surf *and* learn safe food and environmental practices in the Hamptons.

www.ingramcontent.com/pod-product-compliance
Lightning Source LLC
Chambersburg PA
CBHW072058040426
42334CB00040B/1308

9780975383292